Hierarchical Modeling and Inference
in Ecology

Hierarchical Modeling and Inference in Ecology

The Analysis of Data from Populations, Metapopulations and Communities

J. Andrew Royle and Robert M. Dorazio

Amsterdam • Boston • Heidelberg • London
New York • Oxford • Paris • San Diego
San Francisco • Singapore • Sydney • Tokyo

Academic Press is an imprint of Elsevier

ACADEMIC PRESS

Academic Press is an imprint of Elsevier
84 Theobald's Road, London WC1X 8RR, UK
Redarweg 29, PO Box 211, 1000 AE Amsterdam, The Netherlands
Linacre House, Jordan Hill, Oxford OX2 8DP, UK
30 Corporate Drive, Suite 400, Burlington, MA 01803, USA
525 B Street, Suite 1900, San Diego, CA 92101-4495, USA

First edition 2008

Notice
No responsibility is assumed by the publisher for any injury and/or damage to persons
or property as a matter of products liability, negligence or otherwise, or from any use
or operation of any methods, products, instructions or ideas contained in the material
herein. Because of rapid advances in the medical sciences, in particular, independent
verification of diagnoses and drug dosages should be made

ISBN: 978-0-12-374097-7

For information on all Academic Press publications
visit our website at elsevierdirect.com

Transferred to Digital Print 2009

Printed and bound in Great Britain by
CPI Antony Rowe, Chippenham and Eastbourne

Working together to grow
libraries in developing countries

www.elsevier.com | www.bookaid.org | www.sabre.org

ELSEVIER   BOOK AID
            International   Sabre Foundation

CONTENTS

Preface

Statistical models are developed and used to make inferences from observational data in many areas of ecology, such as conservation biology, fisheries and wildlife management, and more theoretical areas, which include macroecology, metapopulation biology, and landscape ecology. The range of applications spans large, continental-scale monitoring programs that often have several competing objectives, and small-scale studies designed to address well-defined, and often more narrowly focused, scientific hypotheses. Regardless of context, statistical modeling is needed to compute inferences about the form or function of ecological systems, to predict the effects of ecological processes, and, in applied problems, to predict the consequences of management actions.

In this book we attempt to provide a modern view of statistical modeling and inference as applied in the analysis of populations, metapopulations, and communities. We rely exclusively on hierarchical models for this purpose. The hierarchical modeling framework that we advocate makes a clear distinction between an observation component of the model - which typically describes nuisance variation in the data – and the process component of the model - which is usually fundamental to the object of inference. Historically, these two components have not usually been distinguished in model formulation. Hierarchical models are sometimes referred to as state-space models. More recently, the term *hierarchical modeling* has been widely adopted and has become synonymous with Bayesian analysis. However, we believe that hierarchical modeling is a conceptual and philosophical approach to doing science, and it has tangible benefits independent of the framework that is adopted for formal inference, i.e., regardless of whether one is "Bayesian" or not. This distinction is not subtle, nor is it a matter of semantics. Hierarchical modeling is a conceptual framework that emphasizes model construction, whereas Bayesian analysis is concerned with technical aspects of inference. Many of our colleagues, when confronted with a complicated modeling problem, declare the need for a Bayesian, hierarchical model, usually with an emphasis on the Bayesian part. However, the Bayesian mode of inference is often selected merely as a computational expedient (owing to the availability of software for fitting these models), not because of any inferential benefits that might be gained from a Bayesian analysis of the problem. Such models can just as well be fitted using classical methods (i.e., maximizing an integrated likelihood). Thus, our book is not about Bayesian analysis per se; instead, we emphasize construction of hierarchical models. That said, we do apply the Bayesian inference paradigm to many problems, especially those whose solution by classical inference methods is incomplete.

A second motivation for this book is that our impression is that the statistical education of most ecologists does not include sufficient exposure to "modeling" – that is, the construction of probability models to describe ecological systems. For example, in modeling occurrence of species almost everyone knows that logistic regression is the preferred modeling framework. Unfortunately, such regression models are usually viewed as *procedures* for inference or prediction with little consideration given to the underlying probability model of species occurrence. In this way ecologists often fail to appreciate or understand what motivates the distributional assumption (independent Bernoulli outcomes) used in logistic regression. In general, we find that many ecologists are not adequately trained in the formulation of probability models for solving ecological inference problems. Our book is intended to help ecologists develop these skills.

We do not attempt comprehensive coverage of any particular topic in this book; however, we think we have achieved relatively broad coverage both in terms of ecological scale (populations, metapopulations, communities) and in the traditional methodological context by which ecological inference problems are classified and rendered into named procedures (e.g., site-occupancy models, capture-recapture, etc.). Moreover, we have tried to cover all of these problems in a cohesive way as opposed to viewing them as distinct entities requiring unique ("black box") solutions. We have tried to describe these problems in a relatively broad context that is unified by their basic formulation and a few common concepts. Our effort is incomplete to be sure, and many interesting and important classes of problems, new or alternative statistical formulations, and novel or distinctive applications remain.

0.1 SOFTWARE, IMPLEMENTATION AND WEBSITE

Another theme of our book is the development of fairly advanced implementation skills. It is our belief that if ecologists don't achieve these implementation skills, they can only do what someone else thinks they should do (say, using some specialized software package). In that regard, we adopt an implementation framework based on the flexible, comprehensive, and freely available software packages **R** and **WinBUGS**. We use these packages to illustrate calculations involved in data analysis and to plot results. If ecologists master these packages, they can do what they want to do. Together these packages provide a flexible computational platform with which any ecologist or data analyst can grow.

R (R Development Core Team, 2004) is an implementation of the S computing language (Becker et al., 1988; Chambers, 1998) and is freely available at the following web site:

http://www.R-project.org.

WinBUGS is an implementation of the BUGS language (Gilks et al., 1994) for specifying models and doing Bayesian analyses. **WinBUGS** is freely available at the following web site:

http://www.mrc-bsu.cam.ac.uk/bugs/welcome.shtml

and can be used alone or accessed remotely from other software packages. We prefer the latter method and use the **R** library, **R2WinBUGS** (Sturtz et al., 2005), to access the **WinBUGS** computational algorithms remotely while working in **R**.

There are easier ways to do some of the analyses described in our book. For example, we ask the reader to write **R** functions to specify and fit logistic regression models, even though such models are implemented in most statistical software packages. (Even **R** has its own generic functions for doing this). However, learning to write such functions will pay dividends over the long term. By learning to do the easy things the hard way, the hard things will become easy.

Complete, functional **R** and **WinBUGS** implementations for most analyses in the book can be accessed from the Web Supplement for the book, which can be found at the URL: http://www.mbr-pwrc.usgs.gov/HMbook/.

0.2 ORGANIZATION OF THIS BOOK

One logical way to organize the book is according to ecological scale of organization: models relevant to populations, models relevant to metapopulations, and models relevant to communities. However, the approach we take is organized more along the lines of technical relationships among the models. For example, after some introductory material (Chapters 1 and 2) we introduce occupancy models (Chapter 3), a metapopulation concept. We then introduce closed population models in their simplest form (Chapter 5). This progression allows us to establish a technical equivalence between the two models via "data augmentation" (Royle et al., 2007a). Though of marginal interest in this particular pair of problems, data augmentation is exploited repeatedly throughout the book as a means of developing useful generalizations or extensions of certain models. For example, in Chapter 6 we use it to develop models of heterogeneity and individual covariates for closed populations. The formulation of these models as "occupancy" models (using data augmentation) yields a generic and flexible solution to these respective problems. In another topical progression, we discuss site-occupancy models for open populations (Chapter 9) and then consider models of population dynamics, including both classical "Jolly-Seber" models (Chapter 10) and models of animal survival (classical "Cormack-Jolly-Seber" models, (Chapter 11)). This progression allows us to show that these two classes of models are, in terms of probability structure, effectively equivalent to a multi-season, site-occupancy model (admitting certain constraints). Thus, an important point of our book is to show that hierarchical formulations of these models establish a formal duality between models of occurrence (closed and open models of site occupancy) and models of individual encounter histories (closed and open models of populations). Such interrelationships would be difficult to emphasize in a different organization of the book.

Acknowledgements

JAR acknowledgements

I would like to begin by acknowledging the intellectual debt this book owes to two of my professional mentors: L. Mark Berliner and Jim Nichols. Working as a post-doc at the National Center for Atmospheric Research under the direction of Mark Berliner, I was first exposed to the philosophy of hierarchical modeling applied to problems in the geophysical sciences. I thank Mark for sharing his forceful vision of hierarchical modeling as a framework for the conduct of science. The profound benefits of hierarchical modeling became especially clear when I came to the Patuxent Wildlife Research Center and had the good fortune to work closely with Jim Nichols, an extraordinary scientist, quantitative ecologist, and intuitive practitioner of the hierarchical modeling philosophy. This interaction focused my interests on the ecological processes of abundance and occurrence, as well as the fundamental relevance of the observation process to the conduct of inference in ecological systems. Nichols had the keen insight that observation and state models translate across ecological systems. The Nichols synthesis recognizes that the basic technical formulations of hierarchical models for populations, metapopulations, communities, and metacommunites are structurally equivalent, and this provides a profound conceptual and methodological unification in statistical ecology.

Special thanks to my great friends and colleagues: Marc Kéry for the enthusiastic and productive collaboration over what is fast becoming many years; Bill Link, for sharing his wisdom and keen intellect; Emily Silverman, for deeply engaging and motivating discussions on statistics, science and the application of statistics to ecological science. Thanks also to my friends and colleagues Emmanuelle Cam, Chris Wikle, Mark Koneff, John Sauer and Tabitha Graves and to my present and former colleagues with USGS and FWS with whom I have had the great fortune of working and collaborating. Finally, I thank my family, Susan, Abigail and Jacob, for their support and understanding, and for bearing the burden of this book for too many months.

RMD acknowledgements

I am thankful to all of my colleagues at the U.S. Geological Survey, especially those with whom I have collaborated on analyses of important ecological inference problems. I'm also thankful to my colleagues in the Department of Statistics at the University of Florida. In particular, I'm grateful for the support and encouragement of Alan Agresti, Malay Ghosh, and Ramon Littell, who have taught me much about the role of statistics in science. Two other significant figures in my training are Brice Carnahan and Jim Wilkes of the University of Michigan (Department of Chemical Engineering), who instilled at an early stage of my career an appreciation for the complex interplay between mathematics and numerical methods of analysis. Finally, I thank my family, and especially my wife Linda, for support of all kinds during the many days that this effort has taken from our time together.

JAR and RMD acknowledgements

We jointly thank Marc Kéry and Beni Schmidt for their efforts at organizing our Zurich workshops (2006, 2007) on Hierarchical Models, and Jim Hines for developing the book website. The following colleagues provided helpful suggestions for improving one or more chapters of the manuscript: Scott Boomer, Ian Fiske, Beth Gardner, Malay Ghosh, Tabitha Graves, Marc Kéry, Jim Nichols, Beni Schmidt, Vivek Roy, Jim Saracco, Michael Schaub, Emily Silverman, Tom Stanley, Chris Wikle, Elise Zipkin. Thanks to Ian Fiske for the graphics in Chapter 9 and computing support. We thank our colleagues who provided data sets used in the book: Beth Hahn (Redstart nest survival), Robin Jung (stream salamanders), Marc Kéry and the Swiss Ornithological Institute (Swiss BBS data, Swiss butterfly data), Ullas Karanth (tigers), Mark Koneff (waterfowl), Cathy Langtimm (manatee counts), Jim Nichols (microtus), John Sauer (NA BBS), Kevin Young (lizards), Bill Zielinski (carnivores), Jim Saracco and Dave DeSante for the MAPS data. Special thanks to Kevin Young for the lizard photographs in Chapter 5.

Thanks to our colleagues and collaborators: Larissa Bailey, Lianne Ball, Jamie Barichivich, Florent Bled, Sarah Converse, Evan Cooch, Ken Dodd, Paul Doherty, Sam Droege, Howard Jelks, Fred Johnson, Frank Jordan, Jeff Keay, Cathy Langtimm, Lynn Lefebvre, Jim Lyons, Bhramar Mukherjee, Franklin Percival, Mike Robblee, Vivek Roy, Mike Runge, Graham Smith, Jennifer Staiger, Linda Weir, and Li Zhang.

We thank Margaret Coons and Jodie Olson for proof reading the manuscript.

J. Andrew Royle
Robert M. Dorazio

1

CONCEPTUAL AND PHILOSOPHICAL CONSIDERATIONS IN ECOLOGY AND STATISTICS

A bird in the hand

is worth two in the bush ....

.... if p = 0.5

Much of contemporary ecological theory, conservation biology, and natural resource management is concerned with variation in the abundance or occurrence of species. Variation may exist over space or time and is often associated with measurable differences in environmental characteristics. Indeed, ecology has been defined as the study of spatial and temporal variation in abundance and distribution of species (Krebs, 2001). In this regard, a variety of problems in ecology require inferences about species abundance or occurrence. Examples include estimation of population size (or density), assessment of species' range and distribution, and identification of landscape-level characteristics that influence occurrence or abundance. Other examples include studies of processes that affect the dynamics of populations, metapopulations, or communities (i.e., survival, recruitment, dispersal, and interactions among species or individuals).

A common problem in the study of populations or communities is that a census (i.e., complete enumeration) of individuals is rarely attainable, given the size of the region that must be surveyed. Consequently, conclusions about populations or communities must be inferred from samples. An additional complication is that individuals exposed to sampling may not be detected. Detection problems are especially acute in surveys of animals in which detection failures can be produced by a variety of sources (e.g., differences in behavior or coloration of individuals, differences in observers' abilities, etc.). In response to these problems, clever sampling protocols and statistical methods of analysis are commonly used in surveys of animal populations. These protocols and analytical methods, whose origins may

1

be traced to the efforts of Dahl (1919), Lincoln (1930) and Leopold (1933) (also see discussions by Hayne (1949) and Le Cren (1965)), include capture–recapture sampling (Cormack, 1964; Jolly, 1965; Seber, 1965), distance sampling (Burnham and Anderson, 1976), and the statistical methods for analyzing such data. One of the first major syntheses of these early efforts is provided by Seber (1982). Updates of his synthesis are still regarded as general references for sampling and inference in animal populations (Seber, 1992; Schwarz and Seber, 1999). A number of books have also been published recently on the subjects of sampling biological populations and statistical modeling and inference in ecology (Borchers et al., 2002; Williams et al., 2002; Gotelli and Ellison, 2004; Buckland et al., 2004b; MacKenzie et al., 2006; Clark and Gelfand, 2006). In addition to these syntheses, countless monographs and review articles on the same topics have appeared in the primary literature.

In the face of this recent proliferation, what more can we possibly offer? We offer a comprehensive treatment of modeling strategies for many different classes of ecological inference problems, ranging from classical populations of individuals to spatially organized community systems (metacommunities). However, our primary conceptual contribution is the development of a principled approach to modeling and inference in ecological systems based on hierarchical models. We adopt a strict focus on the use of parametric inference and probability modeling, which yields a cohesive and generic approach for solving a large variety of problems in population, metapopulation, community and metacommunity systems. Hierarchical models allow an explicit and formal representation of the data into constituent models of the *observations* and of the underlying ecological or *state process*. The model of the ecological process of interest (the 'process model') describes variation (spatial, temporal, etc.) in the ecological process that is the primary object of inference. This process is manifest in a state variable, which is typically unobservable (or partially so). An example would be animal abundance or occurrence at some point in space and time. In contrast, the model of the observations (the 'observation model') contains a probabilistic description of the mechanisms that produce the observable data. In ecology, this often involves an explicit characterization of detection bias.

1.1 SCIENCE BY HIERARCHICAL MODELING

This description of models by explicit observation and state process components is now a fairly conventional approach to statistical modeling in ecology, where the term 'state-space' model is widely used. In fact, the term state-space might be more common in the prevailing ecological literature. Some recent examples include De Valpine and Hastings (2002), Buckland et al. (2004a), Jonsen et al. (2005),Viljugrein et al. (2005), Newman et al. (2006), and Dennis et al. (2006). Our view of hierarchical modeling is that it is not merely a technical approach to model formulation, or a method of variance accounting. Rather, hierarchical modeling is a much broader conceptual framework for doing science. By focusing our thinking on conceptually and scientifically distinct components of a system, it helps clarify the nature of the inference problem in a mathematically and statistically precise way. Thus, while hierarchical models yield a cohesive treatment of many technical issues (components of variance, combining sources of data, 'scale'), they also foster the fundamental (to science) activities of 'model building' and inference. Our colleague L. Mark Berliner advocates this view in application of hierarchical models to geophysical problems. For this reason he has, on several occasions, made a distinction between *scientific modeling* – based on hierarchical models in which the underlying physical or biological process is manifest as one component of the model – and *statistical modeling*, in which this distinction is not made explicitly. The conceptual and practical distinction between these two approaches provides, in large part, the motivation for and content of this book.

Many of the practical benefits of hierarchical modeling are technical – accounting for sources of variance, different kinds of data, 'scales' of observation and many other factors. But the conceptual benefits of hierarchical modeling are both profound and profoundly subtle, and we elucidate and develop supporting arguments for this view in the following chapters of this book, using many classes of models that are widely used in ecology. For example, in Chapter 3 we discuss what is probably the simplest but most widely used class of statistical models in ecology – models for 'occurrence' which are commonly formulated in terms of simple logistic regression. We demonstrate in several subsequent chapters how relatively simple hierarchical models for occurrence provide solutions to problems of fundamental importance in ecology. In Chapter 4, we discuss models of 'occupancy and abundance.' The linkage between occupancy and abundance is expressed naturally using a hierarchical model. In Chapter 12, we show how models of animal community structure are formulated naturally as 'multi-species' models of occurrence. We provide many more examples in other chapters.

1.1.1 Example: Modeling Replicated Counts

We focus briefly on a particular example (which is described in some generality in Chapter 8) that provides an insightful illustration of the profound subtlety alluded to above. This interesting hierarchical model arises in the context of spatially-indexed sampling of a species. Suppose N_i is the (unobserved) 'local abundance' or population size of individuals on spatial sample unit $i = 1, 2, \ldots, M$. Suppose further that y_i is the *observed* count of individuals at sample unit i. A common model for such data is to assume (or assert) that individuals within the population are sampled independently of one another, in which case y_i is binomial with index

N_i and parameter p (commonly interpreted as detection probability). In this case, the N_i 'parameters' are unobserved. As such, it would be natural in many settings to impose a model on them (e.g., thinking of them as random effects). For example, we might suppose that the local abundance parameters are realizations of Poisson random variables, with parameter λ. We have in this case a two-stage hierarchical model composed of a binomial observation model and a Poisson 'process' model, which we denote in compact notation as follows:

$$f(y_i|N_i) = \text{Bin}(y_i|N_i, p)$$
$$g(N_i|\lambda) = \text{Po}(N_i|\lambda)$$

This is a degenerate hierarchical model in the sense that parameters are confounded – that is, the marginal distribution of y is Poisson with mean $p\lambda$, and unique estimates of p and λ cannot be obtained – but it seems like a sensible construction of an ecological sampling problem when sample units are spatially referenced and counts of organisms are obtained on each sample unit. (See Section 8.3 for an application of this model.)

The degeneracy of the model is easily resolved by adding a little bit more information in the form of replicate counts. That is, suppose each local population is sampled $J = 2$ times, yielding counts y_{ij} which we assume, as before, are $\text{Bin}(N_i, p)$ outcomes. In this case, the hierarchical model is not degenerate (Royle, 2004c; see Chapter 8), and we can focus on developing distinct models for the observation process (the binomial sampling model) and for the ecological process (in the form of $g(N|\theta)$).

To illustrate the benefits of this hierarchical construction, let's consider an analysis of counts of harbor seals observed in Prince William Sound, Alaska (PWS). Annual surveys of harbor seals have been conducted at 25 locations throughout PWS following the March 1989 spill of roughly 40 million liters of crude oil by the T/V Exxon Valdez. The surveyed locations include both heavily oiled sites and sites that were less affected by the spill. A primary objective of the survey is to monitor interannual changes in the size of the harbor seal population at these sites and to estimate any trend in seal abundances over time. Harbor seals are counted at each sample location on several days (usually 7 to 10) of each year using low-flying aircraft. Sampling is conducted during the molting season (August to September) when seals spend more time out of the water and are more susceptible to detection. The details of sampling and analyses of the data collected prior to 2000 are reported elsewhere (Frost et al., 1999; Boveng et al., 2003; Ver Hoef and Frost, 2003). In this example, we analyze the counts of harbor seals observed during 1990 to 2002 at 12 sample locations, half of which were heavily oiled sites.

Let y_{ijt} denote the number of harbor seals counted during the jth replicate visit to the ith sample location in year t. A variety of site- and time-specific covariates, such as time of day (`time`), day of year (`day`), and time since low tide (`tide`) are thought to influence the detectability of harbor seals in this survey (Frost et al., 1999; Ver Hoef and Frost, 2003); therefore, in our binomial model of the observed count, $y_{ijt} \sim \text{Bin}(N_{it}, p_{ijt})$, we assume that the detection probability p_{ijt} depends on these covariates as follows

$$\text{logit}(p_{ijt}) = \alpha_{0t} + \alpha_1 \texttt{tide}_{ijt} + \alpha_2 \texttt{tide}_{ijt}^2 + \alpha_3 \texttt{time}_{ijt} + \alpha_4 \texttt{time}_{ijt}^2$$
$$+ \alpha_5 \texttt{day}_{ijt} + \alpha_6 \texttt{day}_{ijt}^2$$

which allows for peaks in detectability with each covariate. We also assume that the average detectability of seals varied among years by specifying random variation in the intercept parameters as follows: $\alpha_{0t} \sim \text{N}(\mu_\alpha, \sigma_\alpha^2)$. No covariates of harbor seal abundance were observed in this survey, so our model of abundance is relatively simple, $N_{it}|\lambda_t \sim \text{Po}(\lambda_t)$, yet it allows seal abundance to vary among sample locations and years. A reduced-parameter version of this model, wherein we assume $\log \lambda_t = \log \lambda_0 + rt$, allows us to specify a trend in population abundance in terms of an intercept parameter λ_0 and a slope parameter r. Thus, we can estimate the rate of exponential growth ($r > 0$) or decline ($r < 0$) in the mean abundance of seals over time.

On fitting the hierarchical model with fixed-year effects on abundance (i.e., a model without explicit trend), we estimated the mean abundance of seals for each year of the survey (Figure 1.1). Evidently, abundance was relatively low in 1990, the year after the oil spill, but then increased markedly in 1991. However, harbor seal abundance appears to have declined steadily during the next 11 years. On fitting the hierarchical model with a trend in abundance, we estimate a rate of decline of $\hat{r} = -0.07$ per year ($SE = 0.0043$). In other words, the mean abundance of harbor seals at these locations is estimated to have decreased by 54 percent between 1991 and 2002.

The seal problem is typical of a much broader class of problems in ecology that involve modeling spatially- and temporally-indexed counts. We have demonstrated that under this common design of spatially and temporally replicate counts of individuals, a very simple hierarchical construction permits a formal rendering of the model into constituent observation and state process components. That is, the mean count is decomposed into two components, one for abundance λ and another for detection p, that have obvious interpretations in the context of the ecological sampling and inference problems. Furthermore, each of these components can be modeled separately allowing auxiliary information (in the form of covariates) to be used in a more meaningful way – i.e., in manner that is consistent with how we think such things distinctly influence *observation* and *process*.

This simple hierarchical model for the counts seems obvious and intuitive, given the context. However, the conventional approach favored by ecologists and

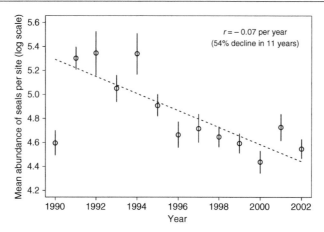

Figure 1.1. Estimates of mean seal abundance per site $(\log \lambda_t)$. The circles are the estimates of the fixed year effects. The dotted line indicates the estimated trend in mean abundance between 1991 and 2002.

statisticians alike is to perform a regression (sometimes a Poisson regression) on the counts y_{ijt}, in which the site effects and replicate effects are all modeled as having the same influence on $\mathrm{E}(y)$, e.g., using a generalized linear mixed model (GLMM). In contrast, the hierarchical model is focused on describing the conditional mean of the observations, say $\mathrm{E}(y|N)$ where N is the abundance process. We provide additional conceptual and technical context of this in Section 1.4.3. What this means practically, in the seal example, is that the mean count of seals is decomposed into an abundance component and a detection component – two components that relate to different 'real' processes. Whereas, in a model for $\mathrm{E}(y)$ (e.g., using a GLMM), the variation due to these distinct processes is considered in the aggregate with no conceptual distinction between them – they both are considered equivalently in models for the mean count of the observations. Some might regard the conventional approach (GLMM) as possessing both an observation and state process model; however, the 'process' is implicit and lacks a scientific interpretation. Does a random site effect correspond to heterogeneity in the count due to detection or due to abundance? Or something else? It's difficult to say, given the ambiguity of the model. While this distinction might seem subtle or esoteric, consider the problem of making predictions of abundance. Using the hierarchical model, we can make a prediction from the abundance component of the hierarchical model which, under the model, provides predictions of abundance that are free of observation effects. Conversely, making a prediction of the marginal mean $\mathrm{E}(y)$ based on a GLMM – a statistical model – then you can predict the expected count that would be observed under whatever sampling method was used to collect the original data. Of what biological relevance is *that*? By ignoring the ecological interpretation and the sampling context induced by viewing the replicate samples as being relevant to the same (local) population of individuals, the conventional approach yields a 'solution' to the inference problem that is more difficult to interpret. We therefore favor a more *science-based* approach to the construction of hierarchical models, as opposed to conventional approaches, which seem to be mostly statistical or descriptive in nature.

1.2 ECOLOGICAL SCALES OF ORGANIZATION

Ecological systems are fundamentally hierarchical systems in which we encounter hierarchies of organization and spatial and temporal scale. These scales often correspond to 'levels' in a hierarchical model. In this book, we address inference at a number of different ecological scales and organizational systems, including populations, metapopulations, communities, and metacommunities (Table 1.1).

Much of quantitative population ecology is concerned with populations of *individuals* – or rather characteristics of populations, such as population size. When the system state is allowed to propagate over time, we introduce population dynamic attributes such as survival and recruitment, and we seek inferences about probabilities or rates or factors that influence them. We can extend the population demographic framework up one hierarchical level to the case of multiple, spatially organized populations. Such systems are commonly referred to as *metapopulations*, the modern conceptual formulation being due to Levins (1969) and Hanski (1998). In metapopulation systems, it is not the size of the population that is relevant so much as the spatial structure in local population attributes and dynamics. Thus, population size becomes spatially indexed, $N(s)$ and one is interested in summaries of $N(s)$ such as mean, variance, and percentiles including $\Pr(N(s) > 0)$, 'occupancy' (see Chapter 3). The metapopulation analogs to individual survival and recruitment are local extinction (analogous to mortality, the complement of survival) and local colonization, which are local population or 'patch-level' attributes, describing transition events from the state $(N(s) = 0)$ to $(N(s) > 0)$, etc. An important focus of applications of metapopulation concepts is the manner in which extinction and colonization dynamics are related to such things as patch area, distance between patches, and other features of the landscape. Metapopulation extinction and colonization are closely related to individual-level dynamics. For example, for a local population to go extinct, all of the individuals have to die. Conversely, for a patch to become colonized, at least one individual must disperse into that patch from a surrounding patch, etc.

Table 1.1. Ecological scales of organization discussed in this book. Each of these systems has some notion of a 'size' parameter, N that is often the quantity of interest in static systems. In dynamic systems, there are analogs of survival and recruitment, which are usually described by extinction and colonization parameters in metapopulation and community systems.

	Static system	Dynamic system
Population of individuals	N = population size	ϕ = survival
		γ = recruitment
Population of populations (metapopulation)	$N(s)$ $\psi(s) = \text{Pr}(N(s) > 0)$	$1 - \phi$ = extinction γ = colonization
Population of species (metacommunity)	N = species richness	ϕ, γ
Population of communities (metacommunity)	$N(s), \psi_i(s)$	ϕ, γ

Population- and metapopulation-level concepts extend up a level of organizational complexity to systems consisting of a population (or populations) of *species*. A single population of species is a community, and spatially organized communities then are metacommunities (Leibold et al., 2004). At the community level, there is also a 'population size' parameter – usually referred to as species richness (see Chapters 6 and 12), which is the number of distinct species in the community. Species richness is a parameter of some concern in conservation biology, biogeography and, naturally, community ecology. Considerable interest exists in using patterns of species richness to inform management or conservation activities. You can examine many ecological journals and find maps of species richness (e.g., Jetz and Rahbek, 2002; Orme et al., 2005). In temporally dynamic community systems, there are also analogs of survival and recruitment that describe changes in the pool of species. These are usually referred to as local extinction (the analog of the complement of survival) and local colonization (Nichols et al., 1998a,b).

Population, metapopulation, community and metacommunity systems often share similar statistical characterizations in the sense that sampling these systems results in replicate binary data on individuals, or sites, or species, or species and sites. Regardless of the system context, we are basically interested in modeling how (or at what rate) 0s become 1s and *vice versa* – as well as how many 'all zeros' there are. Thus, the formulation of models, at least at the level of the observations, is broadly similar and not much more complicated than logistic regression. The equivalence (or perhaps duality) between models of these various systems has been exploited historically in many different contexts to solve new problems using existing methods. For example, Burnham and Overton (1979) used classical models for estimating the size of a closed population of individuals to estimate species richness – the size of a community. Nichols and Karanth (2002) made the linkage between estimating occupancy and closed population size (in the former the zeros are observed, but not so in the latter) which we exploited in Royle et al. (2007a) to aid in the analysis of certain classes of closed population models. Barbraud et al. (2003) exploited this duality between population-level and metapopulation-level systems in the context of an 'open' system – a capture–recapture type model for dynamics of waterbird colonies.

While this technical duality between statistical formulations of models is very exciting to statisticians, the correspondence between ecological levels of organization and the levels of a hierarchical, statistical model is highly relevant to modeling and inference in ecological problems. For example, we might have parameters that are indexed by individual and then a model that governs variation (in parameter values) among individuals within a population; or (as in the seal example) parameters that are indexed by local spatially-indexed populations, combined with a model that governs variation among local populations; or parameters that are indexed by species, which are governed by a model that describes variation among species within a community. Thus, there is a conceptual correspondence between ecological scales of organization and hierarchical levels of organization in the statistical models that we consider in this book.

1.3 SAMPLING BIOLOGICAL SYSTEMS

There are two fundamental considerations common to almost all biological sampling problems regardless of scale or level of organization. These two issues provide considerable motivation for the hierarchical modeling framework that we advocate in this book.

The first issue is that of imperfect detection, or detection bias, which has a number of relevant manifestations. Individual animals in a population are detected imperfectly, species are detected imperfectly, and one might falsely conclude that a species is absent from a site that was surveyed. 'Detectability' of individuals or species is fundamental to assessing rarity and it can be closely linked to demographic processes (see Chapter 4). In contemporary animal sampling, some attention to the issue of detectability is the common thread that ties together a diverse array of sampling methods, protocols, models, and inference strategies. We feel that this issue is important in the context of hierarchical models because it is an important element of the observation process.

The second issue is that sample units are spatially indexed and they almost always represent a subset or sample of all spatial units. Spatial sampling has two consequences that concern hierarchical models. First, it forces us to confront how to combine or aggregate information across space. We confront this issue repeatedly in this book, in which we consistently adopt a model-based approach to solving this problem. Second, spatial attribution induces a component of variation – spatial variation in the state variable – that is directly relevant to most inference problems. This spatial variance component is distinct from variance due to imperfect observation, and it has a profound influence on certain inference problems.

We can see the relevance of imperfect detection and spatial sampling by a formal description of the origin of variance in observations. Suppose that a population is composed of a spatially-indexed collection of subpopulations, of sizes $N_1, N_2, \ldots,$. Suppose we sample the populations and observe n_i individuals at location i. The 'observation variance' is the variation in n_i *conditional* on N_i. That is, the variance in the observed counts under repeated sampling of *the same* population – i.e., holding N_i fixed. This is a function of the sampling method typically, such that the better the method (more money and more time), then the closer n_i is to N_i. One common way in which this source of variation is manifest is in terms of the detection probability parameter of a binomial distribution, as described in the next subsection. The second source of variation that is relevant to some inference problems is the variation in N_i across 'replicate' populations, or spatial sample units. This source of variation is most relevant when the scope of inference is extended to populations beyond those that were sampled, i.e., when making predictions.

1.3.1 Detectability or Detection Bias

The notion of detection bias as a result of incomplete sampling is an important concept that is pervasive in almost all aspects of contemporary animal sampling, and it provides a conceptual unification of many apparently disparate sampling methods, protocols, and models (Seber, 1982; Williams et al., 2002). Detection bias has also been shown to be relevant in the sampling of sessile organisms such as plants (Kéry and Gregg, 2003; Kéry, 2004; Kéry et al., 2006).

The basic problem of imperfect detection is most commonly described using a binomial sampling argument.[1] That is, we suppose that a population composed of N individuals is sampled, and that individuals appear in the sample (i.e., are detected) independently of one another with probability π. Then the number of animals observed in the sample, n, is the outcome of a binomial random variable with sample size N and parameter π. Thus, it must be that $n \leq N$. Binomial sampling gives rise to one of the fundamental equations of animal sampling:

$$\mathrm{E}[n] = \pi \times N.$$

---

[1] This is a natural and concise framework for dealing with this problem, but certainly competing views are reasonable for modeling nuisance variation induced by detectability.

We refer to the general phenomenon of bias in sample quantities (e.g., $E[n] \leq N$) as detection bias. Usually this parameter π is a function of other parameters. It is the probability that an individual appears in the sample at all; thus, if sampling is conducted twice, and detection is independent, then $\pi = 1 - (1-p)^2$ where p is the 'per sample' probability of detection.

One reason that 'detectability' is important is that ecological concepts and scientific hypotheses are formulated in terms of ecological state variables, e.g., abundance, or N, not in terms of the difficulty with which animals are detected. Imperfect detection *induces* a component of variation that is strictly nuisance variation and does not usually correspond to any kind of phenomenon of direct scientific or ecological relevance. This observational variance can obscure our ability to measure ecological processes of interest. Thus, formal attention to variation induced by imperfect detection yields conceptual clarity and strengthens ecological context. A practical reason to be concerned with detectability is that it provides a measure of sampling efficacy – how well animals are being counted. All things being equal, we are inclined to favor a method that counts more individuals. This could be motivated by means of a formal statistical power argument as well. For example, the power to detect differences (e.g., over time) would typically be maximized at $p = 1$.

The importance of detectability is sometimes hotly debated by ecologists and statisticians working in ecology. The approach of basically tossing the issue out the window is as prevalent as the view that detectability must be accounted for or else inferences are completely invalid. Although detectability can be addressed using models that do not explicitly account for a binomial parameter p (models of this nature are sometimes referred to by terms such as relative abundance and abundance indices, etc.), we believe that detection bias, in the form of detection probability, is an important and useful conceptual formulation of the observation process in ecological systems.

In the vast majority of situations, we consider the problem only of 'false absence', in which we fail to observe some individuals or falsely conclude a species doesn't occur. Double counting, misclassification, and 'false positives' are manifestations of imperfect detection that have received considerably less formal attention, but some recent attention to this problem can be found in genetic sampling (Lukacs and Burnham, 2005a,b; Yoshizaki, 2007) and occupancy models (Royle and Link, 2006).

1.3.2 Spatial Sampling and Spatial Variation

That sample units are spatially referenced has important consequences for inference about abundance, occurrence, and other demographic quantities. It forms the other critical consideration motivating the adoption of a hierarchical modeling framework for inference.

We can understand the components of variance clearly by application of standard rules relating marginal and conditional variance (Casella and Berger, 2002, Chapter 4). In particular, the total variance of the observations, $n(s)$, is the sum of two parts: The *observation variance*, which is the variation in what was observed given what you would *like to have observed*, say $\mathrm{Var}(n(s)|N(s))$ (the binomial variance component) and secondly, the *process variance* – the variance of what you would *like to have observed* across replicate spatial sample units, say $\mathrm{Var}_s(N(s))$. Formally, the marginal variance of $n(s)$ is related to the other two quantities according to:

$$\begin{aligned}
\mathrm{Var}(n(s)) &= \mathrm{E}\left\{\mathrm{Var}[n(s)|N(s)]\right\} + \mathrm{Var}\left\{\mathrm{E}[n(s)|N(s)]\right\} \\
&= \mathrm{E}\left\{\pi(1-\pi)N(s)\right\} + \mathrm{Var}\left\{\pi N(s)\right\} \\
&= \pi(1-\pi)\mathrm{E}_s[N(s)] + \pi^2\mathrm{Var}_s[N(s)].
\end{aligned}$$

The first component here is essentially a spatial average of the binomial observation variance – the expected observation variance, whereas the second part is proportional to the spatial variance. Note that as detection improves ($\pi \to 1$), the spatial variance component dominates; whereas, as detection gets worse ($\pi \to 0$), the observation (binomial) variance dominates. Thus, π induces a compromise between these two sources of variation, and we naturally prefer π close to 1. In practice, we have to estimate one or more parameters comprising π that are related to our ability to sample a population or community. As a result, this induces a third component of variance – *estimation variance*.

Thus, in most practical sampling and estimation problems, we have to be concerned with the following components of variance (and sometimes others):

- *Estimation variance* – $\mathrm{Var}(\hat{\pi})$ – variance due to estimating parameters.
- *Observation variance* – $\mathrm{Var}(n|N)$ – variance due to sampling a population of individuals
- *Spatial* or *process variance* – $\mathrm{Var}(N)$ – variance due to sampling a population of populations.

This hierarchical decomposition of variance meshes conceptually with the broader theme of hierarchical models. Regardless of one's view on modeling and inference, importance of detectability, focus on relative abundance or absolute abundance, these different sources of variation exist. Hierarchical models facilitate a formal, model-based accounting for components of variation. However, it is not necessary to adopt a formal model-based solution to properly address the components of variance problem. Classical distance sampling (Buckland et al., 2001) for estimating density provides a good analysis of the components of variance. In the context of distance sampling, it can be shown that the 'total variance' of an estimate of density (or abundance) is the sum of three pieces corresponding to observation variance, process variance, and estimation variance.

The components of variance are often very important in making predictions of the ecological process. Given an estimate $\hat{N}$ of population size for some spatially-indexed population, suppose we wish to predict the population size of a 'similar' population, down the road, in a similar habitat, and in an area of the same size. Estimation usually provides the sampling variance – the variance of the estimator conditional on N – $\mathrm{Var}(\hat{N}|N)$. Then, the same iterated conditional variance formula yields the marginal variance

$$\mathrm{Var}(\hat{N}) = \mathrm{Var}(\hat{N}|N) + \mathrm{Var}(N)$$

which is the variance that applies to a prediction at some 'new' population that was not sampled.

In practice, there are usually other components of variance to consider. For example, in many applications there is also temporal variation that affects inference (Link and Nichols, 1994). We have not addressed this explicitly here because when we are confronted with a temporally dynamic system, our model almost always contains explicit parameters that describe the sources of temporal variation in the system (e.g., survival and recruitment).

1.4 ECOLOGICAL INFERENCE PARADIGMS

Quantitative ecology is practiced by scientists with diverse views and approaches to the analysis of data from ecological systems. In our typology of approaches to the analysis of problems in quantitative ecology, we recognize three more or less distinctive philosophical or conceptual approaches. We refer to these as the observation-driven, process-driven, and the hierarchical view. These are not philosophies, such as Bayesian or frequentist, but rather are somewhat more conceptual views about the analysis of data from ecological systems. On one hand, the observation-driven view focuses on developing models for the data as functions of basic structural parameters. The process-driven view deemphasizes the observation process in favor of building elegant, often complex, models of the underlying ecological process. The hierarchical view that we advocate is a conceptual compromise between these other two views.

1.4.1 The Observation-driven View

In quantitative ecology, there has been a strong focus historically on quantities that are relevant mostly to the sampling apparatus, such as the fraction of the population detected during a sample – quantities that are largely unrelated to any ecological process. This direct focus on modeling and estimation of the observation process characterizes what we call the observation-driven view, and this is the classical view that seems to dominate the segment of statistical ecology concerned with sampling and estimation in biological populations. The key feature of the observation-driven analysis is the prominence of the observation model which, in practical situations, is manifest as an elaborate model of detection probability under a binomial sampling model. Often, there is no attention to the ecological process of interest, except as it relates to an 'adjustment' of sample data based on some estimate of detectability.

As a technical matter, in the observation-driven view, methods are characterized by partial likelihood or conditional estimators of quantities that are relevant. In this approach, the biological quantities of interest are usually formally removed from the model (usually by 'conditioning'), so that attention can be focused on estimating nuisance parameters. As a conceptual example, it is common to treat N as a nuisance parameter by removing it from the likelihood by conditioning on n. This leads to so-called 'conditional likelihood' methods (which we discuss in Chapters 5 and 6) in which the estimator of N is of the form

$$\hat{N} = \frac{n}{\hat{\pi}}.$$

Of course, there is not really anything fundamentally wrong with this estimator of N *per se*. However, as a philosophical matter, removal of the object of inference from the likelihood so that nuisance parameters can be estimated is at odds with statistical norms. Moreover, this approach is philosophically unappealing. More importantly, the basic strategy is ad hoc and becomes self-defeating when the objectives are extended to include not just estimation of N, but developing models, especially predictive models, of the parameters that were removed from the likelihood.

This classical, observation-driven view has been fostered to a large extent by the advent of software that performs certain limited types of analyses, converting sample data to estimates of N or other things. Indeed, it seems that many studies or analyses are motivated largely by the procedures available in existing software. To achieve broader objectives, the practitioner must plug results into a second-stage procedure. The *procedures* are hierarchical (or perhaps 'multi-step' would be a better descriptor), but the *model* is not hierarchical. It is this multi-step approach that gives rise to the phrase 'statistics on statistics', which is almost always an ad hoc attempt at hierarchical modeling. By that, we mean there is usually an improper accounting for variance in the second-stage analysis since the first-stage estimates are treated as data in the second-stage analysis, frequently ignoring error due to estimation.

A typical 'hierarchical procedure' might resemble the following: consider a metacommunity system wherein surveys are carried out at a number of landscape patches. Suppose the goal is to model sources of variation in species richness among patches, e.g., how does the composition of the landscape, patchiness, urbanization, etc. affect species richness? One way in which this has been done (repeatedly) in the literature is to obtain $\hat{N}$ for each patch using some software package, and then conduct a further analysis of the $\hat{N}$'s. For example, fit a regression model relating $\hat{N}$ to covariates. There are a number of problems with this approach – most having to do with improper accounting for uncertainty – components of variance – and the extensibility of the procedure. On the other hand, using hierarchical models, we can achieve a coherent framework for inference – one that properly accounts for sources of variance, and one that is flexible and extensible.

1.4.2 The Process-Driven View

A second philosophy to modeling and inference in ecology is what we refer to as the process-driven view. This is the conceptual opposite of the observation-driven view in the sense that the modeling, estimation, and inference are typically focused on the process component of the model, to the exclusion of any consideration of the observation process. Often, the process-driven approach to a problem involves complex but elegant mathematical models that have been developed based solely on conceptual considerations.

The basic construct is pervasive in models of invasive species and disease systems, which usually involve very complex models of system dynamics, such as invasive spread and density-dependent mechanisms. Another area dominated by the process-driven view are ecosystem models, which are largely 'organizational charts' of trophic levels. A final example can be found in many community ecology analyses, in which counts of species (and other summaries) are viewed as direct measurements of the state process. There was a period in the 1970s and 1980s when considerable research went into 'quantifying' biodiversity, as if species at some point in space and time could simply be tabulated and individuals enumerated. This approach is still the dominant approach in biodiversity research. Another type of problem is the construction of complex models of population dynamics, (e.g., based on Leslie matrix models) that use estimates of various parameters (often of varying quality and relevance) as if they were the truth. This is sometimes justified by an explanation such as 'estimates were obtained from the literature', as if the fact that they had to be estimated from data is irrelevant.

The process-driven view seems to be dominated by mathematicians and engineers, who have a lot of skill with devising models of things but who are less concerned with how observations are obtained in the field. Process-driven applications are characterized by WYSIWYG analyses – analysis of the data as if "what you see is what you get." This can't possibly be the case in practical field situations – nor even in so-called 'field experiments.' In a sense the process-driven view is the natural manner in which scientists want to approach the study of biological or physical processes. The problem is that there can be important effects of the observation process on inference, and this must be considered in the analysis.

1.4.3 The Philosophical Middle Ground: The Hierarchical View

There is a conceptual middle-ground to the observation-driven and process-driven views that is based on hierarchical models. The hierarchical view shares elements with both. Like the observation-driven approach, hierarchical models admit formal consideration of the observation process. Like the process-driven approach, hierarchical models also possess a component model representing the ecological process that is the primary focus of the scientific problem.

We adopt the conceptual definition of a hierarchical model from Berliner (1996). Hierarchical models are composed of a model for the observations – the 'data' – that is conditional on some underlying ecological process that is the focus of inference. A second model – the process model – describes the dynamics of the ecological process. Finally, we may require additional structure to relate the parameters of the observation or process model, in some cases. As a technical, statistical matter, inference is based on the joint distribution, which is the product of submodels. That is (Berliner, 1996)

$$[\text{data}|\text{process, parameters}][\text{process}|\text{parameters}][\text{parameters}]$$

The first component, [data|process, parameters], describes the observation process – the conditional distribution of the data given the state process and parameters. The second component, [process|parameters], describes the state process as we are able to characterize it based on our understanding of the system under study – without regard to 'data' or sampling considerations. Finally, we sometimes express explicit model assumptions about parameters of these two models, and that is the term [parameters].

The existence of the process model is central to the hierarchical modeling view. We recognize two basic types of hierarchical models. First is the hierarchical model that contains an *explicit* process model, which describes realizations of an actual ecological process (e.g., abundance or occurrence) in terms of explicit biological quantities. The second type is a hierarchical model containing an *implicit* process model. The implicit process model is commonly represented by a collection of random effects that are often spatially and or temporally indexed. Usually the implicit process model serves as a surrogate for a real ecological process, but one that is difficult to characterize or poorly informed by the data (or not at all). In this book, we typically deal with hierarchical models having an explicit process model, and we prefer such models when it is possible to characterize them for a given problem.

The distinction was evident in the seal example described in Section 1.1.1. We formulated a model in which the process model described spatial and temporal variation in N, the population size of seals. This spatial model for N represents an explicit process model. We noted in Section 1.1.1 that it is more common in statistical analyses of such data to adopt an analysis based on generalized linear mixed models (GLMMs) or similar methods containing a collection of random effects describing unstructured spatial or temporal variation in the mean of the observations, $E(y)$. We would call this 'random effects' model an implicit process model because the 'process' (the random effects) lack an explicit biological interpretation.

The correspondence (or lack thereof) between implicit and explicit representations comes down to *interpretation* of the implied marginal moment structure of the observations. While the marginal structure may be statistically equivalent (but not necessarily), the interpretation of the parameters is distinctly different between the two approaches. For example, the model for replicate counts (used in the seal example) in which N is Poisson, implies a certain within-group or intra-class correlation (Royle, 2004a). In particular, the covariance between replicate counts of the same local population can be shown to be

$$\mathrm{Cov}(y_{i1}, y_{i2}) = p^2 \lambda,$$

where p is the probability of detection, a component of the observation model, and λ is the mean of N across sites – a component of the process model. In adopting an implicit process model formulation, we would just introduce a variance component into the model to account for this correlation, say $\sigma_{12} = p^2 \lambda$ and estimate σ_{12}. This is essentially what a GLMM does. Conversely, under the explicit hierarchical model, we retain the distinction between the observation and process models and develop distinct models for both p and λ (as we did in the seal example) – both of which have precise interpretations in the context of the sampling and inference problem at hand. In contrast, the marginal covariance parameter σ_{12} is just that regardless of whether we are sampling seal populations in Alaska, taking repeated measurements of pigs in Iowa, or counting dental patients in Ann Arbor. We are not typically concerned about the marginal structure of the observations if our hierarchical model has a coherent formulation that is meaningful in the context of the scientific problem. Where possible, we prefer formulations of hierarchical models in terms of a real ecological process.

1.5 THE ROLE OF PROBABILITY AND STATISTICS

Statistics is a discipline concerned with learning from data. Given a set of observations, we wish to make evidentiary conclusions about some phenomenon or process. The theory of statistics provides the conceptual and methodological framework for doing this. Probability is instrumental to this endeavor in two important respects. First, probability models provide the basis for describing variation in things we can and cannot observe (i.e., variation in observable data and in ecological processes). Second, we use probability to express uncertainty in our conclusions, such as inferences about model parameters and latent variables (things we wish we could observe) of interest.

1.5.1 Probability as the Basis for Modeling Ecological Processes

A model is an abstraction of a phenomenon, process or system. Often, the model takes the form of a mathematical expression, or a probability distribution, but could as well be a graphical description, a pie chart, even a verbal description. Regardless of form, the model represents an abstraction from which we hope to learn about a system or that might elucidate some component of that system. Ecological science is largely concerned with developing models of biological systems and the observation of those systems. In this book, we exclusively adopt probability models as the basis for describing both the observation process, the means by which data are obtained, and of the underlying ecological process.

Classical applied statistics, as taught at most universities, is largely focused on the mechanics of converting data to estimates and p-values. This approach is very procedure-oriented, and there is little attention to broader concepts of model construction – i.e., the use of probability models as the basis for modeling ecological processes. In this book, we try to emphasize the construction of probability models across a broad spectrum of ecological problems.

1.5.2 Probability as the Basis for Inference

Formal inference is the other major use of probability in statistics. By inference, we mean confronting models with data in order to estimate parameters (i.e., fit the model), to carry out some kind of inference (hypothesis test, model selection, model evaluation), to make predictions (Clark et al., 2001), as is done commonly in the construction of species distribution maps (Guisan and Zimmermann, 2000), and even to provide guidance on how to sample the underlying process in an efficient manner (Wikle and Royle, 2005).

While virtually everyone agrees on the use of probability as a tool for building models, there is considerable, sometimes excited, debate over how probability should be used in the conduct of inference. There are at least two schools of thought on this matter. The classical view, which is based on the frequentist idea of a hypothetical collection of repeated samples or experiments, uses probability in many different ways, but never to make direct probability statements about model parameters. In contrast, the Bayesian approach uses probability to make direct probability statements about all unknown quantities in a model.

1.6 STATISTICAL INFERENCE PARADIGMS: BAYESIAN AND FREQUENTIST

The term 'hierarchical modeling' has become almost synonymous with 'Bayesian' in recent years, or at least a good deal of papers on 'Bayesian analysis' also have 'hierarchical model' in the title or key words (e.g. Wikle, 2003; Hooten et al., 2003; Clark, 2003). While hierarchical models are often conveniently analyzed by Bayesian methods, the mode of analysis and inference really stands independent of the formulation of the model. As we have tried to stress, hierarchical modeling is mostly concerned with model *construction*. As such, we freely adopt Bayesian and non-Bayesian analyses of hierarchical models. Although we believe that the analysis of hierarchical models can sometimes be accomplished effectively using classical, non-Bayesian methods, we also believe that Bayesian analysis of hierarchical models is more natural and has some conceptual and practical advantages, which we will describe subsequently and throughout the rest of this book.

The main practical distinction between Bayesian and non-Bayesian treatments of hierarchical models comes down to how latent variables (random effects) are treated. Bayesians put prior distributions on all unknown quantities and use basic probability calculus in conjunction with simulation methods (known as Markov chain Monte Carlo (MCMC), see Link et al., 2002) to characterize the posterior distribution of parameters and random effects by Monte Carlo simulation. The non-Bayesian removes the random effects from the model by integration. This approach works reasonably well in a lot of problems. It does not work as well when the structure of the latent variable model or dependence among latent variables is complex, and it does not always provide a cohesive framework for inference about random effects.

1.6.1 Dueling Caricatures of Bayesianism

We find that much is lost in the transition of classically trained ecologists to the Bayesian paradigm of inference. Or perhaps much confusion is injected into the debate. Ecologists, when first confronted with Bayesianism from an *opposing* view, are often taught that Bayesian analysis is impractical because it is hard to do the calculations, or because model selection is difficult, or because your results depend on the prior. Since you don't usually know anything about the priors, then your inference is not objective. Therefore, Bayesianism is not science. Moreover, priors are not invariant to transformation of the parameters. Therefore, what looks uninformative for θ could very well be informative for $g(\theta)$. Such naive arguments are used by some to explain Bayesianism while, at the same time, debunking it.

Conversely, when confronted with Bayesianism from a *proponent* of Bayesian methods, the ecologist often gets an equal-but-opposite parody in which the profound difference in these two views is reduced to a caricature having to do with Bayesian confidence intervals having a more desirable interpretation, something like this:

> Whereas a frequentist will say "the probability that the interval $[a, b]$ contains the fixed, but unknown, value of θ is 0.95," the Bayesian will say "the probability that θ is in the interval $[a, b]$ is 0.95."

Aside from not being very insightful, it leaves most practical scientists wondering "So what?" Our experience is that ecological scientists do not really care about how confidence intervals should be interpreted. The other part of the pro-Bayesian parody has to do with how great Bayesian methods are for using *prior information* (because, of course, they have prior distributions and the frequentist doesn't!). Unfortunately, prior information isn't a problem that most ecologists have, or even want to have for that matter – or perhaps want to admit. While it is true that the Bayesian framework admits this generalization of the inference problem, it has not proved to be terribly useful in ecology (but see McCarthy and Masters, 2005 for an exception).

In the context of these opposing parodies of Bayesianism, it is useful to reflect on the cynical (but, we believe, accurate) view of Lindley (1986), who remarked "What most statisticians have is a parody of the Bayesian argument, a simplistic view that just adds a woolly prior to the sampling-theory paraphernalia. They look at the parody, see how absurd it is, and thus dismiss the coherent approach as well." While we basically agree with this view, we also emphasize that this book is not about Bayesianism, although we tend to embrace the basic tenets of Bayesianism for analysis and inference in hierarchical models.

1.6.2 Our Bayesian Parody

It would be impossible to elaborate here in a meaningful way on the distinction between classical statistics based on frequentist inference and Bayesian analysis. We don't really want to be guilty of caricaturizing Bayesianism. Several recent books address Bayesian analysis and even Bayesian analysis in Ecology (Clark and Gelfand, 2006; McCarthy, 2007). On the other hand, we probably need to address it briefly, as follows.

In classical statistics, one does not condition on the observed data but rather entertains the notion of replicate realizations (i.e., hypothetical data) and evaluates properties of estimators by averaging over these unobserved things. Conversely, in the Bayesian paradigm, the Bayesian conditions on data, since its the only thing known for certain. So the frequentist will evaluate some procedure, say an estimator, $\hat{\theta}$, that is a function of x, say $\hat{\theta}(x)$, by averaging over realizations of x. The nature of $\hat{\theta}$ becomes somewhat important to the frequentist way of life – there are dozens of rules and procedures for cooking up various flavors of $\hat{\theta}$. On the other hand, the Bayesian will fix x, and base inference on the conditional probability distribution of θ given x, which is called the posterior distribution of θ. For this reason, Bayes is, conceptually, completely objective – inference is *always* based on the posterior distribution. But, therein lies also the conflict. To compute the posterior distribution, the Bayesian has to prescribe a prior distribution for θ, and this is a model *choice*. Fortunately, in practice, this is usually not so difficult to do in a reasonably objective fashion. As such, we view this as a minor cost for being able to exploit probability calculus to yield a coherent framework for modeling and inference in any situation.

There are good philosophical and practical reasons to adopt the Bayesian framework for inference, in general. For example, the Bayesian paradigm is ideal for inference about latent variables and functions of latent variables. Classically, this problem is attacked using a partially Bayesian idea – calculation of the 'conditional posterior distribution' upon which the Best Unbiased Predictor (see Section 2.6) is based. This resolves the main problem (obtaining the point estimate) but creates an additional problem (characterizing uncertainty). A Bayesian formulation of the problem produces a coherent solution to both components of the inference problem. While this point might seem fairly minor, the Bayesian implementation even for very complex models (e.g., non-normal, nonlinear) is conceptually and, using modern methods of computation, practically accessible. An important benefit of Bayesian inference is its relevance to small sample situations (or rather, finite samples). We think that it is under-appreciated by ecologists that frequentist inference is asymptotic, or at least the practical relevance of asymptotic procedures is not often considered. Finally, an important benefit of a Bayesian approach to the analysis of hierarchical models is a transparent accounting for all sources of variation in an estimate or a prediction.

1.7 PARAMETRIC INFERENCE

This book is primarily about model construction, using probability as a basis for modeling and the use of probability as the basis for formal inference. As we have noted, we readily adopt conventional Bayesian and non-Bayesian methods for the analysis of hierarchical models. While the two inference frameworks have important (and profound) technical and conceptual differences, they are unified under the rubric of parametric inference. Thus parametric or 'model-based' inference constitutes one of the overarching themes of this book. That is, our inference is conditional on a prescribed model. This yields a flexible, cohesive framework for inference, and generic procedures having many desirable properties (see Chapter 2).

At many universities, ecologists are taught principles of 'survey sampling' in their statistical curricula. In classical survey sampling or design-based sampling, samples are drawn according to some probabilistic rule, and the properties of estimators are derived from that rule by which the sample is drawn. The actual 'data-generating' process is completely irrelevant. No matter how pathological the data are, the method of sampling *induces* certain desirable properties on parameter estimators. Unfortunately, these desirable properties don't necessarily apply to your data set, obtained under your design. Rather, for example, estimators are unbiased in the sense that, when averaged over all possible samples, the estimator will equal the target population parameter.

A common rationale for reliance on design-based ideas (or at least motivating the use of such methods) is that it will then be 'robust' to parametric model assumptions. However, there is nothing about design-based sampling theory that suggests robustness to arbitrary, unstated model assumptions. Little (2004) gave a good example that turns out to be highly relevant to several procedures used in animal ecology (see Chapter 6). The Horvitz–Thompson estimator (HTE) (Thompson, 2002, Ch. 6) is a widely used procedure for unequal probability sampling. Given a set of observations y_i where observation y_i has probability ψ_i of appearing in the sample, consider estimating the total of a population of N units

$$T = \sum_{i=1}^{N} y_i.$$

Then, the HTE of T based on a sample of size n is $T_{ht} = \sum_{i=1}^{n} y_i/\psi_i$. The HTE has some appealing properties when evaluated from a design-based perspective. Little

(2004) notes that the HTE does have a model-based justification, the estimator arises under the weighted regression model

$$y_i = \beta \psi_i + \psi_i \epsilon_i,$$

where ψ_i is the sample inclusion probability and ϵ_i are *iid* errors with mean 0 and variance σ^2. This leads to $\hat{\beta} = T_{ht}/n$ where n is the sample size. Little concluded: "This analysis suggests that the HTE is likely to be a good estimator when [this model] is a good description of the population, and may be inefficient when it is not." In other words, the HTE is a good (perhaps great?) procedure when inclusion probabilities are relevant to the data-generating mechanism. However, if the inclusion probabilities are not related to the data-generating mechanism, then the HTE may not be that good at all, Little gave an example, the "elephant example," in which they were basically unrelated. This correspondence between design-based and model-based views is not as widely appreciated as we believe it should be.

In classical statistics we are taught elements of finite population sampling. And so, naturally, these notions pervade contemporary statistical ecology as well. But models are more fundamentally relevant to many, if not most, inference problems in ecology. There are always important uncontrollable, unaccounted for sources of variation or practical matters associated with the conduct of field studies that force us to use models to accommodate things that simply could not have been anticipated or controlled for *a priori*. We do have a need for elements of sampling, but often we rely on these things because they give us faith in our models or protect us from severe departures from our model.

1.7.1 Parametric Inference and The Nature of Assumptions

In ecology, we often cannot enumerate sample frames, randomly sample, observe the state variable, or sufficiently control our environment. The consequence of this is that properties of estimators and inference procedures must be, to a large extent, inherited from *parametric inference theory*. What this means, practically, is that we pick a model, we fit that model, and we can compute variances, posterior distributions, p-values, whatever we want under the assumption that the prescribed model is the data-generating model (some would say 'correct' or 'true' model). The unifying conceptual thread of Bayesian and classical frequentist inference is that they are both largely frameworks for the conduct of parametric inference, in which we specify a model, then make an inference that is conditional on the model.

The issue of model-based parametric inference is much bigger than just the distinction between it and other paradigms (i.e., design-based inference). We believe that much of the antagonism between Bayesians and frequentists comes down to anxiety over parametric inference theory vs. these 'other' concepts or philosophies. Biologists favor design-based sampling, at least in part, because the idea of a well-defined set of procedures that can be applied to any problem is reassuring to many. Moreover, one doesn't really have to understand these procedures to apply them. On the other hand, using parametric inference, the specter of 'sensitivity to model' can always be used as a criticism of any analysis. Results are conditional on the model being correctly specified. We accept this as coming with the territory of a generic, flexible, and cohesive framework for inference.

One of the most famous quotes in statistics is that by G.E.P. Box, who remarked: "all models are wrong but some are useful..." which must be the most cited statistical quote of all time[2]. There is a school of thought contending that one must have the right model, or at least the best model achievable, or else you can't do inference. But really these paradigms – parametric inference and model selection – are mutually exclusive to a large extent. You can't carry out a big model selection procedure and then rely on parametric inference in any meaningful way. Consistent with this "all models are wrong ..." sentiment, we view the objective of statistics, at least in part, as being the development of useful models. These (i.e., useful models) exist independent of the quality of data, even of the reasonableness of assumptions since, in our view, the reasonableness of any assumption is completely subjective and debatable. We will further explore model selection and assessment in Chapter 2.

1.7.2 The Hierarchical Rube Goldberg Device

Often, researchers compensate for this reliance on parametric assumptions by building huge elaborate models that obscure what is going on. Now statisticians routinely engage in the development of complex models under the guise of 'hierarchical modeling,' seemingly for the sole purpose of introducing complexity – hierarchical modeling for the sake of hierarchical modeling. Such models are big and conceptually beautiful, and the motives are usually pragmatic, but they can be difficult to understand 'statistically.' It is not easy to criticize such efforts because the mere effort of criticism would be a research project unto itself. The end result is to render a problem unassailable, unrepeatable, unfalsifiable, and beyond comprehension.

In this regard, Dennis Lindley remarked in his book "Understanding Uncertainty" (Lindley, 2006):

---

[2]Besides "Lies, damned lies, and statistics" – by Disraeli.

> There are people who rejoice in the complicated saying, quite correctly, that the real world is complicated and that it is unreasonable to treat it as if it was simple. They enjoy the involved because it is so hard for anyone to demonstrate that what they are saying can be wrong, whereas in a simple argument, fallacies are more easily exposed.

The point being, in the words of our colleague (Link, 2003), "Easily assailable but clearly articulated assumptions ought always to be preferable." Simplicity is a virtue. Not necessarily procedural simplicity, but conceptual simplicity – and clearly assailable assumptions.

1.8 SUMMARY

Hierarchical modeling is something of a growth industry in statistics (and ecology), partially due to the advent of practical Bayesian methods, which we believe foster the adoption of hierarchical models, and partially due to the conceptual advantage of the hierarchical formulation of models in scientific disciplines. Our focus on hierarchical modeling in this book is not about the choice of inference method (Bayesian or frequentist); instead, we focus on providing pragmatic, but principled, solutions to inference problems encountered in ecological studies. Hierarchical models represent a compromise – the conceptual middle ground – between two distinctive approaches to the conduct of ecological science, approaches that we have referred to as the observation and process-driven views. As such, hierarchical models contain explicit representations of both the observation process and also the ecological process of scientific relevance.

Many ecological problems yield naturally to a hierarchical construction because it allows for the formulation of models in terms of the ecological process of interest, the thing that is interesting to most ecologists, while at the same time dealing formally with imperfect observation of the state process. For example, many problems that we encounter have a natural hierarchical structure that is induced by ecological scale: individuals within populations, populations of populations, species within communities, communities of communities. This structure often induces or coincides with components of a hierarchical model, as we will see in subsequent chapters of this book. Secondly, there is almost always a natural and distinct observation component of the model that represents our inability to count individuals. Variation induced in the data by imperfect observation is hardly ever the focus of scientific inquiry, but this source of variation almost always has a profound influence on inference, so must formally be accounted for by statistical procedures. Hierarchical models facilitate and formalize the manner in which the observation process is handled and integrated with the process model. While this clear technical and conceptual distinction between sources of variation in data is a nice feature of hierarchical models, we believe that it is secondary to the main benefit of hierarchical modeling – that it fosters an emphasis on model construction and elucidates the fundamental nature of inference, whether it be about processes, parameters, or predictions. Thus, in our view, hierarchical modeling fosters *scientific modeling*.

2

ESSENTIALS OF STATISTICAL INFERENCE

In the previous chapter we described our philosophy related to statistical inference. Namely, we embrace a model-based view of inference that focuses on the construction of abstract, but hopefully realistic and useful, statistical models of things we can and cannot observe. These models always contain stochastic components that express one's assumptions about the variation in the observed data and in any latent (unobserved) parameters that may be part of the model. However, statistical models also may contain deterministic or structural components that are usually specified in terms of parameters related to some ecological process or theory. Often, one is interested in estimating these parameters given the available data.

We have not yet described how one uses statistical models to estimate parameters, to conduct an inference (hypothesis test, model selection, model evaluation), or to make predictions. These subjects are the focus of this chapter. Inference, by definition, is an inductive process where one attempts to make general conclusions from a collection of specific observations (data). Statistical theory provides the conceptual and methodological framework for expressing uncertainty in these conclusions. This framework allows an analyst to quantify his/her conclusions or beliefs probabilistically, given the evidence in the data.

Statistical theory provides two paradigms for conducting model-based inference: the classical (frequentist) approach and the Bayesian view. As noted in the previous chapter, we find both approaches to be useful and do not dwell here on the profound, foundational issues that distinguish the two modes of inference. Our intent in this chapter is to describe the basis of both approaches and to illustrate their application using inference problems that are likely to be familiar to many ecologists.

We do not attempt to present an exhaustive coverage of the subject of statistical inference. For that, many excellent texts, such as Casella and Berger (2002), are available. Instead, we provide what we regard as essentials of statistical inference in a manner that we hope is accessible to many ecologists (i.e., without using too much mathematics). The topics covered in this chapter are used throughout the book, and it is our belief that they will be useful to anyone engaged in scientific research.

2.1 PRELIMINARIES

2.1.1 Statistical Concepts

Before we can begin a description of model-based inference, we need a foundation for specifying statistical models. That foundation begins with the idea of recognizing that any observation may be viewed as a realization (or outcome) of a stochastic process. In other words, chance plays a part in what we observe.

Perhaps the simplest example is that of a binary observation, which takes one of two mutually exclusive values. For example, we might observe the death or survival of an animal exposed to a potentially lethal toxicant in an experimental setting. Similarly, we might observe the outcomes, 'mated' or 'did not mate', in a study of reproductive behavior. Other binary outcomes common in surveys of animal abundance or occurrence are 'present/absent' and 'detected/not detected.' In all of these examples, there is an element of chance in the observed outcome, and we need a mechanism for specifying this source of uncertainty.

Statistical theory introduces the idea of a *random variable* to represent the role of chance. For example, let Y denote a random variable for a binary outcome, such as death or survival. We might codify a particular outcome, say y, using $y = 0$ for death and $y = 1$ for survival. Notice that we have used uppercase to denote the random variable Y and lowercase to denote an observed value y of that random variable. This notation is a standard practice in the field of statistics. The random variable Y is a theoretical construct, whereas y is real data.

A fully-specified statistical model provides a precise, unambiguous description of its random variables. By that, we mean that the model specifies the probability (or probability density) for *every* observable value of a random variable, i.e., for every possible outcome. Let's consider a model of the binary random variable Y as an example. Suppose $p = \Pr(Y = 1)$ denotes the probability of success (e.g., survival) in a single trial or unit of observation. Since Y is binary, we need only specify $\Pr(Y = 0)$ to complete the model. A fully-specified model requires

$$\Pr(Y = 1) + \Pr(Y = 0) = 1.$$

Given our definition of p, this equation implies $\Pr(Y = 0) = 1 - p$. Thus, we can express a statistical model of the observable outcomes ($y = 0$ or $y = 1$) succinctly as follows:

$$\Pr(Y = y) = p^y (1 - p)^{1-y}, \qquad (2.1.1)$$

where $p \in [0, 1]$ is a formal parameter of the model.

Equation (2.1.1) is an example of a special kind of function in statistics, a *probability mass function* (pmf), which is used to express the probability distribution of a discrete-valued random variable. In fact, Eq. (2.1.1) is the pmf of a Bernoulli distributed random variable. A conventional notation for pmfs is $f(y|\theta)$, which is intended to indicate that the probability of an observed value y depends on the parameter(s) θ used to specify the distribution of the random variable Y. Thus, using our example of a binary random variable, we would say that

$$f(y|p) = p^y(1-p)^{1-y} \tag{2.1.2}$$

denotes the pmf for an observed value of the random variable Y, which has a Bernoulli distribution with parameter p. As summaries of distributions, pmfs honor two important restrictions:

$$f(y|\theta) \geq 0 \tag{2.1.3}$$

$$\sum_y f(y|\theta) = 1, \tag{2.1.4}$$

where the summation is taken over every observable value of the random variable Y.

The probability distribution of a continuous random variable, which includes an infinite set of observable values, is expressed using a *probability density function* (pdf). An example of a continuous random variable might be body weight, which is defined on the set of all positive real numbers. The notation used to indicate pdfs is identical to that used for pmfs; thus, $f(y|\theta)$ denotes the pdf of the continuous random variable Y. Of course, pdfs honor a similar set of restrictions:

$$f(y|\theta) \geq 0 \tag{2.1.5}$$

$$\int_{-\infty}^{\infty} f(y|\theta)\,\mathrm{d}y = 1. \tag{2.1.6}$$

In practice, the above integral need only be evaluated over the range of observable values (or support) of the random variable Y because $f(y|\theta)$ is zero elsewhere (by definition). We describe a variety of pdfs in the next section.

Apart from their role in specifying models, pmfs and pdfs can be used to compute important summaries of a random variable, such as its mean or variance. For example, the mean or *expected value* of a discrete random variable Y with pmf $f(y|\theta)$ is

$$\mathrm{E}(Y) = \sum_y yf(y|\theta).$$

Similarly, the mean of a continuous random variable is defined as

$$\mathrm{E}(Y) = \int_{-\infty}^{\infty} yf(y|\theta)\,\mathrm{d}y.$$

The expectation operator $E(\cdot)$ is actually defined quite generally and applies to *functions of random variables*. Therefore, given a function $g(Y)$, its expectation is computed as

$$E[g(Y)] = \sum_y g(y) f(y|\theta)$$

if Y is discrete-valued and as

$$E[g(Y)] = \int_{-\infty}^{\infty} g(y) f(y|\theta)\, dy$$

if Y is a continuous random variable. One function of particular interest defines the variance,

$$\mathrm{Var}(Y) = E[(Y - E(Y))^2].$$

In other words, the variance is really just a particular kind of expectation. These formulae may seem imposing, but we will see in the next section that the means and variances associated with some common distributions can be expressed in simpler forms. We should not forget, however, that these simpler expressions are actually derived from the general definitions provided above.

2.1.2 Common Distributions and Notation

The construction of fully-specified statistical models requires a working knowledge of some common distributions. In Tables 2.1 and 2.2 we provide a list of discrete and continuous distributions that will be used throughout the book.

For each distribution we provide its pmf (or pdf) expressed as a function of y and the (fixed) parameters of the distribution. We also use these tables to indicate our choice of notation for the remainder of the book. We find it convenient to depart slightly from statistical convention by using lower case to denote both a random variable and its observed value. For example, we use $y \sim N(\mu, \sigma^2)$ to indicate that a random variable Y is normally distributed with mean μ, variance σ^2, and pdf $f(y|\mu, \sigma)$. We also find it convenient to represent pmfs and pdfs using both bracket and shorthand notations. For example, we might represent the pmf of a binomially distributed random variable Y in either of 3 equivalent ways:

- $f(y|N, p)$
- $[y|N, p]$
- $\mathrm{Bin}(y|N, p)$.

Table 2.1. Common distributions for modeling discrete random variables.

Distribution	Notation	Probability mass function	Mean and variance
Poisson	$y \sim \mathrm{Po}(\lambda)$	$f(y\|\lambda) = \exp(-\lambda)\lambda^y/y!$	$\mathrm{E}(y) = \lambda$
	$[y\|\lambda] = \mathrm{Po}(y\|\lambda)$	$y \in \{0, 1, \ldots\}$	$\mathrm{Var}(y) = \lambda$
Bernoulli	$y \sim \mathrm{Bern}(p)$	$f(y\|p) = p^y(1-p)^{1-y}$	$\mathrm{E}(y) = p$
	$[y\|p] = \mathrm{Bern}(y\|p)$	$y \in \{0, 1\}$	$\mathrm{Var}(y) = p(1-p)$
Binomial	$y \sim \mathrm{Bin}(N, p)$	$f(y\|N, p) = \binom{N}{y}p^y(1-p)^{N-y}$	$\mathrm{E}(y) = Np$
	$[y\|N, p]$ $= \mathrm{Bin}(y\|N, p)$	$y \in \{0, 1, \ldots, N\}$	$\mathrm{Var}(y) = Np(1-p)$
Multinomial	$\boldsymbol{y} \sim \mathrm{Multin}(N, \boldsymbol{p})$	$f(\boldsymbol{y}\|N, \boldsymbol{p}) =$ $\binom{N}{y_1\cdots y_k}p_1^{y_1}p_2^{y_2}\cdots p_k^{y_k}$	$\mathrm{E}(y_j) = Np_j$
	$[\boldsymbol{y}\|N, \boldsymbol{p}]$ $= \mathrm{Multin}(\boldsymbol{y}\|N, \boldsymbol{p})$	$\times (1 - p_\bullet)^{N-y_\bullet}$	$\mathrm{Var}(y_j) = Np_j(1-p_j)$
		$y_j \in \{0, 1, \ldots, N\}$	$\mathrm{Cov}(y_i, y_j) = -Np_ip_j$
Negative-binomial	$y \sim \mathrm{NegBin}(\lambda, \alpha)$	$f(y\|\lambda, \alpha) =$ $\frac{\Gamma(y+\alpha)}{y!\,\Gamma(\alpha)}\left(\frac{\lambda}{\alpha+\lambda}\right)^y\left(\frac{\alpha}{\alpha+\lambda}\right)^\alpha$	$\mathrm{E}(y) = \lambda$
	$[y\|\lambda, \alpha]$ $= \mathrm{NegBin}(y\|\lambda, \alpha)$	$y \in \{0, 1, \ldots\}$	$\mathrm{Var}(y) = \lambda + \lambda^2/\alpha$
Beta-binomial	$y \sim \mathrm{BeBin}(N, \alpha, \beta)$	$f(y\|N, \alpha, \beta)$ $= \binom{N}{y}\frac{\Gamma(\alpha+y)\Gamma(N+\beta-y)}{\Gamma(\alpha+\beta+N)}$	$\mathrm{E}(y) = N\alpha/(\alpha+\beta)$
	$[y\|N, \alpha, \beta]$ $= \mathrm{BeBin}(y\|N, \alpha, \beta)$	$\times \frac{\Gamma(\alpha+\beta)}{\Gamma(\alpha)\Gamma(\beta)}$	$\mathrm{Var}(y)$ $= N\frac{\alpha\beta(\alpha+\beta+N)}{(\alpha+\beta)^2(\alpha+\beta+1)}$
		$y \in \{0, 1, \ldots, N\}$	

The bracket and shorthand notations are useful in describing hierarchical models that contain several distributional assumptions. The bracket notation is also useful when we want to convey probabilities or probability densities without specific reference to a particular distribution. For example, we might use $[y|\theta]$ to denote the probability of y given a parameter θ without reference to a particular distribution. Similarly, we might use $[y]$ to denote the probability of y without specifying a particular distribution or its parameters.

We adhere to the common practice of using a regular font for scalars and a bold font for vectors or matrices. For example, in our notation $\boldsymbol{y} = (y_1, y_2, \ldots, y_n)'$ indicates a $n \times 1$ vector of scalars. A prime symbol is used to indicate the transpose of a matrix or vector. There is one exception in which we deviate from this notational convention. We sometimes use θ with regular font to denote a model parameter

Table 2.2. Common distributions for modeling continuous random variables.

Distribution	Notation	Probability density function	Mean and variance
Normal	$y \sim \mathrm{N}(\mu, \sigma^2)$	$f(y|\mu, \sigma)$ $= \frac{1}{\sqrt{2\pi}\sigma} \exp\left(-\frac{(y-\mu)^2}{2\sigma^2}\right)$	$\mathrm{E}(y) = \mu$
	$[y|\mu, \sigma] = \mathrm{N}(y|\mu, \sigma^2)$	$y \in \mathbb{R}$	$\mathrm{Var}(y) = \sigma^2$
Multivariate	$\boldsymbol{y} \sim \mathrm{N}(\boldsymbol{\mu}, \boldsymbol{\Sigma})$	$f(\boldsymbol{y}|\boldsymbol{\mu}, \boldsymbol{\Sigma})$ $= (2\pi)^{-p/2}|\boldsymbol{\Sigma}|^{-1/2}$	$\mathrm{E}(\boldsymbol{y}) = \boldsymbol{\mu}$
normal	$[\boldsymbol{y}|\boldsymbol{\mu}, \boldsymbol{\Sigma}] = \mathrm{N}(\boldsymbol{y}|\boldsymbol{\mu}, \boldsymbol{\Sigma})$	$\times \exp\left(-\frac{1}{2}(\boldsymbol{y} - \boldsymbol{\mu})'\boldsymbol{\Sigma}^{-1}\right.$ $\left.(\boldsymbol{y} - \boldsymbol{\mu})\right) \boldsymbol{y} \in \mathbb{R}^p$	$\mathrm{Var}(\boldsymbol{y}) = \boldsymbol{\Sigma}$
Uniform	$y \sim \mathrm{U}(a, b)$	$f(y|a, b) = 1/(b - a)$	$\mathrm{E}(y) = (a + b)/2$
	$[y|a, b] = \mathrm{U}(y|a, b)$	$y \in [a, b]$	$\mathrm{Var}(y) = (b - a)^2/12$
Beta	$y \sim \mathrm{Be}(\alpha, \beta)$	$f(y|\alpha, \beta) = \frac{\Gamma(\alpha+\beta)}{\Gamma(\alpha)\Gamma(\beta)}y^{\alpha-1}$ $\times (1 - y)^{\beta-1}$	$\mathrm{E}(y) = \alpha/(\alpha + \beta)$
	$[y|\alpha, \beta] = \mathrm{Be}(y|\alpha, \beta)$	$y \in [0, 1]$	$\mathrm{Var}(y) = \alpha\beta/\{(\alpha + \beta)^2(\alpha + \beta + 1)\}$
Dirichlet	$\boldsymbol{y} \sim \mathrm{Dir}(\boldsymbol{\alpha})$	$f(\boldsymbol{y}|\boldsymbol{\alpha})$ $= \frac{\Gamma(\alpha_1+\cdots+\alpha_k)}{\Gamma(\alpha_1)\cdots\Gamma(\alpha_k)}y_1^{\alpha_1-1}\cdots$	$\mathrm{E}(y_j)$ $= \alpha_j/\sum_{l=1}^{k}\alpha_l$
	$[\boldsymbol{y}|\boldsymbol{\alpha}] = \mathrm{Dir}(\boldsymbol{y}|\boldsymbol{\alpha})$	$y_k^{\alpha_k-1} y_j \in [0, 1];$ $\sum_{j=1}^{k} y_j = 1$	$\mathrm{Var}(y_j)$ $= \frac{\alpha_j(-\alpha_j+\sum_l \alpha_l)}{(\sum_l \alpha_l)^2(1+\sum_l \alpha_l)}$
Gamma	$y \sim \mathrm{Gamma}(\alpha, \beta)$	$f(y|\alpha, \beta)$ $= \frac{\beta^\alpha}{\Gamma(\alpha)}y^{\alpha-1}\exp(-\beta y)$	$\mathrm{E}(y) = \alpha/\beta$
	$[y|\alpha, \beta] = \mathrm{Gamma}(y|\alpha, \beta)$	$y \in \mathbb{R}^+$	$\mathrm{Var}(y) = \alpha/\beta^2$

that may be a scalar *or* a vector depending on the context. To avoid confusion, in these cases, we state explicitly that θ is possibly vector-valued.

2.1.3 Probability Rules for Random Variables

Earlier we noted that a statistical model is composed of one or more random variables. In fact, in most inferential problems it would be highly unusual to observe the value of only one random variable because it is difficult to learn much from a sample of size one. Therefore, we need to understand the 'rules' involved in modeling multiple outcomes.

Since outcomes are modeled as observed values of random variables, it should come as no surprise that the *laws of probability* provide the foundation for modeling multiple outcomes. To illustrate, let's consider the joint distribution of only two random variables. Let (y, z) denote a vector of two discrete random variables. The *joint pmf* of (y, z) is defined as follows:

$$f(y, z) = \Pr(Y = y, Z = z),$$

where we suppress the conditioning on the parameter(s) needed to characterize the distribution of (y, z). The notation for the *joint pdf* of two continuous random variables is identical (i.e., $f(y, z)$). Now suppose we want to calculate a *marginal pmf* or *marginal pdf* for each random variable. If y and z are discrete-valued, the marginal pmf is calculated by summation:

$$f(y) = \sum_z f(y, z)$$

$$f(z) = \sum_y f(y, z).$$

If y and z are continuous random variables, their marginal pdfs are computed by integration:

$$f(y) = \int_{-\infty}^{\infty} f(y, z)\, \mathrm{d}z$$

$$f(z) = \int_{-\infty}^{\infty} f(y, z)\, \mathrm{d}y$$

Statistical models are often formulated in terms of *conditional* outcomes; therefore, we often need to calculate conditional probabilities, such as $\Pr(Y = y | Z = z)$ or $\Pr(Z = z | Y = y)$. Fortunately, *conditional pmfs* and *conditional pdfs* are easily calculated from joint and marginal distribution functions. In particular, the conditional pmf (or pdf) of y given z is

$$f(y|z) = \frac{f(y, z)}{f(z)}.$$

Likewise, the conditional pmf (or pdf) of z given y is

$$f(z|y) = \frac{f(y, z)}{f(y)}.$$

The above formulae may not seem useful now, but they will be used extensively in later chapters, particularly in the construction of hierarchical models. However, one immediate use is in evaluating the consequences of independence. Suppose random variables y and z are assumed to be *independent*; thus, knowing the value of z gives us no additional information about the value of y and vice versa. Then, by definition, the joint pmf (or pdf) of the the vector (y, z) equals the product of the marginal pmfs (pdfs) as follows:

$$f(y, z) = f(y)f(z).$$

Now, recalling the definition of the conditional pmf (pdf) of y given z and substituting the above expression yields

$$
\begin{aligned}
f(y|z) &= \frac{f(y, z)}{f(z)} \\
&= \frac{f(y)f(z)}{f(z)} \\
&= f(y).
\end{aligned}
$$

Therefore, the conditional probability (or probability density) of y given z is identical to the marginal probability (or probability density) of y. This result confirms that knowledge of z provides no additional information about y when the random variables y and z are independent.

One well-known application of the laws of probability is called the *law of total probability*. To illustrate, let's consider two discrete random variables, y and z, and suppose z has n distinct values, $z_1, z_2, \ldots, z_n$. The law of total probability corresponds to the expression required for calculating the marginal probability $f(y)$ given only the conditional and marginal pmfs, $f(y|z)$ and $f(z)$, respectively. We know how to calculate $f(y)$ from the joint pmf $f(y, z)$:

$$f(y) = \sum_{z=z_1}^{z_n} f(y, z)$$

By definition of conditional probability, $f(y, z) = f(y|z)f(z)$. Therefore, substituting this expression into the right-hand side of the above equation yields

$$
\begin{aligned}
f(y) &= \sum_{z=z_1}^{z_n} f(y|z)f(z) \\
&= f(y|z_1)f(z_1) + f(y|z_2)f(z_2) + \cdots + f(y|z_n)f(z_n)
\end{aligned}
$$

which is the law of total probability. In this equation $f(y)$ can be thought of as a weighted average (or expectation) of the conditional probabilities where the weights correspond to the marginal probabilities $f(z)$. The law of total probability is often used to remove latent random variables from hierarchical models so that the parameters of the model can be estimated. We will see several examples of this type of marginalization in subsequent chapters.

2.2 THE ROLE OF APPROXIMATING MODELS

Inference begins with the data observed in a sample. These data may include field records from an observational study (a survey), or they may be outcomes of a designed experiment. In either case the observed data are manifestations of at least two processes: the sampling or experimental procedure, i.e., the process used to collect the data, *and* the ecological process that we hope to learn about.

Proponents of model-based inference base their conclusions on one or more *approximating models* of the data. These models should account for the observational (or data-gathering) process and for the underlying ecological process. Let's consider a simple example. Suppose we randomly select n individual animals from a particular location and measure each animal's body mass y_i $(i = 1, \ldots, n)$ for the purpose of estimating the mean body mass of animals at that location. In the absence of additional information, such as the age or size of an individual, we might be willing to entertain a relatively simple model of each animal's mass, e.g., $y_i \sim N(\mu, \sigma^2)$. This model has two parameters, μ and σ, and we cannot hope to estimate them from the body mass of a single individual; therefore, an additional modeling assumption is required. Because animals were selected at random to obtain a representative sample of those present, it seems reasonable to assume *mutual independence* among the n measurements of body mass. Given this additional assumption, an approximating model of the sample data is

$$[y_1, y_2, \ldots, y_n | \mu, \sigma] = \prod_{i=1}^{n} N(y_i | \mu, \sigma^2). \tag{2.2.1}$$

Therefore, the joint pdf of observed body masses is modeled as a product of identical marginal pdfs (in this case $N(y|\mu, \sigma^2)$). We describe such random variables as *independent and identically distributed* (abbreviated as *iid*) and use the notation, $y_i \overset{iid}{\sim} N(\mu, \sigma^2)$, as a compact summary of these assumptions.

Now that we have formulated a model of the observed body masses, how is the model used to estimate the mean body mass of animals that live at the sampled locations? After all, that is the real inferential problem in our example. To answer this question, we must recognize the connection between the parameters of the model and the scientifically relevant estimand, the mean body mass of animals in the population. In our model-based view each animal's body mass is a random variable, and we can prove that $E(y_i) = \mu$ under the assumptions of the approximating model. Therefore, in estimating μ we solve the inference problem.

This equivalence may seem rather obvious; however, if we had chosen a *different* approximating model of body masses, the mean body mass would necessarily be specified in terms of *that* model's parameters. For example, suppose the sample of n animals includes both males and females but we don't observe the sex of each individual. If we know that males and females of similar ages have different average masses and if we have reason to expect an uneven sex ratio, then we might consider a mixture of 2 normal distributions as an approximating model:

$$y_i \overset{iid}{\sim} p\mathrm{N}(\mu_m, \sigma^2) + (1-p)\mathrm{N}(\mu_f, \sigma^2),$$

where p denotes the unknown proportion of males in the population and μ_m and μ_f denote the mean body masses of males and females, respectively. Under the assumptions of this model we can show that $\mathrm{E}(y_i) = p\mu_m + (1-p)\mu_f$; therefore, to solve the inference problem of estimating mean body mass, we must estimate the model parameters, p, μ_m, and μ_f.

We have used this example to illustrate the crucial role of modeling in inference problems. The parameters of a model specify the theoretical properties of random variables, and we can use those properties to deduce how a model's parameters are related to one or more scientifically relevant estimands. Often, but not always, these estimands may be formulated as summaries of observations, such as the sample mean. In these cases it is important to remember that such summary statistics may be related to the parameters of a model, but the two are not generally equivalent.

2.3 CLASSICAL (FREQUENTIST) INFERENCE

In the previous section we established the role of modeling in inference. Here we describe classical procedures for estimating model parameters and for using the estimates to make some kind of inference or prediction.

Let $\boldsymbol{y} = (y_1, \ldots, y_n)$ denote a sample of n observations. Suppose we develop an approximating model of $\boldsymbol{y}$ that contains a (possibly vector-valued) parameter θ. The model is a formal expression of the processes that are assumed to have produced the observed data. In classical inference the model parameter θ is assumed to have a fixed, but unknown, value. The observed data $\boldsymbol{y}$ are regarded as a single realization of the stochastic processes specified in the model. Similarly, any summary of $\boldsymbol{y}$, such as the sample mean $\bar{y}$, is viewed as a random outcome.

Now suppose we have a procedure or method for estimating the value of θ given the information in the sample, i.e, given $\boldsymbol{y}$. In classical statistics such procedures are called *estimators* and the result of their application to a particular data set yields an *estimate* $\hat{\theta}$ of the fixed parameter θ. Of course, different estimators can produce different estimates given the same set of data, and considerable statistical theory has been developed to evaluate the operating characteristics of different estimators (e.g., bias, mean squared error, etc.) in different inference problems. However, regardless of the estimator chosen for analysis, classical inference views the estimate $\hat{\theta}$ as a random outcome because it is a function of $\boldsymbol{y}$, which also is regarded as a random outcome.

To make inferences about θ, classical statistics appeals to the idea of hypothetical outcomes under *repeated sampling*. In other words, classical statistics views $\hat{\theta}$ as a single outcome that belongs to a distribution of estimates associated with hypothetical repetitions of an experiment or survey. Under this view, the fixed value θ and the assumptions of the model represent a mechanism for generating a random, hypothetical sequence of data sets and parameter estimates:

$$(\boldsymbol{y}_1, \hat{\theta}_1), (\boldsymbol{y}_2, \hat{\theta}_2), (\boldsymbol{y}_3, \hat{\theta}_3), \ldots$$

Therefore, probability statements about θ (i.e., inferences) are made with respect to the distribution of estimates of θ that could have been obtained in repeated samples.

For this reason those who practice classical statistics are often referred to as *frequentists*. In classical statistics the role of probability in computing inferences is based on the relative frequency of outcomes in repeated samples (experiments or surveys). Frequentists never use probability directly as an expression of degrees of belief in the magnitude of θ. Probability statements are based entirely on the hypothetical distribution of $\hat{\theta}$ generated under the model and repeated sampling. We will have more to say about the philosophical and practical differences that separate classical statistics and Bayesian statistics later in this chapter.

2.3.1 Maximum Likelihood Estimation

We have not yet described an example of a model-based estimator, i.e., a procedure for estimating θ given the observed data $\boldsymbol{y}$. One of the most widely adopted examples in all of classical statistics is the *maximum likelihood estimator* (MLE), which can be traced to the efforts of Daniel Bernoulli and Johann Heinrich Lambert in the 18th century (Edwards, 1974). However, credit for the MLE is generally awarded to the brilliant scientist, Ronald Aylmer Fisher, who in the early 20th century fully developed the MLE for use in inference problems (Edwards, 1992). Among his many scientific contributions, Fisher invented the concept of *likelihood* and described its application in point estimation, hypothesis testing, and other inference problems. The concept of likelihood also has connections to Bayesian inference. Therefore, we have chosen to limit our description of classical statistics to that of likelihood-based inference.

Let's assume, without loss of generality, that the observed data $\boldsymbol{y}$ are modeled as continuous random variables and that $f(\boldsymbol{y}|\theta)$ denotes the joint pdf of $\boldsymbol{y}$ given a model indexed by the parameter θ. In many cases the observations are mutually independent so that the joint pdf can be expressed as a product of individual pdfs as follows:

$$f(\boldsymbol{y}|\theta) = \prod_{i=1}^{n} g(y_i|\theta).$$

However, the concept of likelihood applies equally well to samples of dependent observations, so we will not limit our notation to cases of independence.

To define the MLE of θ, Fisher viewed the joint pdf of $\boldsymbol{y}$ as a function of θ,

$$L(\theta|\boldsymbol{y}) \equiv f(\boldsymbol{y}|\theta),$$

which he called the *likelihood function*. The MLE of θ, which we denote by $\hat{\theta}$, is defined as the particular value of θ that maximizes the likelihood function $L(\theta|\boldsymbol{y})$. Heuristically, one can think of $\hat{\theta}$ as the value of θ that is most likely given the data because $\hat{\theta}$ assigns the highest chance to the observations in $\boldsymbol{y}$. Fisher always intended the likelihood $L(\theta|\boldsymbol{y})$ to be interpreted as a measure of *relative* support for different hypotheses (i.e., different values of θ); he never considered the likelihood function to have an absolute scale, such as a probability. This distinction provides an important example of the profound differences between the classical and Bayesian views of inference, as we shall see later (Section 2.4).

The MLE of θ is, by definition, the solution of an optimization problem. In some cases this problem can be solved analytically using calculus, which allows the MLE to be expressed in closed form as a function of $\boldsymbol{y}$. If this is not possible, numerical methods of optimization must be used to compute an approximation of $\hat{\theta}$; however, these calculations can usually be done quite accurately and quickly with modern computers and optimization algorithms.

2.3.1.1 Example: estimating the probability of occurrence (analytically)

Suppose a survey is designed to estimate the average occurrence of an animal species in a region. As part of the survey, the region is divided into a lattice of sample units of uniform size and shape, which we will call 'locations', for simplicity. Now imagine that n of these locations are selected at random and that we are able to determine with certainty whether the species is present ($y = 1$) or absent ($y = 0$) at each location.[1]

On completion of the survey, we have a sample of binary observations $\boldsymbol{y} = (y_1, \ldots, y_n)$. Because the sample locations are selected at random, it seems

---

[1]Observations of animal occurrence are rarely made with absolute certainty when sampling natural populations; however, we assume certainty in this example to keep the model simple.

reasonable to assume that the observations are mutually independent. If we further assume that the probability of occurrence is identical at each location, this implies a rather simple model of the data:

$$y_i \overset{iid}{\sim} \text{Bern}(\psi), \tag{2.3.1}$$

where ψ denotes the probability of occurrence.

We could derive the MLE of ψ based on Eq. (2.3.1) alone; however, we can take a mathematically equivalent approach by considering the implications of Eq. (2.3.1) for a summary of the binary data. Let $v = \sum_{i=1}^{n} y_i$ denote the total number of sample locations where the species is present. This definition allows the information in $\boldsymbol{y}$ to be summarized as the frequency of ones, v, and the frequency of zeros, $n - v$. Given the assumptions in Eq. (2.3.1), the probability of any particular set of observations in $\boldsymbol{y}$ is

$$\psi^v (1 - \psi)^{n-v} \tag{2.3.2}$$

for $v = 0, 1, \ldots, n$. However, to formulate the model in terms of v, we must account for the total number of ways that v ones and $n - v$ zeros could have been observed, which is given by the combinatorial

$$\binom{n}{v} = \frac{n!}{v!(n-v)!}. \tag{2.3.3}$$

Combining Eqs. (2.3.2) and (2.3.3) yields the total probability of observing v ones and $n - v$ zeros independent of the ordering of the binary observations in $\boldsymbol{y}$:

$$f(v|\psi) = \binom{n}{v} \psi^v (1 - \psi)^{n-v}. \tag{2.3.4}$$

The astute reader will recognize that $f(v|\psi)$ is just the pmf of a binomial distribution with index n and parameter ψ (see Table 2.1).

This simple model provides the likelihood of ψ given v (and the sample size n)

$$L(\psi|v) = \psi^v (1 - \psi)^{n-v}, \tag{2.3.5}$$

where the combinatorial term has been omitted because $\binom{n}{v}$ does not involve ψ. (Note that the combinatorial term may be ignored given Fisher's definition of likelihood because $\binom{n}{v}$ does not involve ψ and only contributes a multiplicative constant to the likelihood.) The MLE of ψ is the value of ψ that maximizes

$L(\psi|v)$. Because $L(\psi|v)$ is non-negative for admissible values of ψ, the MLE of ψ also maximizes

$$\log L(\psi|v) = v \log \psi + (n - v) \log(1 - \psi). \qquad (2.3.6)$$

This results stems from the fact that the natural logarithm is a one-to-one, monotone function of its argument. The MLE of ψ is the solution of either of the following equations,

$$\frac{\mathrm{d}L(\psi|z)}{\mathrm{d}\psi} = 0 \quad \text{or} \quad \frac{\mathrm{d}\log L(\psi|z)}{\mathrm{d}\psi} = 0$$

and equals $\hat{\psi} = v/n$. Therefore, the MLE of ψ is equivalent to the sample mean of the binary observations: $\bar{y} = (1/n) \sum_{i=1}^{n} y_i = v/n$.

2.3.1.2 Example: estimating the probability of occurrence (numerically)

In the previous example we were able to derive $\hat{\psi}$ in closed form; however, in many estimation problems the MLE cannot be obtained as the analytic solution of a differential equation (or system of differential equations for models with ≥ 2 parameters). In such problems the MLE must be estimated numerically.

To illustrate, let's consider the previous example and behave as though we could not have determined that $\hat{\psi} = v/n$. Suppose we select a random sample of $n = 5$ locations and observe a species to be present at only $v = 1$ of those locations. How do we compute the MLE of ψ numerically?

One possibility is a brute-force calculation. Because ψ is bounded on the interval $[0, 1]$, we can evaluate the log-likelihood function in Eq. (2.3.6) for an arbitrarily large number of ψ values that span this interval (e.g., $\epsilon, 2\epsilon, \ldots, 1 - \epsilon$, where $\epsilon > 0$ is an arbitrarily small, positive number). Then, we identify $\hat{\psi}$ as the particular value of ψ with the highest log-likelihood (see Figure 2.1). Panel 2.1 contains the **R** code needed to do these calculations and yields $\hat{\psi} = 0.2$, which is assumed to be correct within $\pm 10^{-6}$ $(= \epsilon)$. In fact, the answer is exactly correct, since $v/n = 1/5 = 0.2$.

Another possibility for computing a numerical approximation of $\hat{\psi}$ is to employ an optimization algorithm. For example, **R** contains two procedures, `nlm` and `optim`, which can be used to find the *minima* of functions of arbitrary form. Of course, these procedures can also be used to find maxima by simply changing the sign of the objective function. For example, *maximizing* a log-likelihood function is equivalent to *minimizing* a *negative* log-likelihood function. In addition to defining the function to be minimized, **R**'s optimization algorithms require a starting point. Ideally, the starting point should approximate the solution. Panel 2.2 contains an example of **R** code that uses `optim` to do these calculations and produces an estimate of $\hat{\psi}$ that is very close to the correct answer. Note, however, that **R** also reports a warning message upon fitting the model. During the optimization `optim` apparently

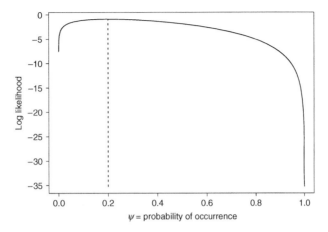

Figure 2.1. Log likelihood for a binomial outcome ($v = 1$ successes in $n = 5$ trials) evaluated over the admissible range of ψ values. Dashed vertical line indicates the MLE.

```
> v=1
> n=5
> eps=1e-6
> psi = seq(eps, 1-eps, by=eps)
> logLike = v*log(psi) + (n-v)*log(1-psi)
> psi[logLike==max(logLike)]
[1] 0.2
```

Panel 2.1. R code for brute-force calculation of MLE of ψ.

attempted to evaluate the function **dbinom** using values of ψ that were outside the interval $[0, 1]$. In this case the message can be ignored, but it's an indication that computational improvements are possible. We will return to this issue later.

2.3.1.3 Example: estimating parameters of normally distributed data

In Section 2.2 we described an example where body mass was measured for each of n randomly selected animals. As a possible model, we assumed that the body masses in the sample were independent and identically distributed as follows: $y_i \overset{iid}{\sim} N(\mu, \sigma^2)$. Here, we illustrate methods for computing the MLE of the parameter vector (μ, σ^2).

Given our modeling assumptions, the joint pdf of the observed body masses $\boldsymbol{y}$ is a product of identical marginal pdfs, $N(y_i|\mu, \sigma^2)$, as noted in Eq. (2.2.1). Therefore, the likelihood function for these data is

$$L(\mu, \sigma^2|\boldsymbol{y}) = (2\pi\sigma^2)^{-n/2} \exp\left(-\frac{1}{2\sigma^2} \sum_{i=1}^{n} (y_i - \mu)^2\right). \qquad (2.3.7)$$

To find the MLE, we need to solve the following set of simultaneous equations:

$$\frac{\partial \log L(\mu, \sigma^2)}{\partial \mu} = 0$$

$$\frac{\partial \log L(\mu, \sigma^2)}{\partial \sigma^2} = 0.$$

It turns out that an analytical solution exists in closed form. The MLE is $(\hat{\mu}, \hat{\sigma}^2) = (\bar{y}, \frac{n-1}{n}s^2)$, where $\bar{y}$ and s^2 denote the sample mean and variance, respectively, of the observed body masses. Note that $\hat{\sigma}^2$ is strictly less than s^2, the usual (unbiased) estimator of σ^2; however, the difference becomes negligible as sample size n increases.

Suppose we could not have found the MLE of (μ, σ^2) in closed form. In this case we need to compute a numerical approximation. We could try the brute-force approach used in the earlier example, but in this example we would need to evaluate the log-likelihood function over 2 dimensions. Furthermore, the parameter space is unbounded $(\mathbb{R} \times \mathbb{R}^+)$, so we would need to restrict evaluations of the log-likelihood to be in the vicinity of the MLE, which we do not know!

It turns out that the brute-force approach is seldom feasible, particularly as the number of model parameters becomes large. Therefore, numerical methods of

---

```
> v=1
> n=5
> negLogLike = function(psi) -dbinom(v, size=n, prob=psi, log=TRUE)
> fit = optim(par=0.5, fn=negLogLike, method='BFGS')
Warning messages: 1: NaNs produced in: dbinom(x, size, prob, log) 2:
NaNs produced in: dbinom(x, size, prob, log)
>
> fit$par
[1] 0.2000002
>
```

Panel 2.2. R code for numerically maximizing the likelihood function to approximate $\hat{\psi}$.

optimization are often used to compute an approximation of the MLE. To illustrate, suppose we observe the following body masses of 10 animals:

$$\boldsymbol{y} = (8.51, 4.03, 8.20, 4.19, 8.72, 6.15, 5.40, 8.66, 7.91, 8.58)$$

which have sample mean $\bar{y} = 7.035$ and sample variance $s^2 = 3.638$. Panel 2.3 contains **R** code for approximating the MLE and yields the estimates $\hat{\mu} = 7.035$ and $\hat{\sigma}^2 = 3.274$, which are the correct answers.

2.3.2 Properties of MLEs

Maximum likelihood estimators have several desirable properties. In this section we describe these properties and illustrate their consequences in inference problems. In particular, we show how MLEs are used in the construction of confidence intervals.

2.3.2.1 Invariance to reparameterization

If $\hat{\theta}$ is the MLE of θ, then for any one-to-one function $\tau(\theta)$, the MLE of $\tau(\theta)$ is $\tau(\hat{\theta})$. This invariance to reparameterization can be extremely helpful in computing

```
> y = c(8.51, 4.03, 8.20, 4.19, 8.72, 6.15, 5.40, 8.66, 7.91, 8.58)
>
> neglogLike = function(param) {
+ mu = param[1]
+ sigma = exp(param[2])
+-sum(dnorm(y,mean=mu,sd=sigma, log=TRUE))
+}
>
> fit = optim(par=c(0,0), fn=neglogLike, method='BFGS')
> fit$par
[1] 7.0350020 0.5930949
>
> exp(fit$par[2])^2
[1] 3.274581
>
```

Panel 2.3. **R** code for numerically maximizing the likelihood function to approximate $(\hat{\mu}, \log \hat{\sigma})$.

MLEs by numerical approximation. For example, let $\gamma = \tau(\theta)$ and suppose we can compute $\hat{\gamma}$ that maximizes $L_1(\gamma|\boldsymbol{y})$ easily; then, by the property of invariance we can deduce that $\hat{\theta} = \tau^{-1}(\hat{\gamma})$ maximizes $L_2(\theta|\boldsymbol{y})$ without actually computing the solution of $\mathrm{d}L_2(\theta|\boldsymbol{y})/\mathrm{d}\theta = 0$, which may involve numerical difficulties.

To illustrate, let's consider the problem of estimating the probability of occurrence ψ that we described in Section 2.3.1.1. The *logit* function, a one-to-one transformation of ψ, is defined as follows:

$$\mathrm{logit}\,(\psi) = \log\left(\frac{\psi}{1-\psi}\right)$$

and provides a mapping from the domain of ψ ($[0,1]$) to the entire real line. Let $\theta = \mathrm{logit}(\psi)$ denote a reparameterization of ψ. We can maximize the likelihood of θ given v to obtain $\hat{\theta}$ and then calculate $\hat{\psi}$ by inverting the transformation as follows:

$$\hat{\psi} = \mathrm{logit}^{-1}(\hat{\theta})$$
$$= 1/(1 + \exp(-\hat{\theta})).$$

We will use this inversion often; therefore, in the remainder of this book we let $\mathrm{expit}(\theta)$ denote the function $\mathrm{logit}^{-1}(\theta)$ as a matter of notational convenience.

Panel 2.4 contains **R** code for computing $\hat{\theta}$ by numerical optimization and for computing $\hat{\psi}$ by inverting the transformation. Notice that the definition of the **R** function `neglogLike` is identical to that used earlier (Panel 2.2) except that we have substituted `theta` for `psi` as the function's argument and `expit(theta)` for `psi` in the body of the function. Therefore, the extra coding required to compute ψ on the logit scale is minimal. Notice also in Panel 2.4 that in maximizing the likelihood function of θ, **R** did not produce the somewhat troubling warning messages that appeared in maximizing the likelihood function of ψ (cf. Panel 2.2). The default behavior of **R**'s optimization functions, `optim` and `nlm`, is to provide an *unconstrained* minimization wherein no constraints are placed on the magnitude of the argument of the function being minimized. In other words, if the function's argument is a vector of p components, their value is assumed to lie anywhere in $\mathbb{R}^p$. In our example the admissible values of θ include the entire real line; in contrast, the admissible values of ψ are confined to a subset of the real line ($[0,1]$).

The lesson learned from this example is that when using *unconstrained* optimization algorithms to maximize a likelihood function of p parameters, one should typically try to formulate the likelihood so that the parameters are defined in $\mathbb{R}^p$. The invariance of MLEs always allows us to back-transform the parameter estimates if that is necessary in the context of the problem.

2.3.2.2 Consistency

Suppose the particular set of modeling assumptions summarized in the joint pdf $f(\boldsymbol{y}|\theta)$ is true, i.e., the approximating model of the data $\boldsymbol{y}$ correctly describes the process that generated the data. Under these conditions, we can prove that $\hat{\theta}$, the MLE of θ, converges to θ as the sample size n increases, which we denote mathematically as follows:

$$\hat{\theta} \to \theta \quad \text{as } n \to \infty.$$

Although the assumptions of an approximating model are unlikely to hold exactly, it is reassuring to know that with enough data, the MLE is guaranteed to provide the 'correct' answer.

```
> v=1
> n=5
> expit = function(x) 1/(1+exp(-x))
>
> neglogLike = function(theta) -dbinom(v, size=n, prob=expit(theta), log=TRUE)
> fit = optim(par=0, fn=neglogLike, method='BFGS')
>
> fit$par
[1] -1.386294
>
> expit(fit$par)
[1] 0.2
```

Panel 2.4. R code for numerically maximizing the likelihood function to estimate $\hat{\theta} = \text{logit}(\hat{\psi})$.

2.3.2.3 Asymptotic normality

As in the previous section, suppose the particular set of modeling assumptions summarized in the joint pdf $f(\boldsymbol{y}|\theta)$ is true. If, in addition, a set of 'regularity conditions' that have to do with technical details[2] are satisfied, we can prove the following limiting behavior of the MLE of θ:

In a hypothetical set of repeated samples with θ fixed and with $n \to \infty$,

$$(\hat{\theta} - \theta) \mid \theta \sim \text{N}(0, [I(\hat{\theta})]^{-1}), \qquad (2.3.8)$$

where $I(\hat{\theta}) = -\frac{\partial^2 \log L(\theta|\boldsymbol{y})}{\partial \theta \partial \theta}\big|_{\theta=\hat{\theta}}$ is called the *observed information*.

---

[2]Such as identifiability of the model's parameters and differentiability of the likelihood function. See page 516 of Casella and Berger (2002) for a complete list of conditions.

If θ is a vector of p parameters, then $I(\hat{\theta})$ is a $p \times p$ matrix called the *observed information matrix*.

According to Eq. (2.3.8), the distribution of the discrepancy, $\hat{\theta}-\theta$, obtained under repeated sampling is approximately normal with mean zero as $n \to \infty$. Therefore, $\hat{\theta}$ is an *asymptotically unbiased* estimator of θ. Similarly, Eq. (2.3.8) implies that the inverse of the observed information provides the estimated *asymptotic variance* (or *asymptotic covariance matrix*) of $\hat{\theta}$.

The practical utility of asymptotic normality is evident in the construction of $100(1 - \alpha)$ percent *confidence intervals* for θ. For example, suppose θ is scalar-valued; then in repeated samples, the *random* interval

$$\hat{\theta} \pm z_{1-\alpha/2}([I(\hat{\theta})]^{-1})^{1/2} \tag{2.3.9}$$

'covers' the fixed value θ $100(1 - \alpha)$ percent of the time, provided n is sufficiently large. Here, $z_{1-\alpha/2}$ denotes the $(1-\alpha/2)$ quantile of a standard normal distribution. Note that Eq. (2.3.9) does *not* imply that any *individual* confidence interval includes θ with probability $1 - \alpha$. This misinterpretation of the role of probability is an all-too-common occurrence in applications of statistics. An individual confidence interval either includes θ or it doesn't. A correct probability statement (or inference) refers to the proportion of confidence intervals that include θ in a hypothetical, infinitely long sequence of repeated samples. In this sense $1 - \alpha$ is the probability (relative frequency) that an interval constructed using Eq. (2.3.9) includes the fixed value θ.

Example: estimating the probability of occurrence
As an illustration, let's compute a 95 percent confidence interval for ψ, the probability of occurrence, that was defined earlier in an example (Section 2.3.1.1). The information is easily derived using calculus:

$$\frac{\mathrm{d}^2 \log L(\psi|v)}{\mathrm{d}\psi^2} = I(\psi) = \frac{n}{\psi(1 - \psi)}.$$

The model has only one parameter ψ; therefore, we simply take the reciprocal of $I(\psi)$ to compute its inverse. Substituting $\hat{\psi}$ for ψ yields the 95 percent confidence interval for ψ:

$$\hat{\psi} \pm 1.96\sqrt{\frac{\hat{\psi}(1 - \hat{\psi})}{n}}.$$

Suppose we had not been able to derive the observed information or to compute its inverse analytically. In this case we would need to compute a numerical approximation of $[I(\hat{\psi})]^{-1}$. Panel 2.5 contains the **R** code for computing the MLE of

ψ and a 95 percent confidence interval having observed only $v = 1$ occupied site in a sample of $n = 5$ sites. As before, we estimate $\hat{\psi} = 0.20$. A numerical approximation of $I(\hat{\psi})$ is computed by adding `hessian=TRUE` to the list of `optim`'s arguments. After rounding, our **R** code yields the following 95 percent confidence interval for ψ: $[-0.15, 0.55]$. This is the correct answer, but it includes negative values of ψ, which don't really make sense because ψ is bounded on $[0, 1]$ by definition.

One solution to this problem is to compute a confidence interval for $\theta = \text{logit}(\psi)$ and then transform the upper and lower confidence limits back to the ψ scale (see Panel 2.6). This approach produces an asymmetrical confidence interval for ψ ($[0.027, 0.691]$), but the interval is properly contained in $[0, 1]$.

Another solution to the problem of nonsensical confidence limits is to use a procedure which produces limits that are invariant to reparameterization. We will describe such procedures in the context of hypothesis testing (see Section 2.5). For now, we simply note that confidence limits computed using these procedures and those computed using Eq. (2.3.9) are asymptotically equivalent. The construction of intervals based on Eq. (2.3.9) is far more popular because the confidence limits are relatively easy to compute. In contrast, the calculation of confidence limits based on alternative procedures is more challenging in many instances.

Before leaving our example of interval estimation for ψ, let's examine the influence of sample size. Suppose we had examined a sample of $n = 50$ randomly selected

```
> v=1
> n=5
> neglogLike = function(psi) -dbinom(v, size=n, prob=psi, log=TRUE)
> fit = optim(par=0.5, fn=neglogLike, method='BFGS', hessian=TRUE)
Warning messages: 1: NaNs produced in: dbinom(x, size, prob, log) 2:
NaNs produced in: dbinom(x, size, prob, log)
>
> fit$par
[1] 0.2000002
>
> psi.mle = fit$par
> psi.se = sqrt(1/fit$hessian)
> zcrit = qnorm(.975)
> c(psi.mle-zcrit*psi.se,  psi.mle+zcrit*psi.se)
[1] -0.1506020  0.5506024
>
```

Panel 2.5. R code for computing a 95 percent confidence interval for ψ.

locations (a tenfold increase in sample size) and had observed the species to be present at $v = 10$ of these locations. Obviously, our estimate of ψ is unchanged because $\hat{\psi} = 10/50 = 0.2$. But how has the uncertainty in our estimate changed? Earlier we showed that $I(\psi) = n/(\psi(1 - \psi))$. Because $\hat{\psi}$ is identical in both samples, we may conclude that the observed information in the sample of $n = 50$ locations is ten times higher than that in the sample of $n = 5$ locations, as shown in Table 2.3. In fact, this is easily illustrated by plotting the log-likelihood functions for each sample (Figure 2.2). Notice that the curvature of the log-likelihood function in the vicinity of the MLE is greater for the larger sample $(n = 50)$. This is consistent with the differences in observed information because $I(\hat{\psi})$ is the negative of the second derivative of $\log L(\psi|v)$ evaluated at $\hat{\psi}$, which essentially measures the curvature of $\log L(\psi|v)$ at the MLE. The log-likelihood function decreases with distance from $\hat{\psi}$ more rapidly in the sample of $n = 50$ locations than in the sample of $n = 5$ locations; therefore, we might expect the estimated precision of $\hat{\psi}$ to be higher in the larger sample. This is exactly the case; the larger sample yields a narrower confidence interval for ψ. Table 2.3 and Figure 2.2 also illustrate the effects of parameterizing the log-likelihood in terms of $\theta = \text{logit}(\psi)$. The increase in observed information associated with the larger sample is the same (a tenfold increase), and this results in a narrower confidence interval. The confidence limits for ψ based on the asymptotic normality of $\hat{\theta}$ are not identical to those based on the asymptotic normality of $\hat{\psi}$, as mentioned earlier; however, the discrepancy between the two confidence intervals is much lower in the larger sample.

```
> v=1
> n=5
> expit = function(x) 1/(1+exp(-x))
>
> neglogLike = function(theta) -dbinom(v, size=n, prob=expit(theta), log=TRUE)
> fit = optim(par=0, fn=neglogLike, method='BFGS', hessian=TRUE)
>
> theta.mle = fit$par
> theta.se = sqrt(1/fit$hessian)
> zcrit = qnorm(.975)
> expit(c(theta.mle-zcrit*theta.se, theta.mle+zcrit*theta.se))
[1] 0.02718309 0.69104557
>
```

Panel 2.6. **R** code for computing a 95 percent confidence interval for ψ by back-transforming the lower and upper limits of $\theta = \text{logit}(\psi)$.

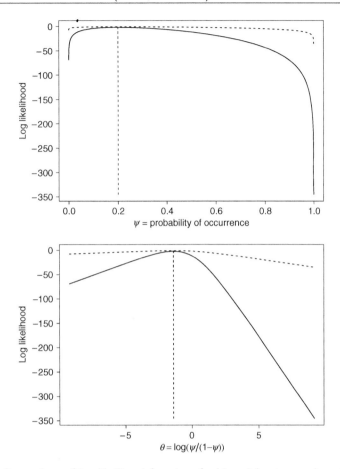

Figure 2.2. Comparison of log-likelihood functions for binomial outcomes based on different sample sizes, $n = 5$ (dashed line) and $n = 50$ (solid line), and different parameterizations, ψ (upper panel) and $\text{logit}(\psi)$ (lower panel). Dashed vertical line indicates the MLE, which is identical in both samples.

Table 2.3. Effects of sample size n and parameterization on 95 percent confidence intervals for ψ.

n	$I(\hat{\psi})$	$[I(\hat{\psi})]^{-1}$	95% C.I. for ψ	$I(\hat{\theta})$	$[I(\hat{\theta})]^{-1}$	95% C.I. for $\psi = \text{expit}(\theta)$
5	31.25	0.0320	[-0.15, 0.55]	0.8	1.250	[0.03, 0.69]
50	312.50	0.0032	[0.09, 0.31]	8.0	0.125	[0.11, 0.33]

Example: estimating parameters of normally distributed data

We conclude this section with a slightly more complicated example to illustrate the construction of confidence intervals when the model contains two or more parameters. In an earlier example, the body weights of n animals were modeled as $y_i \overset{iid}{\sim} \mathrm{N}(\mu, \sigma^2)$, and we determined that the MLE is $(\hat{\mu}, \hat{\sigma}^2) = (\bar{y}, \frac{n-1}{n} s^2)$, where $\bar{y}$ and s^2 denote the sample mean and variance, respectively, of the observed body weights.

Suppose we want to compute a 95 percent confidence interval for the model parameter μ. To do this, we rely on the asymptotic normality of MLEs stated in Eq. (2.3.8). Let $\theta = (\mu, \sigma^2)$. It turns out that $[I(\hat{\theta})]^{-1}$ may be expressed in closed form

$$[I(\hat{\theta})]^{-1} = \begin{bmatrix} \frac{s^2}{n} \cdot \frac{n-1}{n} & 0 \\ 0 & \frac{s^4}{n^2} \cdot \frac{2(n-1)^2}{n} \end{bmatrix}.$$

Therefore, the asymptotic normality of MLEs justifies the following approximation

$$\begin{bmatrix} \hat{\mu} - \mu \\ \hat{\sigma}^2 - \sigma^2 \end{bmatrix} \sim \mathrm{N}\left(\begin{bmatrix} 0 \\ 0 \end{bmatrix}, \begin{bmatrix} \frac{s^2}{n} \cdot \frac{n-1}{n} & 0 \\ 0 & \frac{s^4}{n^2} \cdot \frac{2(n-1)^2}{n} \end{bmatrix} \right)$$

which implies that the 95 percent confidence interval for μ may be computed as follows:

$$\bar{y} \pm 1.96 \sqrt{\frac{s^2}{n} \cdot \frac{n-1}{n}}.$$

Applying this formula to the sample of $n = 10$ body weights yields a confidence interval of $[5.913, 8.157]$.

If we had not been able to derive $[I(\hat{\theta})]^{-1}$ in closed form, we still could have computed it by numerical approximation. For example, Panel 2.7 contains the **R** code for computing the MLE of μ and its 95 percent confidence interval. **R** contains several functions for inverting matrices, including `chol2inv` and `solve`. In Panel 2.7 we use the function `chol` to compute the Cholesky decomposition of $I(\hat{\theta})$ and then `chol2inv` to compute its inverse. This procedure is particularly accurate because $I(\hat{\theta})$ is a positive-definite symmetric matrix (by construction).

2.4 BAYESIAN INFERENCE

In this section we describe the Bayesian approach to model-based inference. To facilitate comparisons with classical inference procedures, we will apply the Bayesian approach to some of the same examples used in Section 2.3.

Let $\boldsymbol{y} = (y_1, \ldots, y_n)$ denote a sample of n observations, and suppose we develop an approximating model of $\boldsymbol{y}$ that contains a (possibly vector-valued) parameter θ. As in classical statistics, the approximating model is a formal expression of the processes that are assumed to have produced the observed data. However, in the Bayesian view the model parameter θ is treated as a random variable and the approximating model is elaborated to include a probability distribution for θ that specifies one's beliefs about the magnitude of θ *prior to having observed the data.* This elaboration of the model is therefore called the *prior distribution.*

In the Bayesian view, computing an inference about θ is fundamentally just a probability calculation that yields the probable magnitude of θ given the assumed

```
> y = c(8.51, 4.03, 8.20, 4.19, 8.72, 6.15, 5.40, 8.66, 7.91, 8.58)
>
> neglogLike = function(param) {
+ mu = param[1]
+ sigma = exp(param[2])
+-sum(dnorm(y,mean=mu,sd=sigma, log=TRUE))
+}
>
> fit = optim(par=c(0,0), fn=neglogLike, method='BFGS', hessian=TRUE)
>
> fit$hessian
              [,1]            [,2]
[1,]   3.053826e+00  -1.251976e-05
[2,]  -1.251976e-05   2.000005e+01
>
> covMat = chol2inv(chol(fit$hessian))
> covMat
              [,1]            [,2]
[1,] 3.274581e-01  2.049843e-07
[2,] 2.049843e-07  4.999987e-02
>
> mu.mle = fit$par[1]
> mu.se = sqrt(covMat[1,1])
> zcrit = qnorm(.975)
>
> c(mu.mle-zcrit*mu.se, mu.mle+zcrit*mu.se)
[1] 5.913433 8.156571
```

Panel 2.7. R code for computing a 95 percent confidence interval for μ.

prior distribution and given the evidence in the data. To accomplish this calculation, the observed data $\boldsymbol{y}$ are assumed to be *fixed* (once the sample has been obtained), and all inferences about θ are made with respect to the fixed observations $\boldsymbol{y}$. Unlike classical statistics, Bayesian inferences do not rely on the idea of hypothetical repeated samples or on the asymptotic properties of estimators of θ. In fact, probability statements (i.e., inferences) about θ are *exact* for any sample size under the Bayesian paradigm.

2.4.1 Bayes' Theorem and the Problem of 'Inverse Probability'

To describe the principles of Bayesian inference in more concrete terms, it's convenient to begin with some definitions. Let's assume, without loss of generality, that the observed data $\boldsymbol{y}$ are modeled as continuous random variables and that $f(\boldsymbol{y}|\theta)$ denotes the joint pdf of $\boldsymbol{y}$ given a model indexed by the parameter θ. In other words, $f(\boldsymbol{y}|\theta)$ is an approximating model of the data. Let $\pi(\theta)$ denote the pdf of an assumed *prior distribution* of θ. Note that $f(\boldsymbol{y}|\theta)$ provides the probability of the data given θ. However, once the data have been collected the value of $\boldsymbol{y}$ is known; therefore, to compute an inference about θ, we really need the probability of θ given the evidence in the data, which we denote by $\pi(\theta|\boldsymbol{y})$.

Historically, the question of how to compute $\pi(\theta|\boldsymbol{y})$ was called the 'problem of inverse probability.' In the 18th century Reverend Thomas Bayes (1763) provided a solution to this problem, showing that $\pi(\theta|\boldsymbol{y})$ can be calculated to update one's prior beliefs (as summarized in $\pi(\theta)$) using the laws of probability[3]:

$$\pi(\theta|\boldsymbol{y}) = \frac{f(\boldsymbol{y}|\theta)\pi(\theta)}{m(\boldsymbol{y})}, \tag{2.4.1}$$

where $m(\boldsymbol{y}) = \int f(\boldsymbol{y}|\theta)\pi(\theta)\,d\theta$ denotes the marginal probability of $\boldsymbol{y}$. Eq. (2.4.1) is known as Bayes' theorem (or Bayes' rule), and $\theta|\boldsymbol{y}$ is called the *posterior distribution* of θ to remind us that $\pi(\theta|\boldsymbol{y})$ summarizes one's beliefs about the magnitude of θ *after having observed the data*. Bayes' theorem provides a coherent, probability-based framework for inference because it specifies how prior beliefs about θ can be converted into posterior beliefs in light of the evidence in the data.

Close ties obviously exist between Bayes' theorem and likelihood-based inference because $f(\boldsymbol{y}|\theta)$ is also the basis of Fisher's likelihood function (Section 2.3.1). However, Fisher was vehemently opposed to the 'theory of inverse probability',

---

[3]Based on the definition of conditional probability, we know $[\theta|\boldsymbol{y}] = [\boldsymbol{y},\theta]/[\boldsymbol{y}]$ and $[\boldsymbol{y}|\theta] = [\boldsymbol{y},\theta]/[\theta]$. Rearranging the second equation yields the joint pdf, $[\boldsymbol{y},\theta] = [\boldsymbol{y}|\theta][\theta]$, which when substituted into the first equation produces Bayes' rule: $[\theta|\boldsymbol{y}] = ([\boldsymbol{y}|\theta][\theta])/[\boldsymbol{y}]$.

as applications of Bayes' theorem were called in his day. Fisher sought inference procedures that did not rely on the specification of a prior distribution, and he deliberately used the term 'likelihood' for $f(\boldsymbol{y}|\theta)$ instead of calling it a probability. Therefore, it is important to remember that although the likelihood function is present in both inference paradigms (i.e., classical and Bayesian), dramatic differences exist in the way that $f(\boldsymbol{y}|\theta)$ is used and interpreted.

2.4.1.1 Example: estimating the probability of occurrence

Let's reconsider the problem introduced in Section 2.3.1.1 of computing inferences about the probability of occurrence ψ. Our approximating model of v, the total number of sample locations where the species is present, is given by the binomial pmf $f(v|\psi)$ given in Eq. (2.3.4). A prior density $\pi(\psi)$ is required to compute inferences about ψ from the posterior density

$$\pi(\psi|v) = \frac{f(v|\psi)\pi(\psi)}{m(v)},$$

where $m(v) = \int_0^1 f(v|\psi)\pi(\psi)\,\mathrm{d}\psi$. It turns out that $\pi(\psi|v)$ can be expressed in closed form if the prior $\pi(\psi) = \mathrm{Be}(\psi|\mathrm{a},\mathrm{b})$ is assumed, where the values of a and b are fixed (by assumption). To be specific, this choice of prior implies that the posterior distribution of ψ is $\mathrm{Be}(\mathrm{a} + v, \mathrm{b} + n - v)$. Thus, the prior and posterior distributions belong to the same class of distributions (in this case, the class of beta distributions). This equivalence, known as *conjugacy*, identifies the beta distribution as the *conjugate prior* for the success parameter of a binomial distribution. We will encounter other examples of conjugacy throughout this book. For now, let's continue with the example.

Suppose we assume prior indifference in the magnitude of ψ. In other words, before observing the data, we assume that all values of ψ are equally probable. This assumption is specified with a $\mathrm{Be}(1,1)$ prior ($\equiv \mathrm{U}(0,1)$ prior) and implies that the posterior distribution of ψ is $\mathrm{Be}(1 + v, 1 + n - v)$. It's worth noting that the mode of this distribution equals v/n, which is equivalent to the MLE of ψ obtained in a classical, likelihood-based analysis. Now suppose a sample of $n = 5$ locations contains only $v = 1$ occupied site; then the $\mathrm{Be}(2,5)$ distribution, illustrated in Figure 2.3, may be used to compute inferences for ψ. For example, the posterior mean and mode of ψ are 0.29 and 0.20, respectively. Furthermore, we can compute the $\alpha/2$ and $1 - \alpha/2$ quantiles of the $\mathrm{Be}(2,5)$ posterior and use these to obtain a $100(1 - \alpha)$ percent *credible interval* for ψ. (Bayesians use the term, 'credible interval', to distinguish it from the frequentist concept of a confidence interval.) For example, the 95 percent credible interval for ψ is $[0.04, 0.64]$.

We use this example to emphasize that a Bayesian credible interval and a frequentist confidence interval have completely different interpretations. The Bayesian

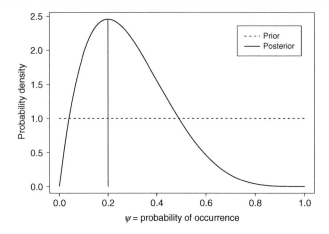

Figure 2.3. Posterior distribution for the probability of occurrence assuming a uniform prior. Vertical line indicates the posterior mode.

credible interval is the result of a probability calculation and reflects our posterior belief in the probable range of ψ values given the evidence in the observed data. Thus, we might choose to summarize the analysis by saying, "the probability that ψ lies in the interval $[0.04, 0.64]$ is 0.95." In contrast, the probability statement associated with a confidence interval corresponds to the proportion of confidence intervals that contain the fixed, but unknown, value of ψ in an infinite sequence of hypothetical, repeated samples (see Section 2.3.2). The frequentist's interval therefore requires considerably more explanation and is far from a direct statement of probability. Unfortunately, the difference in interpretation of credible intervals and confidence intervals is often ignored in practice, much to the consternation of many statisticians.

2.4.2 Pros and Cons of Bayesian Inference

Earlier we mentioned that one of the virtues of Bayesian inference is that probability statements about θ are *exact* for any sample size. This is especially meaningful when one considers that a Bayesian analysis yields the entire posterior pdf of θ, $\pi(\theta|\boldsymbol{y})$, as opposed to a single point estimate of θ. Therefore, in addition to computing summaries of the posterior, such as its mean $\mathrm{E}(\theta|\boldsymbol{y})$ or variance $\mathrm{Var}(\theta|\boldsymbol{y})$, *any function* of θ can be calculated while accounting for all of the posterior uncertainty in θ. The benefits of being able to manage errors in estimation in this way are especially evident in computing inferences for latent parameters of hierarchical models, as we will illustrate in Section 2.6, or in computing predictions that depend on the estimated value of θ.

Specification of the prior distribution may be perceived as a benefit or as a disadvantage of the Bayesian mode of inference. In scientific problems where prior information about θ may exist or can be elicited (say, from expert opinion), Bayes' theorem reveals precisely how such information may be used when computing inferences for θ. In other (or perhaps most) scientific problems, little may be known about the probable magnitude of θ in advance of an experiment or survey. In these cases an *objective* approach would be to use a prior that places equal (or nearly equal) probability on all values of θ. Such priors are often called 'vague' or 'non-informative.' A problem with this approach is that priors are not invariant to transformation of the parameters. In other words a prior that is 'non-informative' for θ can be quite informative for $g(\theta)$, a one-to-one transformation of θ.

One solution to this problem is to develop a prior that is both non-informative *and* invariant to transformation of its parameters. A variety of such 'objective priors', as they are currently called (see Chapter 5 of Ghosh et al. (2006)), have been developed for models with relatively few parameters. Objective priors are often improper (that is, $\int \pi(\theta) \, d\theta = \infty$); therefore, if an objective prior is to be used, the analyst must prove that that the resulting posterior distribution is proper (that is, $\int f(\boldsymbol{y}|\theta)\pi(\theta) \, d\theta < \infty$). Such proofs often require considerable mathematical expertise, particularly for models that contain many parameters.

A second solution to the problem of constructing a non-informative prior is to identify a particular parameterization of the model for which a uniform (or nearly uniform) prior makes sense. Of course, this approach is possible only if we are able to assign scientific relevance and context to the model's parameters. We have found this approach to be useful in the analysis of ecological data, and we use this approach throughout the book.

Specification of the prior distribution can be viewed as the 'price' paid for the exactness of inferences computed using Bayes' theorem. When the sample size is low, the price of an exact inference may be high. As the size of a sample increases, the price of an exact inference declines because the information in the data eventually exceeds the information in the prior. We will return to this tradeoff in the next section, where we describe some asymptotic properties of posteriors.

2.4.3 Asymptotic Properties of the Posterior Distribution

We have noted already that the Bayesian approach to model-based inference has several appealing characteristics. In this section we describe additional features that are associated with computing inferences from large samples.

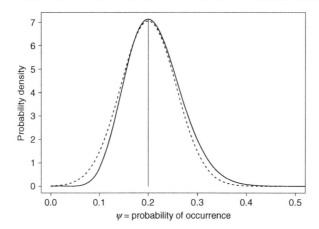

Figure 2.4. A normal approximation (dashed line) of the posterior distribution of the probability of occurrence (solid line). Vertical line indicates the posterior mode.

2.4.3.1 Approximate normality

Let $[\theta|\boldsymbol{y}]$ denote the posterior distribution of θ given an observed set of data $\boldsymbol{y}$. If a set of 'regularity conditions' that have to do with technical details, such as identifiability of the model's parameters and differentiability of the posterior density function $\pi(\theta|\boldsymbol{y})$, are satisfied, we can prove that as sample size $n \to \infty$,

$$(\theta - \hat{\theta}) \mid \boldsymbol{y} \sim \mathrm{N}(0, [I(\hat{\theta})]^{-1}), \qquad (2.4.2)$$

where $\hat{\theta}$ is the posterior mode and $I(\hat{\theta}) = -\frac{\partial^2 \log \pi(\theta|\boldsymbol{y})}{\partial\theta\partial\theta}|_{\theta=\hat{\theta}}$ is called the *generalized observed information* (Ghosh et al., 2006). The practical utility of this limiting behavior is that the posterior distribution of θ can be approximated by a normal distribution $\mathrm{N}(\hat{\theta}, [I(\hat{\theta})]^{-1})$ if n is sufficiently large. In other words, when n is large, we can expect the posterior to become highly concentrated around the posterior mode $\hat{\theta}$.

Example: estimating the probability of occurrence

Recall from Section 2.4.1.1 that the posterior mode for the probability of occurrence was $\hat{\psi} = v/n$ when a $\mathrm{Be}(1,1)$ prior was assumed for ψ. It is easily proved that $[I(\hat{\psi})]^{-1} = \hat{\psi}(1 - \hat{\psi})/n$ given this choice of prior; therefore, according to Eq. (2.4.2) we can expect a $\mathrm{N}(\hat{\psi}, \hat{\psi}(1-\hat{\psi})/n)$ distribution to approximate the true posterior, a $\mathrm{Be}(1+v, 1+n-v)$ distribution, when n is sufficiently large. Figure 2.4 illustrates that the approximation holds very well for a sample of $n = 50$ locations, of which $z = 10$ are occupied.

The asymptotic normality of the posterior (indicated in Eq. (2.4.2)) is an important result because it establishes formally that the relative importance of the prior distribution must decrease with an increase in sample size. To see this, note that $I(\theta)$ is the sum of two components, one due to the likelihood function $f(\boldsymbol{y}|\theta)$ and another due to the prior density $\pi(\theta)$:

$$
\begin{aligned}
I(\theta) &= -\frac{\partial^2 \log \pi(\theta|\boldsymbol{y})}{\partial\theta\partial\theta} \\
&= -\frac{\partial^2 \log f(\boldsymbol{y}|\theta)}{\partial\theta\partial\theta} - \frac{\partial^2 \log \pi(\theta)}{\partial\theta\partial\theta}.
\end{aligned} \tag{2.4.3}
$$

As n increases, only the magnitude of the first term on the right-hand side of Eq. (2.4.3) increases, whereas the magnitude of the second term, which quantifies the information in the prior, remains constant. An important consequence of this result is that we can expect inferences to be insensitive to the choice of prior if we have enough data. On the other hand, if the sample size is relatively small, the prior distribution may be a critical part of the model specification.

2.4.4 Modern Methods of Bayesian Computation

Thus far, we have illustrated the Bayesian approach to inference using a rather simple model (binomial likelihood and conjugate prior), where the posterior density function $\pi(\theta|\boldsymbol{y})$ could be expressed in closed form. However, in many (perhaps most) cases of scientific interest, an approximating model of the data will be more complex, often involving many parameters and multiple levels of parameters. In such cases it is often difficult or impossible to calculate the normalizing constant $m(\boldsymbol{y})$ accurately because the calculation requires a p-dimensional integration if the model contains p distinct parameters. Therefore, the posterior density is often known only up to a constant of proportionality

$$
\pi(\theta|\boldsymbol{y}) \propto f(\boldsymbol{y}|\theta)\pi(\theta).
$$

This computational impediment is one the primary reasons why the Bayesian approach to inference was not widely used prior to the late 20th century. Opposition by frequentists, many of whom strongly advocated classical inference procedures, is another reason. In 1774, Pierre Simon Laplace developed a method for computing a large-sample approximation of the normalizing constant $m(\boldsymbol{y})$, but this procedure is applicable only in cases where the unnormalized posterior density is a smooth function of θ with a sharp maximum at the posterior mode (Laplace, 1986). Refinements of Laplace's method have been developed (Ghosh et al., 2006,

Section 4.3.2), but these refinements, as with Laplace's method, lack generality in their range of application.

A recent upsurge in the use of Bayesian inference procedures can be attributed to the widespread availability of fast computers and to the development of efficient algorithms, known collectively as *Markov chain Monte Carlo* (MCMC) samplers. These include the Gibbs sampler, the Metropolis–Hastings algorithm, and others (Robert and Casella, 2004). The basic idea behind these algorithms is to compute an arbitrarily large sample from the posterior distribution $\theta|\boldsymbol{y}$ without actually computing its normalizing constant $m(\boldsymbol{y})$. Given an arbitrarily large sample of the posterior, the posterior density $\pi(\theta|\boldsymbol{y})$ can be approximated quite accurately (say, using a histogram or a kernel-density smoother). In addition, any function of the posterior, such as the marginal mean, median, or standard deviation for an individual component of θ, can be computed quite easily without actually evaluating the integrals implied in the calculation.

2.4.4.1 Gibbs sampling

One of the most widely used algorithms for sampling posterior distributions is called the *Gibbs sampler*. This algorithm is a special case of the *Metropolis–Hastings* algorithm, and the two are often used together to produce efficient hybrid algorithms that are relatively easy to implement.

The basic idea behind Gibbs sampling (and other MCMC algorithms) is to produce a random sample from the *joint* posterior distribution of ≥ 2 parameters by drawing random samples from a sequence of *full conditional* posterior distributions. The motivation for this idea is that while it may be difficult or impossible to draw a sample directly from the joint posterior, drawing a sample from the full conditional distribution of each parameter is often a relatively simple calculation.

Let's illustrate the Gibbs sampler for a model that includes 3 parameters: θ_1, θ_2, and θ_3. Given a set of data $\boldsymbol{y}$, we wish to compute an arbitrarily large sample from the joint posterior $[\theta_1, \theta_2, \theta_3|\boldsymbol{y}]$. We assume that the full-conditional distributions

$$[\theta_1|\theta_2, \theta_3, \boldsymbol{y}]$$
$$[\theta_2|\theta_1, \theta_3, \boldsymbol{y}]$$
$$[\theta_3|\theta_1, \theta_2, \boldsymbol{y}]$$

of the model's parameters are relatively easy to sample. To begin the Gibbs sampler, we assign an arbitrary set of initial values to each parameter, i.e., $\theta_1 = \theta_1^{(0)}$, $\theta_2 = \theta_2^{(0)}$, $\theta_3 = \theta_3^{(0)}$, where the superscript in parentheses denotes the order in the sequence of random draws. The Gibbs sampling algorithm proceeds as follows:

Step 1 Draw $[\theta_1^{(1)} \sim \theta_1|\theta_2^{(0)}, \theta_3^{(0)}, \boldsymbol{y}]$.

Step 2 Draw $\theta_2^{(1)} \sim [\theta_2 | \theta_1^{(1)}, \theta_3^{(0)}, \boldsymbol{y}]$.

Step 3 Draw $\theta_3^{(1)} \sim [\theta_3 | \theta_1^{(1)}, \theta_2^{(1)}, \boldsymbol{y}]$.

This completes one iteration of the Gibbs sampler and generates a new set of parameter values, say $\boldsymbol{\theta} = \boldsymbol{\theta}^{(1)}$ where $\boldsymbol{\theta} = (\theta_1, \theta_2, \theta_3)$. Steps 1 to 3 are then repeated using the values of $\boldsymbol{\theta}$ from the previous iteration to obtain $\boldsymbol{\theta}^{(2)}, \boldsymbol{\theta}^{(3)}, \ldots$. Typically several thousand iterations of the Gibbs sampler will be required to obtain an accurate sample from the joint posterior $\boldsymbol{\theta} | \boldsymbol{y}$.

The Gibbs sampling algorithm produces a Markov chain, $\boldsymbol{\theta}^{(0)}, \boldsymbol{\theta}^{(1)}, \boldsymbol{\theta}^{(2)}, \boldsymbol{\theta}^{(3)}, \ldots$, whose stationary distribution is equivalent to the joint posterior $\boldsymbol{\theta} | \boldsymbol{y}$, provided a set of technical, regularity conditions are satisfied. Consequently, to obtain a sample from $\boldsymbol{\theta} | \boldsymbol{y}$, the beginning of the Markov chain, which is often called the *burn-in*, is typically discarded. To obtain a random sample of approximately independent draws from $\boldsymbol{\theta} | \boldsymbol{y}$, the remainder of the Markov chain must be subsampled (say, choosing every kth draw where $k > 1$) because successive draws $\boldsymbol{\theta}^{(t)}$ and $\boldsymbol{\theta}^{(t+1)}$ are not independent. Alternatively, one may retain only the last draw of the Markov chain and repeat the entire process n times (to obtain a posterior sample of size n) by using n different starting values for $\boldsymbol{\theta}^{(0)}$. This alternative procedure, though originally proposed by Gelfand and Smith (1990), is seldom used or even necessary. An extensive literature is devoted to assessing the convergence of Markov chains to stationarity. Because proofs of convergence are difficult to establish for complex models, a variety of diagnostics have been developed to assess convergence empirically by subsampling a few independently initialized Markov chains (Robert and Casella, 2004, Chapter 12). Contemporary Bayesian analyses routinely use such diagnostics to assess whether an MCMC sampler has been run long enough to produce an accurate sample from the joint posterior.

2.4.4.2 Example: estimating survival of larval fishes

We extend an example described by Arnold (1993) to illustrate the Gibbs sampler. Suppose an experiment is conducted to estimate the daily probability of survival for a single species of larval fish. In the experiment larval fishes are added to each of $n = 5$ replicate containers by pouring the contents (i.e., water and fishes) of a common source into each container. In practice, the contents of the source are carefully mixed, and equal volumes are added to each of the 5 replicate containers. This procedure is used because handling of individual fishes is thought to reduce their survival.

An important aspect of the experiment is that the number of fishes added to each replicate container is not known precisely, and only the survivors in each replicate can be enumerated. Fortunately, the number of fishes in the source container is known to be 250, so we may assume that the average number of fishes initially added to each replicate container is 50 (=250/5). At the end of a 24-hour incubation period, the numbers of surviving fishes in the 5 replicate containers are $\boldsymbol{y} = (15, 11, 12, 5, 12)$.

To formulate a model of these counts, we define a latent parameter N_i for the number of fishes initially added to the ith replicate container; then we assume $y_i | N_i, \phi \sim \text{Bin}(N_i, \phi)$, where ϕ denotes the daily probability of survival of each fish. Given the design of the experiment, it is reasonable to assume $N_i \sim \text{Po}(\lambda)$, where $\lambda = 50$ denotes the average number of fishes initially added to each replicate container. These modeling assumptions imply that the marginal pmf of the observed counts can be expressed in closed form because

$$[y_i | \phi] = \sum_{N_i = y_i}^{\infty} \text{Bin}(y_i | N_i, \phi) \, \text{Po}(N_i | \lambda)$$
$$= \text{Po}(y_i | \lambda \phi)$$

(proof omitted). To complete the model for a Bayesian analysis, we assume a non-informative $\text{Be}(a, b)$ prior for ϕ (wherein $a = b = 1$) and independence among replicate observations. These assumptions yield the following posterior density function

$$[\phi | \boldsymbol{y}] = \frac{\left(\prod_{i=1}^{n} [y_i | \phi] \right) \text{Be}(\phi | a, b)}{m(\boldsymbol{y})}, \tag{2.4.4}$$

where $m(\boldsymbol{y})$ is the normalizing constant. After a bit of algebra, it can be shown that the posterior pdf in Eq. (2.4.4) is equivalent to

$$[\phi | \boldsymbol{y}] = c^{-1} \phi^{a - 1 + \sum_{i=1}^{n} y_i} (1 - \phi)^{b-1} \exp(-n\lambda\phi), \tag{2.4.5}$$

where

$$c = \int_0^1 \phi^{a - 1 + \sum_{i=1}^{n} y_i} (1 - \phi)^{b-1} \exp(-n\lambda\phi) \, d\phi \tag{2.4.6}$$

is the constant of integration needed to make $[\phi | \boldsymbol{y}]$ a proper probability density function. However, the definite integral in Eq. (2.4.6) cannot be evaluated in closed form; therefore, c must be computed by numerical integration.

In this example c can be computed quite accurately because the integral is only one-dimensional. However, as an alternative, let's examine how Gibbs sampling may be used to compute a sample from the marginal posterior of $\phi | \boldsymbol{y}$. Instead of eliminating $\boldsymbol{N} = (N_1, \dots, N_n)$ from the model, we will use the Gibbs algorithm to compute a sample from the joint posterior distribution of $\boldsymbol{N}$ and ϕ; then we simply ignore the simulated values of $\boldsymbol{N}$ to obtain a sample from $\phi | \boldsymbol{y}$. In this

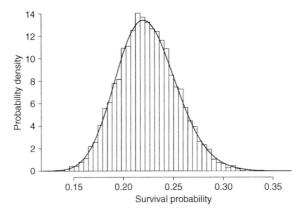

Figure 2.5. Posterior distribution of survival probability ϕ (solid line) and an approximation of the distribution obtained by Gibbs sampling (histogram).

way, the marginal posterior distribution of ϕ is obtained implicitly without actually integrating the joint posterior density $[\boldsymbol{N}, \phi | \boldsymbol{y}]$.

The full conditional distributions needed to compute a Gibbs sample of the joint posterior $[\boldsymbol{N}, \phi | \boldsymbol{y}]$ are given by

$$\phi | \boldsymbol{N}, \boldsymbol{y} \sim \mathrm{Be}\left(a + \sum_{i=1}^{n} y_i, b + \sum_{i=1}^{n} N_i - y_i\right)$$
$$N_i | \phi, y_i \sim y_i + \mathrm{Po}(\lambda(1 - \phi)) \quad i = 1, \dots, n$$

both of which are standard distributions that are easy to sample. In fact, we computed a sample of 10000 random draws from the joint posterior by running the Gibbs sampler for 100000 iterations and then retaining every fifth draw of the last 50000 iterations. A histogram of the simulated values of ϕ provides an excellent approximation of the posterior density computed using Eq. (2.4.5) (Figure 2.5).

Now suppose the full conditional distributions had been more difficult to sample or that we simply did not want to bother with developing an efficient algorithm. In either case we might choose to use the **WinBUGS** software, which provides an implementation of the Gibbs sampler that does not require the user to specify a set of full conditional distributions. In fact, all that **WinBUGS** requires is a statement of the distributional assumptions used in the model. For example, the section of Panel 2.8 labeled `native WinBUGS code` is all that **WinBUGS** needs to compute the joint posterior of $\boldsymbol{N}$ and ϕ in our example. We prefer to access **WinBUGS** remotely while working in **R**. The remaining sections of Panel 2.8 provide **R** code for assigning values to $\boldsymbol{y}$ and to the prior parameters and for using the **R** library, **R2WinBUGS** (Sturtz et al., 2005), to direct the calculations. We will use **WinBUGS** extensively throughout the book to illustrate how Bayesian methods may be used to fit models of varying complexity.

2.5 HYPOTHESIS TESTING

The development of formal rules for testing scientific hypotheses has a long history in classical statistics. The application of these rules is especially useful in making evidentiary conclusions (inferences) from *designed experiments*, where the scientist exercises some degree of control over the relevant sources of uncertainty in the observed outcomes. In a well-designed experiment, hypothesis testing can be used to establish *causal relationships* between experimental outcomes and the systematic sources of variation in those outcomes that are manipulated as part of the design.

Hypothesis testing also can be applied in *observational studies* (surveys), where the scientist does *not* have direct control over the sources of uncertainty being tested. In such studies hypothesis testing may be used to assess whether estimated levels of association (or correlation) between one or more observable outcomes are 'statistically significant' (i.e., are unlikely to have occurred by chance given a particular significance level α of the test). However, in observational studies hypothesis testing cannot be used to determine whether a significant association between outcomes is the result of a coincidence of events or of an underlying causal relationship. Therefore, it can be argued that hypothesis testing is more useful scientifically in the analysis of designed experiments.

We find hypothesis testing to be useful because it provides a general context for the description of some important inference problems, including model selection, construction of confidence intervals, and assessment of model adequacy. We describe and illustrate these topics in the following subsections.

2.5.1 Model Selection

In our model-based approach to statistical inference, the classical problem of testing a scientific hypothesis is equivalent to the problem of selecting between two *nested* models of the data. By nested, we mean that the parameters of one model are a restricted set of the parameters of the other model. To illustrate, suppose we are interested in selecting between two linear regression models, one which simply contains an intercept parameter α and another which contains both intercept and slope parameters, α and β, respectively. By defining the parameter vector $\theta = (\alpha, \beta)$, we can specify the restricted model as $\theta = (\alpha, 0)$ and the full model as

```
--------------data------------------
> y=c(15,11,12,5,12)
> lambda=50
> a=1
> b=1
--------------arguments for R2WinBUGS--------------
> data = list(n=length(y), y=y, lambda=lambda, a=a, b=b)
> params = list('phi','N')
>
> inits = function() {
+   phi = rbeta(1,a,b)
+   N = rpois(length(y),lambda)
+   list(phi=phi, N=N)
+   }
>
--------------native WinBUGS code------------
> sink('MarginalDensity.txt')
> cat('
 model {
  phi ~ dbeta(a,b)
  for (i in 1:n) {
     N[i] ~ dpois(lambda)
     y[i] ~ dbin(phi, N[i])
  }
  }
   ', fill=TRUE)
> sink()
>
--------------call bugs() to fit model----------------
> library(R2WinBUGS)
> fit = bugs(data, inits, params,
  model.file='MarginalDensity.txt',
  debug=F, n.chains=1, n.iter=100000, n.burnin=50000, n.thin=5)
>
> phi = fit$sims.matrix[,'phi']
>
> summary(phi)
  Min. 1st Qu.  Median    Mean 3rd Qu.    Max.
 0.1198  0.2033  0.2227  0.2242  0.2435  0.3541
```

Panel 2.8. **R** and **WinBUGS** code for sampling the posterior distribution of survival probability ϕ.

$\theta = (\alpha, \beta)$. Therefore, a decision to select the full model over the restricted model is equivalent to rejecting the null hypothesis that $\beta = 0$ in favor of the alternative hypothesis that $\beta \neq 0$, given that α is present in both models.

The connection between model selection and hypothesis testing can be specified quite generally. Let H_0 and H_1 denote two complementary hypotheses which represent the *null* and *alternative* hypotheses of a testing problem. In addition, assume that both of these hypotheses can be specified in terms of a (possibly vector-valued) parameter θ, which lies in the parameter space Θ. Given these definitions, the null and alternative hypotheses can be represented as follows:

$$H_0 : \theta \in \Theta_0$$
$$H_1 : \theta \in \Theta_0^c,$$

where Θ_0 is a subset of the parameter space Θ and Θ_0^c is the complement of Θ_0. The model selection problem is basically equivalent to a two-sided hypothesis test wherein

$$H_0 : \theta = \theta_0$$
$$H_1 : \theta \neq \theta_0.$$

Thus, if we reject H_0, we accept the more complex model for which $\theta \neq \theta_0$; otherwise, we accept the simpler model for which $\theta = \theta_0$.

In classical statistics the decision to accept or reject H_0 is based on the asymptotic distribution of a test statistic. Although a variety of test statistics have been developed, we describe only two, the Wald statistic and the likelihood-ratio statistic, because they are commonly used in likelihood-based inference.

2.5.1.1 Wald test

The Wald test statistic is derived from the asymptotic normality of MLEs. Recall from Eq. (2.3.8) that the distribution of the discrepancy $\hat{\theta} - \theta$ is asymptotically normal

$$(\hat{\theta} - \theta) \mid \theta \sim \mathrm{N}(0, [I(\hat{\theta})]^{-1}).$$

Suppose we have a single-parameter model and want to test $H_0 : \theta = \theta_0$. Under the assumptions of the null model, the asymptotic normality of $\hat{\theta}$ implies

$$[I(\hat{\theta})]^{1/2}(\hat{\theta} - \theta_0) \mid \theta_0 \sim \mathrm{N}(0, 1)$$

or, equivalently,

$$\frac{\hat{\theta} - \theta_0}{\mathrm{SE}(\hat{\theta})} \sim \mathrm{N}(0, 1), \tag{2.5.1}$$

where $\text{SE}(\hat{\theta}) = [I(\hat{\theta})]^{-1/2}$ is the asymptotic standard error of $\hat{\theta}$ computed by fitting the alternative model H_1. The left-hand side of Eq. (2.5.1) is called the Wald test statistic and is often denoted by z. Based on Eq. (2.5.1), we reject H_0 when $|z| > z_{1-\alpha/2}$, where $z_{1-\alpha/2}$ denotes the $(1 - \alpha/2)$ quantile of a standard normal distribution.

The distribution of the square of a standard normal random variable is a chi-squared distribution with 1 degree of freedom, which we denote by $\chi^2(1)$; therefore, an equivalent test of H_0 is to compare z^2 to the $(1 - \alpha)$ quantile of a $\chi^2(1)$ distribution. We mention this only to provide a conceptual link to tests of hypotheses that involve multiple parameters. For example, if several parameters are held fixed under the assumptions of H_0, the multi-parameter version of the Wald test statistic and its asymptotic distribution are

$$(\hat{\theta} - \theta_0)' I(\hat{\theta})(\hat{\theta} - \theta_0) \sim \chi^2(\nu),$$

where ν denotes the rank of $I(\hat{\theta})$ (i.e., the number of parameters to be estimated under the alternative model H_1).

2.5.1.2 Likelihood ratio test

The likelihood ratio test is rooted in the notion that the likelihood function $L(\theta|\boldsymbol{y})$ provides a measure of *relative* support for different values of the parameter θ. Therefore, in the model selection problem the ratio

$$\Lambda = \frac{L(\hat{\theta}_0|\boldsymbol{y})}{L(\hat{\theta}|\boldsymbol{y})} \tag{2.5.2}$$

provides the ratio of likelihoods obtained by computing the MLE of θ_0 (the parameters of the model associated with H_0) and the MLE of θ (the parameters of the model associated with H_1). Because θ_0 is a restricted version of θ, $L(\hat{\theta}|\boldsymbol{y}) > L(\hat{\theta}_0|\boldsymbol{y})$ (by definition), the likelihood ratio must be a fraction (i.e., $0 < \Lambda < 1$). Thus, lower values of Λ lend greater support to H_1.

Under the assumptions of the null model H_0, the asymptotic distribution of the statistic $-2 \log \Lambda$ is chi-squared with ν degrees of freedom

$$-2 \log(\Lambda) \sim \chi^2(\nu), \tag{2.5.3}$$

where ν equals the difference in the number of free parameters to be estimated under H_0 and H_1. The left-hand side of Eq. (2.5.3), which is called the *likelihood ratio statistic*, is strictly positive. In practice, we reject H_0 for values of $-2 \log \Lambda$ that exceed $\chi^2_{1-\alpha}(\nu)$, the $(1 - \alpha)$ quantile of a chi-squared distribution with ν degrees of freedom.

The likelihood ratio test can be used to evaluate the *goodness of fit* of a model of counts provided the sample is sufficiently large. In this context H_1 corresponds to a 'saturated' model in which the number of parameters equals the sample size n. We cannot learn anything new from a saturated model because its parameters essentially amount to a one-to-one transformation of the counts $\boldsymbol{y}$; however, a likelihood ratio comparison between the saturated model and an approximating model H_0 *can* be used to assess the goodness of fit of H_0. For example, suppose the approximating model contains k free parameters to be estimated; then, the value of the likelihood ratio statistic $-2 \log \Lambda$ can be compared to $\chi^2_{1-\alpha}(n-k)$ to determine whether H_0 is accepted at the α significance level. If H_0 is accepted, we may conclude that the approximating model provides a satisfactory fit to the data. In the context of this test the likelihood ratio statistic provides a measure of discrepancy between the counts in $\boldsymbol{y}$ and the approximating model's estimate of $\boldsymbol{y}$; consequently, $-2 \log \Lambda$ is often called the *deviance* test statistic, or simply the deviance, in this setting. We will see many applications of the deviance test statistic in later chapters.

2.5.1.3 Example: Mortality of moths exposed to cypermethrin

We illustrate model selection and hypothesis testing using data observed in a dose-response experiment involving adults of the tobacco budworm (*Heliothis virescens*), a moth species whose larvae are responsible for damage to cotton crops in the United States and Central and South America (Collett, 1991, Example 3.7). In the experiment, batches of 20 moths of each sex were exposed to a pesticide called cypermethrin for a period of 72 hours, beginning two days after the adults had emerged from pupation. Both sexes were exposed to the same range of pesticide doses: 1, 2, 4, 8, 16, and 32 μg cypermethrin. At the end of the experiment the number of moths in each batch that were either knocked down (movement of moth was uncoordinated) or dead (moth was unable to move and was unresponsive to a poke from a blunt instrument) was recorded.

The experiment was designed to test whether males and females suffered the same mortality when exposed to identical doses of cypermethrin. The results are shown in Figure 2.6 where the empirical logit of the proportion of moths that died in each batch is plotted against $\log_2$(dose), which linearizes the exponential range of doses. The *empirical logit*, which is defined as follows

$$\log \left(\frac{y_i + 0.5}{N - y_i + 0.5} \right)$$

(wherein y_i ($i = 1, \ldots, 12$) denotes the number of deaths observed in the ith batch of $N = 20$ moths per batch), is the least biased estimator of the true logit of the proportion of deaths per batch (Agresti, 2002). We use the empirical logit because it allows outcomes of 100 percent mortality ($y = N$) or no mortality ($y = 0$) to be plotted on the logit scale along with the other outcomes of the experiment.

The empirical logits of mortality appear to increase linearly with $\log_2$(dose) for both sexes (Figure 2.6); therefore, we consider logistic regression models as a reasonable set of candidates for finding an approximating model of the data. Let x_i denote the $\log_2$(dose) of cypermethrin administered to the ith batch of moths that contained either males ($z_i = 1$) or females ($z_i = 0$). A logistic-regression model containing 3 parameters is

$$y_i|N, p_i \sim \text{Bin}(N, p_i)$$
$$\text{logit}(p_i) = \alpha + \beta x_i + \gamma z_i,$$

where α is the intercept, β is the effect of cypermethrin and γ is the effect of sex.

Let $\theta = (\alpha, \beta, \gamma)$ denote a vector of parameters. The relevant hypotheses to be examined in the experiment are

$$H_0 : \theta = (\alpha, \beta, 0)$$
$$H_1 : \theta = (\alpha, \beta, \gamma),$$

where $\gamma \neq 0$ in H_1. In other words, the test of H_0 amounts to selecting between two models of the data: the null model, wherein only α and β are estimated, and the alternative model, wherein all 3 parameters are estimated.

To conduct a Wald test of H_0, we need only fit the alternative model, which yields the MLE $\hat{\theta} = (-3.47, 1.06, 1.10)$. We obtain $\text{SE}(\hat{\gamma}) = 0.356$ from the inverse of the observed information matrix, and we compute a Wald test statistic of $z = (\hat{\gamma} - 0)/\text{SE}(\hat{\gamma}) = 3.093$. Because $|z| > 1.96$, we reject H_0 at the 0.05 significance level and select the alternative model of the data in favor of the null model.

To conduct a likelihood ratio test of H_0, we must compute MLEs for the parameters of both null and alternative models. Using these estimates, we obtain $\log L(\hat{\alpha}, \hat{\beta}|\boldsymbol{y}) = -23.547$ for the null model and $\log L(\hat{\alpha}, \hat{\beta}, \hat{\gamma}|\boldsymbol{y}) = -18.434$ for the alternative model. Therefore, the likelihood ratio statistic is $-2\log(\Lambda) = -2\{\log L(\hat{\alpha}, \hat{\beta}|\boldsymbol{y}) - \log L(\hat{\alpha}, \hat{\beta}, \hat{\gamma}|\boldsymbol{y})\} = 10.227$. The number of parameters estimated under the null and alternative models differ by $\nu = 3 - 2 = 1$; therefore, to test H_0 we compare the value of the likelihood ratio statistic to $\chi^2_{0.95}(1) = 3.84$. Since $10.227 > 3.84$, we reject H_0 at the 0.05 significance level and select the alternative model of the data in favor of the null model.

The null hypothesis is rejected regardless of whether we use the Wald test or the likelihood ratio test; therefore, we may conclude that the difference in mortality of male and female moths exposed to the same dose of cypermethrin in the experiment is statistically significant, a result which certainly appears to be supported by the data in Figure 2.6.

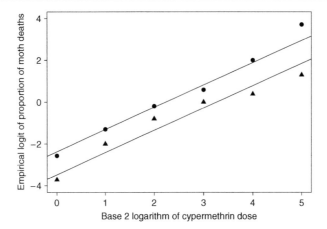

Figure 2.6. Mortality of male (circle) and female (triangle) moths exposed to various doses of cypermethrin. Lines indicate the fit of a logistic regression model with dose ($\log_2$ scale) and sex as predictors.

We may also use a likelihood ratio test to assess the goodness of fit of the model that we have selected as a basis for inference. In this test the parameter estimates of the alternative ('saturated') model H_1 correspond to the observed proportions of moths that died in the 12 experimental batches, i.e., $\hat{\theta} = (\hat{p}_1, \hat{p}_2, \ldots, \hat{p}_{12}) = (y_1/N, y_2/N, \ldots, y_{12}/N)$. For this model $\log L(\hat{p}_1, \ldots, \hat{p}_{12}|\boldsymbol{y}) = -15.055$; therefore the likelihood ratio statistic for testing goodness of fit is $-2\{\log L(\hat{\alpha}, \hat{\beta}, \hat{\gamma}|\boldsymbol{y}) - \log L(\hat{p}_1, \ldots, \hat{p}_{12}|\boldsymbol{y}) = 6.757\}$. To test H_0, we compare this value to $\chi^2_{0.95}(9) = 16.92$. Since $6.757 \leq 16.92$, we accept H_0 and conclude that the model with parameters (α, β, γ) cannot be rejected for lack of fit at the 0.05 significance level.

2.5.2 Inverting Tests to Estimate Confidence Intervals

In many studies a formal test of the statistical significance of an effect (e.g., a treatment effect in a designed experiment) is less important scientifically than an estimate of the magnitude of the effect. This is particularly true in observational studies where the estimated level of association between one or more observable outcomes is of primary scientific interest. In these cases the main inference problem is to estimate a parameter and to provide a probabilistic description of the uncertainty in the estimate.

In classical statistics the solution to this problem involves the construction of a confidence interval for the parameter. However, a variety of procedures have been developed for constructing confidence intervals. For example, in Section 2.3.2

we described how the asymptotic normality of MLEs provides a $(1 - \alpha)$ percent confidence interval of the form

$$\hat{\theta} \pm z_{1-\alpha/2} \, \text{SE}(\hat{\theta}) \tag{2.5.4}$$

for a scalar-valued parameter θ wherein $\text{SE}(\hat{\theta}) = [I(\hat{\theta})]^{-1/2}$ (cf. Eq. (2.3.9)). It turns out that this confidence interval may be constructed by inverting the Wald test described in Section 2.5.1. To see this, note that the null hypothesis $H_0 : \theta = \theta_0$ is accepted if

$$\left| \frac{\hat{\theta} - \theta_0}{\text{SE}(\hat{\theta})} \right| \leq z_{1-\alpha/2}.$$

Simple algebra can be used to prove that the range of θ_0 values that satisfy this inequality are bounded by the confidence limits given in Eq. (2.5.4); consequently, there is a direct correspondence between the acceptance region of the null hypothesis (that is, the values of θ for which $H_0 : \theta = \theta_0$ is accepted) and the $(1 - \alpha)$ percent confidence interval for θ.

Confidence intervals also may be constructed by inverting the likelihood ratio test described in Section 2.5.1. To see this, note that the null hypothesis $H_0 : \theta = \theta_0$ is accepted if

$$-2\{\log L(\hat{\theta}_0|\boldsymbol{y}) - \log L(\hat{\theta}|\boldsymbol{y})\} \leq \chi_{1-\alpha}^2(\nu). \tag{2.5.5}$$

Therefore, the range of the fixed parameters in $\hat{\theta}_0$ that satisfy this inequality provide a $(1-\alpha)\%$ confidence region for those parameters. Such confidence regions are often more difficult to calculate than those based on inverting the Wald test because the free parameters in θ_0 must be estimated by maximizing $L(\theta_0|\boldsymbol{y})$ for each value of the parameters in θ_0 that are fixed. For this reason $L(\theta_0|\boldsymbol{y})$ is called the *profile likelihood* function of θ_0. If a confidence interval is required for only a single parameter, a numerical root-finding procedure may be used to calculate the values of that parameter that satisfy Eq. (2.5.5).

In sufficiently large samples, confidence intervals computed by inverting the Wald test or the likelihood ratio test will be nearly identical. An obvious advantage of intervals based on the Wald test is that they are easy to compute. However, in small samples, such intervals can produce undesirable results, as we observed in Table 2.3, where an interval for the probability of occurrence ψ includes negative values. Intervals based on inverting the likelihood ratio test can be more difficult to calculate, but an advantage of these intervals is that they are *invariant to transformation* of a model's parameters. Therefore, regardless of whether we parameterize the probability of occurrence in terms of ψ or logit(ψ), the confidence intervals for ψ will be the same. Table 2.4 illustrates the small-sample benefits of computing intervals for ψ by inverting the likelihood ratio test.

Table 2.4. Comparison of 95 percent confidence intervals for ψ based on inverting a Wald test and a likelihood ratio test. All intervals have the same MLE ($\hat{\psi} = 0.2$), but sample size n differs.

n	Wald test	Likelihood ratio test
5	[-0.15, 0.55]	[0.01, 0.63]
50	[0.09, 0.31]	[0.11, 0.32]

2.5.2.1 Example: Mortality of moths exposed to cypermethrin

As an additional comparison of procedures for constructing confidence intervals, we compute 95 percent confidence intervals for γ, the parameter of the logistic regression model that denotes the effect of sex on mortality of moths exposed to cypermethrin (see Section 2.5.1.3). Recall that the MLE for the logit-scale effect of sex is $\hat{\gamma} = 1.10$, and an estimate of its uncertainty is $\text{SE}(\hat{\gamma}) = 0.356$. Therefore, the 95 percent confidence interval for γ based on inverting the Wald test is $[0.40, 1.80]$ ($= 1.10 \pm 1.96 * 0.356$).

To compute a 95 percent confidence interval for γ based on inverting the likelihood ratio test, we must find the values of γ_0 that satisfy the following inequality

$$-2\{\log L(\hat{\alpha}, \hat{\beta}, \gamma_0 | \boldsymbol{y}) - \log L(\hat{\alpha}, \hat{\beta}, \hat{\gamma} | \boldsymbol{y})\} \leq 3.84. \qquad (2.5.6)$$

Using a numerical root-finding procedure, we find that $[0.42, 1.82]$ is the 95 percent confidence interval for γ. In Figure 2.7 we plot the left-hand side of Eq. (2.5.6) against fixed values of γ_0 to show that we have calculated this interval correctly.

2.5.3 A Bayesian Approach to Model Selection

In this section we develop a Bayesian approach to the problem of selecting between two nested models, which was described in Section 2.5.1. Recall that H_0 and H_1 denote two complementary hypotheses that can be specified in terms of a (possibly vector-valued) parameter θ:

$$H_0 : \theta \in \Theta_0$$
$$H_1 : \theta \in \Theta_0^c,$$

where Θ_0 is a subset of the parameter space Θ and Θ_0^c is the complement of Θ_0. The problem is to select either the null model represented by H_0 or the alternative (more complex) model represented by H_1.

A Bayesian analysis of the problem requires a prior for each model's parameters. Let $\pi(\theta|H_0)$ and $\pi(\theta|H_1)$ denote prior density functions for the parameters of null

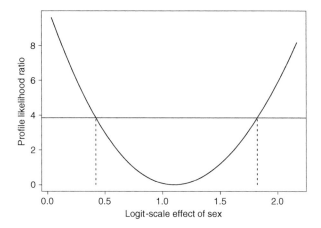

Figure 2.7. Profile likelihood ratio for the logit-scale effect of sex on mortality of moths exposed to cypermethrin. Dashed vertical lines indicate 95% confidence limits. Horizontal line is drawn at $\chi^2_{0.95}(1)$.

and alternative models, respectively. Given these priors, the posterior densities are defined in the usual way (using Bayes' rule) and have normalizing constants

$$m(\boldsymbol{y}|H_0) = \int_{\Theta_0} f(\boldsymbol{y}|\theta)\pi(\theta|H_0)\,\mathrm{d}\theta$$

and

$$m(\boldsymbol{y}|H_1) = \int_{\Theta_0^c} f(\boldsymbol{y}|\theta)\pi(\theta|H_1)\,\mathrm{d}\theta,$$

where $f(\boldsymbol{y}|\theta)$ specifies a model-specific likelihood of the observed data $\boldsymbol{y}$. An additional requirement for Bayesian analysis is that we assign prior probabilities to H_0 and H_1. Let $q_0 = \mathrm{Pr}(H_0) \equiv \mathrm{Pr}(\theta \in \Theta_0)$ quantify our prior opinion of whether the null model should be accepted. Since only two models are involved in the comparison, $\mathrm{Pr}(H_1) \equiv \mathrm{Pr}(\theta \in \Theta_0^c) = 1 - q_0$. As an example, we might use $q_0 = 0.5$ to specify an impartial prior for the two models.

A Bayesian solution of the model selection problem is obtained by computing the *posterior* probability of H_0 given $\boldsymbol{y}$

$$\mathrm{Pr}(H_0|\boldsymbol{y}) = \frac{\mathrm{Pr}(H_0)\,m(\boldsymbol{y}|H_0)}{\mathrm{Pr}(H_0)\,m(\boldsymbol{y}|H_0) + \mathrm{Pr}(H_1)\,m(\boldsymbol{y}|H_1)} \qquad (2.5.7)$$

$$= \frac{q_0\,m(\boldsymbol{y}|H_0)}{q_0\,m(\boldsymbol{y}|H_0) + (1 - q_0)\,m(\boldsymbol{y}|H_1)} \qquad (2.5.8)$$

which follows from a single application of Bayes' rule. Because there are only two models involved, the posterior probability of H_1 must be $\Pr(H_1|\boldsymbol{y}) = 1 - \Pr(H_0|\boldsymbol{y})$. It's worth mentioning that Eq. (2.5.8) could have been derived by specifying the prior density of θ as a finite mixture of model-specific priors as follows: $\pi(\theta) = q_0\pi(\theta|H_0) + (1 - q_0)\pi(\theta|H_1)$.

How should a Bayesian use posterior model probabilities in model selection? One possibility is to reject the null model (and accept the alternative) if $\Pr(H_0|\boldsymbol{y}) < \Pr(H_1|\boldsymbol{y})$ or, equivalently, if $\Pr(H_0|\boldsymbol{y}) < 0.5$. On the other hand, if a Bayesian wants to guard against falsely rejecting the null model, he may wish to reject H_0 under a more conservative (lower) threshold (e.g., $\Pr(H_0|\boldsymbol{y}) < 0.1$).

The Bayesian approach to model selection possesses some appealing features. Unlike the classical approach, the analyst does not have to choose a particular statistic for testing. Furthermore, decisions to reject or accept H_0 do not require samples to be large enough for the application of asymptotic distributions. That said, the Bayesian approach to model selection is not without its problems. For one thing, the posterior model probabilities can be sensitive to the priors assumed for each model's parameters. In addition, the model-specific normalizing constants needed for computing posterior model probabilities can be difficult to calculate when models contain many parameters. Recall that this difficulty impeded routine use of Bayesian statistics for many years and motivated the development of MCMC sampling algorithms.

In Section 2.5.4 we will examine some alternative Bayesian methods of model selection. For now, however, we illustrate a clever approach developed by Kuo and Mallick (1998), which avoids the calculation of normalizing constants but is equivalent conceptually to the approach described earlier in this section. Kuo and Mallick introduce a latent, binary *inclusion parameter*, say w, that is used as a multiplier on those parameters which are present in the alternative model and absent from the null model. By assuming $w \sim \text{Bern}(\psi)$, the prior probability of the alternative model is specified implicitly because

$$\Pr(H_1) \equiv \Pr(\theta \in \Theta_0^c) = \frac{\Pr(w = 1)}{\Pr(w = 0) + \Pr(w = 1)}$$

$$= \frac{\psi}{1 - \psi + \psi}$$

$$= \psi.$$

Therefore, we have the identity $q_0 = 1 - \psi$. To specify prior impartiality in the selection of null and alternative models, we simply assume $\psi = 0.5$.

Kuo and Mallick showed that Gibbs sampling may be used to fit this elaboration of the alternative model. The Gibbs output yields a sample from the marginal posterior distribution of $w|\boldsymbol{y}$, which provides a direct solution to the model selection problem because the posterior probability of the null model $\Pr(H_0|\boldsymbol{y}) = \Pr(w = 0|\boldsymbol{y})$ is easy to estimate from the sample.

The model-selection methodology developed by Kuo and Mallick (1998) applies equally well to much more difficult problems. For example, suppose selection is required among regression models that may include as many as p distinct predictors. In this problem a $p \times 1$ vector $\boldsymbol{w}$ of inclusion parameters may be used to select among the 2^p possible regression models by computing posterior probabilities for the 2^p values of $\boldsymbol{w}$. Alternatively, these posterior probabilities can be used to compute some quantity of scientific interest, by averaging over model uncertainty. Thus, model-averaging can be done at practically no additional computational expense once the posterior is calculated. We conclude this section with an illustration of this approach based on a comparison of two logistic regression models described in a previous example (see Section 2.5.1.3).

2.5.3.1 Example: Mortality of moths exposed to cypermethrin

Recall that the logistic-regression model of moth mortalities contains 3 parameters, an intercept α, the effect of cypermethrin β, and the effect of sex γ. In the test of H_0 we wish to compare a null model for which $\gamma = 0$ against an alternative for which $\gamma \neq 0$. To apply the approach of Kuo and Mallick, we elaborate the alternative model using the binary inclusion parameter w as follows:

$$y_i|N, p_i \sim \text{Bin}(N, p_i)$$
$$\text{logit}(p_i) = \alpha + \beta x_i + w\,\gamma z_i$$
$$w \sim \text{Bern}(0.5)$$
$$\alpha \sim \text{N}(0, \sigma^2)$$
$$\beta \sim \text{N}(0, \sigma^2)$$
$$\gamma \sim \text{N}(0, \sigma^2),$$

where σ is chosen to be sufficiently large to specify vague priors for the regression parameters.

We fit this model using **WinBUGS** and estimated the posterior probability of the null model $\Pr(w = 0|\boldsymbol{y})$ to be 0.13. Since $0.13 < 0.5$, we reject H_0 and accept the alternative model. We can estimate the parameters of the alternative model by computing summary statistics from the portion of the Gibbs output for which $w = 1$. This yields the following posterior means and standard errors (in parentheses) of the regression parameters: $\hat{\alpha} = -3.53\,(0.47)$, $\hat{\beta} = 1.08\,(0.13)$, and $\hat{\gamma} = 1.11\,(0.36)$. Given the sample size and our choice of priors for this model's parameters, it is not surprising that the posterior means are similar in magnitude to the MLEs that we reported earlier.

2.5.4 Assessment and Comparison of Models

In this chapter we have described both classical and Bayesian approaches for conducting model-based inference. In applications of either approach, the role of the approximating model of the data is paramount, as noted in Section 2.2. The approximating model provides an unambiguous specification of the data-gathering process and of one or more ecological processes that are the targets of scientific interest.

Statistical theory assures us that inferences computed using either classical or Bayesian approaches are valid, *provided* the model correctly specifies the processes that have generated the data (i.e., nature and sampling). However, since we never know all of the processes that might have influenced the data, we are never really able to determine the accuracy of an approximating model. What we *can* do is define, in precise terms, the operating characteristics of an 'acceptable' model. This definition of acceptability is necessarily subjective, but it provides an honest basis for assessing a candidate model's adequacy. This approach is also useful in the comparison and selection of alternative approximating models. For example, if we are asked questions such as, "Is model A better than model B?" or "Should model A be selected in favor of model B?", a clear definition is needed for what 'better than' means in the context of the scientific problem.

In defining the operating characteristics of an acceptable model, one clearly must consider how the model is to be used. Will the model be used simply to compute inferences for a single data set, or will the model be used to make decisions that depend on the model's predictions? In either case, decision theory may be used to define acceptability formally in terms of a utility function that specifies the benefits and costs of accepting a particular model (Draper, 1996). Similarly, utility functions may be used to specify the benefits and costs of selecting a particular model in favor of others. Such functions provide a basis for deciding whether a model should be simplified, expanded, or left alone (that is, assuming the model is acceptable as is) (Spiegelhalter, 1995; Key et al., 1999). In ecology a popular method of model selection involves the comparison of a scalar-valued criterion, such as Akaike information criteria (AIC) (Burnham and Anderson, 2002) or deviance information criteria (DIC) (Spiegelhalter et al., 2002). This approach is an application of decision theory, though the utility function on which the criterion is based is not always stated explicitly.

The need for a unified theory or set of procedures for assessing and comparing statistical models seems to be more important than ever, given the complexity of models that can be fitted with today's computing algorithms (e.g., MCMC samplers) and software. Unfortunately, no clear solution or consensus seems to have emerged regarding this problem. There does seem to be a growing appreciation among many statisticians, that while inference and prediction are best conducted using Bayes' theorem, the evaluation and comparison of alternative models are best accomplished from a frequentist perspective (Box, 1980; Rubin, 1984; Draper, 1996; Gelman et al., 1996; Little, 2006). The idea here is that if a model's inferences and predictions are to be well-calibrated, they should have good operating characteristics in a sequence of hypothetical repeated samples.

In the absence of a unified theory for assessing and comparing statistical models, a variety of approaches are currently practiced. We do not attempt to provide an exhaustive catalog of these approaches because recent reviews (Claeskens and Hjort, 2003; Kadane and Lazar, 2004) and books (Zellner et al., 2001; Burnham and Anderson, 2002; Miller, 2002) on the subject are available. Instead we list a few of the more commonly used methods, noting some of their advantages and disadvantages.

In Section 2.5.1 we described the likelihood ratio test for comparing nested models and for assessing the goodness of fit of models of counts. These are frequentist procedures based on the calculation of MLEs and on their asymptotic distributions in repeated samples. We also described a Bayesian procedure for model selection that is based on the calculation of posterior model probabilities. An advantage of the Bayesian procedure is that it can be used to select among *non-nested* models; however, the posterior model probabilities can be *extremely sensitive* to the form of priors assumed for model parameters when such priors are intended to convey little or no information (Kass and Raftery, 1995; Kadane and Lazar, 2004). Several remedies have been proposed for this deficiency (e.g., intrinsic Bayes factors, fractional Bayes factors, etc.), but none is widely used or is without criticism, as noted by Kadane and Lazar (2004).

Some approaches to model selection involve the comparison of an omnibus criterion, which typically values a model's goodness of fit and penalizes a model's complexity in the interest of achieving parsimony. Examples of such criteria include the Akaike, the Bayesian, and the deviance information criteria (i.e., AIC, BIC, and DIC) (Burnham and Anderson, 2002; Spiegelhalter et al., 2002), but there are many others (Claeskens and Hjort, 2003).

Alternative approaches to model selection recognize that some components of a model's specification may not be adequately summarized in a single omnibus criterion. For example, scientific interest may be focused on a particular quantity that requires predictions of the model. In this instance, the operating characteristics of a model's predictions of the scientifically relevant estimand should be used to define a basis for comparing models. Examples of this approach include the use of posterior-predictive checks (Laud and Ibrahim, 1995; Gelman et al., 1996; Gelfand and Ghosh, 1998; Chen et al., 2004) and the focused information criterion (Claeskens and Hjort, 2003).

At this point, the reader may realize that the list of approaches for assessing and comparing statistical models is long. In fact, the published literature on this subject is vast. That the field of Statistics has not produced a unified theory or set of procedures for assessing and comparing models may seem disconcerting; however, given the wide variety of problems to which models are applied, the absence of unified procedures is perhaps understandable. At present, the construction and evaluation of statistical models necessarily include elements of subjectivity, and perhaps that is as it should be. One cannot automate clear thinking and the subjective inputs required for a principled analysis of data.

2.6 HIERARCHICAL MODELS

In Chapter 1 we argued that hierarchical modeling provides a framework for building models of ecological processes to solve a variety of inference problems. We defined the term *hierarchical model* only conceptually without using mathematics. Here, we define hierarchical models in more concrete terms and provide a number of examples as illustrations. We also illustrate the inferential implications of fitting these models with classical or Bayesian methods.

2.6.1 Modeling Observations and Processes

As with any approximating model of data, hierarchical models must account for the observational (or data-gathering) process and for one or more underlying ecological processes. An interesting and useful feature of hierarchical models is that these processes are specified separately using constituent models of observable and unobservable quantities.

To illustrate, let's consider a typical hierarchical model of 3 components. The first component corresponds to the data y, which are assumed to have been generated by an observation process $f(y|z, \theta_y)$ that depends on some state variable z and parameter(s) θ_y. The state variable z is often the thing we would like to know about (and compute inferences for), especially in situations where z is the outcome of an ecological process. Unfortunately, z is unobserved (or only partially so), and inferences about z can only be achieved by including it as a second component in the hierarchy. This requires a model $g(z|\theta_z)$ whose parameters θ_z are used to specify the variation in z. In many situations g is a model of an underlying ecological process. It's worth noting that if the state variable z had been observed, the model g could have been fitted directly (i.e., ignoring y and the observation process) using statistical methods appropriate for the particular form of g. The third component of the hierarchy corresponds to the parameters θ_z. We typically make a distinction between parameters that appear in the observation model but not the state model – nuisance parameters – and those appearing in the state model that are often of some intrinsic interest.

By providing an explicit representation of the model into its constituents (*observations*, *state variables*, and ecological *processes*), a complex modeling problem often can be rendered into smaller, and simpler, pieces. For example, if we were to fit the model using Bayesian methods, we would need to specify the joint probability density $[y, z, \theta_y, \theta_z]$. The hierarchical formulation of the model provides the following *hierarchical factorization* of the joint density:

$$[y, z, \theta_y, \theta_z] = f(y|z, \theta_y)\, g(z|\theta_z)\, [\theta_y, \theta_z],$$

where $[\theta_y, \theta_z]$ denotes a prior density for the parameters. In practice, it can be very difficult to conceptualize or describe $[y, z, \theta_y, \theta_z]$ directly, but each of the components (mainly $f(y|z, \theta_y)$ and $g(z|\theta_z)$) are often relatively simple to construct. Thus, hierarchical modeling allows an analyst to focus on more manageable components of the problem.

Hierarchical models also provide conceptual clarity by allowing factors that influence observations to be decoupled from those that influence the ecological process. For example, although we cannot observe the state variable z directly, hierarchical models allow us to proceed with model construction as if we *could* have observed it. Thus, by pretending as if we had observed the state variable, the formulation of model components f and g is greatly simplified.

2.6.1.1 Example: estimating survival of larval fishes

We have already encountered an example of a hierarchical model in Section 2.4.4.2, though we did not identify it as such. In this example the number of surviving larval fishes y_i in the ith replicate container is the observation. In the first stage of the model, y_i is assumed to depend on the survival probability ϕ (the parameter of scientific interest) and on the unknown number of fishes N_i initially added to the ith container. In this model N_i corresponds to the state variable, but it is not of any scientific interest (i.e., N_i is just a nuisance parameter in the observation model). The second stage of the hierarchical model expresses variation in N_i among replicate containers in terms of a known parameter λ, the average number of fishes added to each container.

2.6.2 Versatility of Hierarchical Models

The general approach to hierarchical modeling that we described above can be used to solve many inference problems. In fact, hierarchical models appear almost everywhere in contemporary statistics (Hobert, 2000). Such models are commonly used to specify overdispersion and to account for correlated outcomes that arise by design, as with 'repeated measurements' taken on the same individual or at the same location. In modeling, overdispersion hierarchical models sometimes give rise to standard, 'named' distributions. For example, the beta-binomial and negative-binomial distributions, which are often used to model overdispersed counts, can be generated by marginalizing the binomial-beta and Poisson-gamma mixtures, respectively. Both mixtures are examples of hierarchical models.

Although various synonyms have been used for hierarchical models (e.g., *mixed-effects models* and *multi-level models*), conceptually there is no difference between these models. For historical reasons, statisticians sometimes refer to mid-level parameters as '*random effects*' and to upper-level parameters as '*fixed effects*'. However, in this book we do not focus on the labels that have been attached to different kinds of model parameters or on differences in terminology.

We are more concerned with specifying hierarchical models in terms of quantities that have a recognizable, scientific interpretation — what we have referred to as explicit process models in Chapter 1. In any given problem a scientifically relevant quantity may correspond to an unobserved state variable, a mid-level parameter, an upper-level parameter, a prediction, or perhaps some function of these different model components. That said, we also are interested in selecting an inference method (classical or Bayesian) that allows us to make conclusions (about quantities of interest) that are useful in the context of the problem. The choice of inference method can be especially important in small samples and in problems where predictions are needed. In subsequent chapters of this book we provide several examples where the choice of inference method can be crucial.

For now, we would like to illustrate a few of the differences associated with fitting hierarchical models by different inference methods. In this illustration we use an example from the class of *linear mixed-effects models* (Demidenko, 2004), which are widely applied in biology, agriculture, and other disciplines. We also use this example to introduce some terms that are often used in association with hierarchical modeling.

2.6.2.1 Example: a normal-normal mixture model

Let $\boldsymbol{y}_i = (y_{i1}, y_{i2}, \ldots, y_{im})$ denote a set of replicate measurements made on each of $i = 1, 2, \ldots, n$ subjects. We consider a simple observation model in which y_{ij} is assumed to be normally distributed with a mean that depends on a subject-specific parameter α_i as follows:

$$y_{ij} \sim \mathrm{N}(\alpha_i, \sigma^2). \tag{2.6.1}$$

In practice, m might be relatively small, in which case this model contains many parameters relative to the number of observations in the sample ($=nm$). A common modification is to impose additional structure on α_i, and in many cases this additional structure makes sense in the context of the problem. For example, suppose the subjects in the sample are selected to be representative of a larger population of individuals. Then we might reasonably assume α_i itself to be a realization of a random variable, say,

$$\alpha_i \sim N(\mu, \sigma_\alpha^2). \tag{2.6.2}$$

The model implied by combining the assumptions in Eqs. (2.6.1) and (2.6.2) is quite common. In classical statistics the subject-level parameters $\{\alpha_i\}$ are often called *random effects*, whereas the parameters, μ, σ^2, and σ_α^2, that are assumed to be fixed among all subjects in the sample (and population) are called *fixed effects*. Since the model contains both kinds of parameters it qualifies as a *mixed-effects model.*

The issue we examine here is "how does a Bayesian analysis of this model compare to a non-Bayesian analysis?" The answer to that question depends, in part, on the objective of the analysis. Suppose interest is focused primarily on computing inferences for the random effect α_i. The classical (non-Bayesian) solution to this problem requires two steps. In the first step the fixed effects are estimated by removing the random effects from the likelihood; then in the second step the random effects are estimated conditional on the estimated values of the fixed effects. To be more specific, in the first step the MLEs of the fixed effects are computed by maximizing a *marginal* or *integrated likelihood* function that does not involve the random effects,

$$L(\mu, \sigma^2, \sigma_\alpha^2 \mid \boldsymbol{y}_1, \ldots, \boldsymbol{y}_n) = \prod_{i=1}^{n} [\boldsymbol{y}_i \mid \mu, \sigma^2, \sigma_\alpha^2],$$

where

$$[\boldsymbol{y}_i \mid \mu, \sigma^2, \sigma_\alpha^2] = \int_{-\infty}^{\infty} \prod_{j=1}^{m} [y_{ij} \mid \alpha_i, \sigma^2] [\alpha_i \mid \mu, \sigma_\alpha^2] \, \mathrm{d}\alpha_i \tag{2.6.3}$$

denotes the marginal probability density of $\boldsymbol{y}_i$ given the fixed effects. This step regards $\{\alpha_i\}$ as a set of nuisance parameters to be eliminated by integration (i.e., marginalization over the joint density of $\boldsymbol{y}_i$ and α_i) (Berger et al., 1999). In the normal-normal mixture the integral in Eq. (2.6.3) can be evaluated in closed form to establish that the vector $\boldsymbol{y}_i$ has a multivariate normal distribution with mean $\mu \boldsymbol{1}$ and covariance matrix having diagonal elements $\sigma^2 + \sigma_\alpha^2$ and off-diagonal elements σ_α^2; therefore, it is relatively straightforward to compute the MLE of the fixed effects by maximizing the integrated likelihood function.

The second step of the classical solution involves estimating α_i *conditional* on the data and on estimates of the fixed effects. Estimates of α_i are based on the conditional distribution of $\alpha_i | \boldsymbol{y}_i, \mu, \sigma^2, \sigma_\alpha^2$, which is normal with mean

$$\mathrm{E}(\alpha_i | \cdot) = \frac{\sigma_\alpha^2}{\sigma_\alpha^2 + \sigma^2/m} \left(\bar{y}_i - \mu \right) \qquad (2.6.4)$$

where $\bar{y}_i = (1/m) \sum_{j=1}^m y_{ij}$ denotes the sample mean of the measurements taken on the ith subject. In classical statistics Eq. (2.6.4) is known as the *best linear unbiased predictor* (BLUP) of α_i (Robinson, 1991; Searle et al., 1992). To compute an estimate of the BLUP (sometimes called an *empirical* BLUP), the MLE computed in the first step is substituted for the fixed effects in Eq. (2.6.4).

The classical solution to the estimation of random effects is also known as *parametric empirical Bayes* (Morris, 1983) because it stops short of a fully Bayesian analysis by not specifying a prior distribution for the fixed effects. However, many Bayesians find this approach to be objectionable, primarily because it uses the data twice (first to compute the MLE as a surrogate for the unspecified prior, and then to compute the posterior of the random effect on which the BLUP is based). Carlin and Louis (2000) provide some colorful historical quotations of some prominent Bayesians (de Finetti, Lindley, and Savage), who were strongly opposed to the empirical Bayes approach.

The main problem with the empirical BLUP (or empirical Bayes estimator) of α_i is that by conditioning on the MLE of the fixed effects, it fails to account for the uncertainty in the MLE. Therefore, estimates of variation in α_i based on $[\alpha_i | \boldsymbol{y}_i, \hat{\mu}, \hat{\sigma}^2, \hat{\sigma}_\alpha^2]$ will generally be negatively biased. Laird and Louis (1987) developed a method for correcting for this bias (at least approximately), but the method requires a considerable amount of additional calculation and the analysis of a large number of parametric bootstrap samples.

So how does a Bayesian compute inferences for the random effect α_i? The answer is simple if one adopts modern methods of Bayesian computation. After specifying a prior distribution for the fixed effects, MCMC methods may be used to compute an arbitrarily large sample from the joint posterior distribution, whose unnormalized density is

$$[\mu, \sigma^2, \sigma_\alpha^2, \alpha_1, \ldots, \alpha_n \mid \boldsymbol{y}_1, \ldots, \boldsymbol{y}_n] \propto [\mu, \sigma^2, \sigma_\alpha^2] \prod_{i=1}^n \left\{ [\alpha_i \mid \mu, \sigma_\alpha^2] \prod_{j=1}^m [y_{ij} \mid \alpha_i, \sigma^2] \right\}.$$

Inferences for the random effect α_i are based on its marginal posterior distribution. A sample from this distribution may be obtained without any additional calculations. We simply ignore the other parameters in the sample of the joint posterior to obtain a sample from $[\alpha_i \mid \boldsymbol{y}_1, \ldots, \boldsymbol{y}_n]$. In doing so, we implicitly integrate over all parameters (except α_i) of the joint posterior and thereby account for the uncertainty in estimating those parameters. The posterior sample of α_i values may be used to compute an estimate, such as $\mathrm{E}(\alpha_i|\boldsymbol{y}_1,\ldots,\boldsymbol{y}_n)$, a credible interval for α_i, or any other summary deemed to be useful in the context of the problem.

2.6.3 Hierarchical Models in Ecology

Hierarchical models can be used to solve many common inference problems in ecology. The canonical example is probably that of estimating the occurrence or distribution of a species using 'presence/absence' data collected by many different observers in a standardized survey. Variables that describe habitat or landscape typically influence the occurrence of a species, but there may be other variables, such as sampling effort or weather, that primarily influence one's ability to observe species and hence prevent an error-free assessment of the true occurrence state. We address this type of inference problem in detail in Chapter 3. Often, this problem is attacked using logistic regression to build complex models of the mean response, without regard to the distinction between occurrence and its observation. We provide a hierarchical formulation of the problem that provides a good illustration of the sensibility and ease with which hierarchical models may be fashioned from simpler, component models. In this case a hierarchical model for observations contaminated by false-negative errors consists of a compound, logistic regression model – a logistic regression model for the observations conditional on the true occurrence state, and a logistic regression model for the true occurrence state.

Hierarchical models also can be used to estimate abundance from spatially referenced counts of individual animals. For example, in Chapter 1 we described a model wherein the local abundance of harbor seals at each of n spatial sample units was estimated from a set of independent counts at each unit while accounting for the imperfect detectability of the seals. The same modeling strategy can be adopted with other types of counts, which arise when other sampling protocols, such as capture–recapture or double-observer sampling, are used. In this setting hierarchical modeling can be used to produce maps of the spatial distribution of abundance estimates. We describe these types of hierarchical models in Chapter 8.

A final example of hierarchical models that we wish to introduce involves inference about the structure of a biological community. There are competing views as to how to solve this inference problem: the observation-driven view and the process-driven view. Both views are pervasive in ecology. Advocates of the observation-driven view regard the community of species as a classical closed population, wherein species play the role of individuals. Such models may be used to estimate summaries of community structure, such as species richness, but these summaries fail to preserve species identity and are less likely to meet the needs of conservation and management. Although the observation error is formally accounted for in this view, solving the actual inference problem often requires a second analysis, wherein the estimates of species richness are treated as though they were data. The process-driven view essentially ignores the sampling process and regards the collection of species in the sample as *the* community of interest. Inference about these species proceeds accordingly. Our view is that inference problems associated with community structure are naturally formulated in terms of occurrence of individual species. In a recent series of papers (Dorazio and Royle, 2005a; Dorazio et al., 2006; Kéry and Royle, 2008a,b) we have described a hierarchical, multi-species formulation of occupancy models that accommodates imperfect detection of species and allows species-level models of occurrence to be developed. We demonstrated how such models can be used to estimate many important summaries of community structure. Despite the technical complexity of these models, the multi-species occupancy model stems from a simple hierarchical construction and is relatively easy to implement in **WinBUGS**. We provide a more detailed description of this type of hierarchical model in Chapter 12.

3

MODELING OCCUPANCY AND OCCURRENCE PROBABILITY

In this chapter, we consider models for the occurrence of species. Occurrence is a binary state variable, indexed by space (or space and time), so that $z_i = 1$ if a patch 'i' is occupied, and $z_i = 0$ if a patch is not occupied by the species of interest. (i is the index to a finite collection of spatial units, patches, or sites.) *Occurrence probability* is the parameter $\psi = \Pr(z_i = 1)$. The distinction between the binary state variable, *occurrence* or *occupancy*, which is the outcome of a stochastic process (the realization of a random variable, z), and its expected value, occurrence *probability*, can be relevant in certain inference problems (see Section 3.7). Another term in widespread use is *proportion of area occupied*, or PAO. This is often regarded as being synonymous with the parameter ψ. However, implicit in the term ('proportion') is that it relates to the *realization* of some stochastic process, i.e., PAO is the proportion of some specified collection of spatially-indexed units that are presently occupied. That is, it is a function of z.

Interest in occurrence-based summaries is widespread in ecology. Occupancy is a fundamental concept in macroecology (Brown and Maurer, 1989), landscape ecology, and metapopulation ecology (Levins, 1969; Hanski, 1998). Metapopulation biology is largely concerned with the dynamics of occupancy in patchy landscapes – e.g., how occupancy is influenced by patch size, configuration, and other characteristics. There is some interest in occupancy that is due to its relationship to abundance (He and Gaston, 2000a,b; Royle and Nichols, 2003; MacKenzie and Nichols, 2004; Dorazio, 2007; Royle, 2007), which is the topic of Chapter 4. From a practical perspective, modeling distribution or range of a species, i.e., the extent of occurrence, is important in many management and conservation problems, and probability of occurrence is a natural summary of (habitat) suitability (Boyce and McDonald, 1999). There is also practical interest in the proportion or number of occupied sites as an 'index' of population status when it is not possible to monitor populations of individuals effectively. In the context of monitoring, it is logistically simple and economically efficient (time, money) to obtain presence/absence data informative about occupancy instead of individual-level 'encounter histories' or counts of individuals based on systematic surveys. Moreover, for some species, conventional population sampling methods might not be readily adaptable (e.g., rare or elusive species (Thompson, 2004)). The extension of occurrence models to dynamic systems broadens substantially the scope of their relevance (see Chapter 9). Models with explicit consideration of extinction and colonization have direct relevance to modeling invasive and imperiled species.

One of the most important issues relevant to modeling species occurrence is that of imperfect detection. This problem has a long history throughout wildlife sampling (e.g., capture–recapture, distance sampling, etc.), but has only recently been considered in the context of modeling species occurrence (Bayley and Peterson, 2001; Kéry, 2002; Moilanen, 2002; MacKenzie et al., 2002). There is a vast literature in landscape ecology and biogeography that has historically ignored the issue altogether. Conversely, a relatively smaller segment of literature adopts a distinct focus on the problem of modeling detectability. We believe that imperfect detection is a fundamental component of the observation process and that it is difficult to make a coherent inference about occurrence probability without taking detectability into consideration. For many species, especially rare, cryptic, or elusive species, the existence of 'false negatives' can be especially problematic. From a practical perspective, the probability of species detection, say p, is essentially a measure of the efficacy of your sampling apparatus. Thus, it provides a natural gauge for calibrating survey methods and programs. Finally, we note that false positives are a distinctively different kind of problem – that being related to an observer's fundamental ability to identify (or not) the species in question, whereas false negatives are primarily effort/expense driven. Absence of false positives is assumed in most of what we do here (but see Section 3.8.1).

Models of occupancy or occurrence probability, especially that take into account imperfect detection, are a growth industry, with much recent attention having been focused on developing theory and methods associated with modeling occurrence probability, occupancy, and related summaries of metapopulation structure from observational data. Bayley and Peterson (2001), MacKenzie et al. (2002), and Moilanen (2002) address the conduct of inference about occupancy from survey data that are subject to imperfect detection of individuals. Basic interest in the topic was also shown by Moilanen (2002), Tyre et al. (2003), Gu and Swihart (2004), and Wintle et al. (2004). Methodological extensions include Barbraud et al. (2003), MacKenzie et al. (2003), Royle and Nichols (2003), and Dorazio and Royle (2005a). Such models have been applied to a myriad of taxa including anurans (Mazerolle et al., 2005; Weir et al., 2005; Pellet and Schmidt, 2005; Schmidt, 2005; Schmidt and Pellet, 2005; Brander et al., 2007), salamanders (Bailey et al., 2004; McKenny et al., 2006), snakes (Kéry, 2002), the dreaded Otago skink (Roughton and Seddon, 2006), insects (Sileshi et al., 2006), various bird species (Olson et al., 2005; Dreitz et al., 2006; Winchell and Doherty , 2008), small mammals (Stanley and Royle,

2005), and carnivores (Sargeant et al., 2005; O'Connell et al., 2006). This frenzy culminated in the recent synthetic treatment by MacKenzie et al. (2006).

Given this explosive growth in the topic, what could we possibly offer here? And, how is it distinct from existing treatments of the topic? As with the rest of the material covered in this book, our primary contribution is a hierarchical or state-space rendering of models of occurrence. In this regard, we make a formal and explicit distinction, in the model, between the latent occupancy state variable z and the observations y. Under this formulation, methods that address inference about occurrence in the context of observation error are closely related to simple logistic regression models for the underlying occurrence process. We provide classical and Bayesian formulations of models of occurrence and, in keeping with our 'teach a man to fish' philosophy, we provide implementations in **R** and **WinBUGS** which we hope will enable the reader to grow. Logistic regression is a major conceptual and technical building-block. It is relevant to a large segment of methods and applications in statistical ecology. The basic models are binomial models (for modeling 1s and 0s) with structure on the parameters of the binomial distribution. The **R/WinBUGS** implementation of those models forms the basis of many more complicated problems, including capture–recapture models for closed *and* open populations, models of animal community structure, and many other widely used classes of models.

Thus, we begin by introducing logistic regression models such as might be used for modeling species distribution. Classical and Bayesian analysis of these models is addressed, and various extensions are considered. One important extension is to the case where occurrence is not observed perfectly. This extension yields probably the simplest true hierarchical model that we consider in this book. The resulting model consists of two distinct logistic regression models – one for the occurrence state variable (which is only partially observed) and one for the observations of detection/non-detection. This formulation allows us to focus our modeling efforts on these distinct components of the problem.

3.1 LOGISTIC REGRESSION MODELS OF OCCURRENCE

Logistic regression is one of the most widely used statistical procedures in ecology, and most ecologists have been exposed to it at some level. Fewer realize that these models are formalized as a binomial probability model for the observations. If you suggested to someone that a logistic regression might be suitable for their data, they would be comfortable with that. If, on the other hand, you suggested that $z_i \sim \text{Bin}(J, \pi_i)$ seemed reasonable, its relevance to logistic regression would be unclear to many.

In the prototypical logistic regression application in ecology, suppose that $i = 1, 2, \ldots, M$ spatial units are sampled (surveyed) for the species in question. Sample units are sites, points, quadrats, patches etc., and we use those terms interchangeably except where the specific context dictates otherwise. Sampling yields observations of species' presence or absence, which we regard as binary observations according to

$$z_i = \begin{cases} 1 \text{ if present} \\ 0 \text{ if absent.} \end{cases}$$

We will suppose here, as is typical of logistic regression applications, that actual presence or absence is observed without error.[1] The logistic regression model begins with the assumption of a binomial distribution for the observations

$$z_i \sim \text{Bin}(1, \pi_i).$$

In some cases there might be repeated independent observations per sample unit. In this case, if z_i is the total number of detections, and J_i represents the number of observations for unit i, then $z_i \sim \text{Bin}(J_i, \pi_i)$. Typically, interest is focused on modeling variation in the binomial success probability, e.g., how π_i changes in response to some covariate, say x_i. This is accomplished by specifying a functional relationship between π_i and the covariates, most commonly a linear relationship between the logit-transformed π_i and the covariate x_i, according to:

$$\text{logit}(\pi_i) = \beta_0 + \beta_1 x_i, \tag{3.1.1}$$

where β_0 and β_1 are parameters to be estimated. Other link functions are possible, but the logit link is customary. Choice of link function is seldom considered as an interesting or useful model-selection problem in ecology, although the problem has been addressed in some contexts (Collett, 1991; Morgan, 1992). In the context of modeling occurrence, there are some instances where there is a natural choice of link function other than the logit (see Section 4.5).

Under the logistic regression model, when the observations are independent of one another, the joint pmf (see Chapter 2) of the observations is the product of M binomial probability mass functions. Thus, the likelihood for the observations is

$$L(\boldsymbol{\beta}|\mathbf{z}) \propto \prod_{i=1}^{M} \pi_i(\boldsymbol{\beta})^{z_i} (1 - \pi_i(\boldsymbol{\beta}))^{J_i - z_i}.$$

Here, the dependence of the cell probabilities on parameters $\boldsymbol{\beta}$ via Eq. (3.1.1) is indicated explicitly by the notation $\pi(\boldsymbol{\beta})$. $L(\boldsymbol{\beta}|\mathbf{y})$ can be maximized effectively

---

[1] Strictly speaking, we might prefer to qualify our analyses by noting that 'presence' here means *observed* or *apparent* presence but we are temporarily avoiding the conceptual and practical implications of that.

using standard numerical methods, e.g., as implemented in **R** using a number of nonlinear optimization routines, as we illustrate in the following section.

3.1.1 Modeling Species Distribution: Swiss Breeding Birds

We consider data from the Swiss Monitoring Häufige Brutvögel (MHB) – the Swiss Survey of Common Breeding Birds[2]. We will refer to this program as the Swiss Breeding Bird Survey, or Swiss BBS in some instances.

The Swiss BBS is an annual survey conducted during the breeding season. Sample units are 1 km^2 quadrats that are sampled 2 or 3 times each by volunteer observers. Each observer surveys on foot a route of variable length (among quadrats) but consistent for all surveys of a given quadrat. Observers can also choose the intensity of sampling, i.e., the amount of time they spend surveying their chosen route. Thus, route selection, length, and situation within the quadrat are largely up to each observer. Data are collected using a method of territory mapping (Bibby et al., 1992) that involves significant interpretation on the part of the biologist, but yields a rich data structure that is amenable to a number of different analytic renderings. Routes aim to cover as large a proportion of a quadrat as possible and remain the same from year to year. During each survey, an observer marks every visual or acoustic contact with a potential breeding species on a large-scale map and notes additional information such as sex, behavior, territorial conflicts, location of pairs, or simultaneous observations of individuals from different territories. Date and time are also noted for each survey.

Detailed descriptions of the survey, the resulting data, and a number of different analyses can be found in Kéry and Schmid (2004), Kéry et al. (2005), Royle et al. (2005), Kéry and Schmid (2006), Royle et al. (2007b), Royle and Kéry (2007), and Kéry (2008). Here we use data on the willow tit (*Parus montanus*) from 237 quadrats. Table 3.1 shows a subset of the data with the number of observed territories for each of 3 sample periods. For the present purposes, we have quantized the counts in Table 3.1 to simple presence/absence data, so that if the count is 1 or greater, it was set to 1 while observations of 0 remain so. For any model fitting, the covariates elevation and forest were standardized to have mean 0 and unit variance.

We will address a number of features of the data relevant to various analyses in this and subsequent chapters. First, the counts were typically made on different days for each quadrat. Second, both route length and survey effort (duration

---

[2] We are immensely grateful to Hans Schmid and Marc Kéry of the Swiss Ornithological Institute for making these data available to us.

Table 3.1. Swiss bird survey data consisting of 3 replicate quadrat counts of the willow tit (*Parus montanus*) during the breeding season and covariates elevation (meters above sea level) and forest cover (percent). Only a subset of quadrat counts are shown here. The symbol 'NA' indicates a missing value in an **R** data set.

rep1	rep2	rep3	elevation	forest
0	0	0	910	21
0	0	0	540	11
0	0	0	450	29
0	0	0	470	55
0	0	0	470	54
0	1	NA	1880	32
1	0	1	1400	32
0	0	0	470	3
0	0	0	830	58
0	1	2	1950	36
0	0	0	1660	86
0	1	1	1210	75
1	0	0	380	23
0	0	NA	1840	0
1	0	NA	1910	18
1	2	2	1630	33
0	0	0	1540	21
2	4	3	1340	39
2	3	1	1410	52
14	13	9	2030	36
6	6	8	1880	66

of survey) vary among quadrats, as they are selected by individual observers. In addition to these variables that might explain the likelihood of observing a $y = 1$, we have auxiliary covariates of biological relevance, elevation (Figure 3.1) and forest cover (Figure 3.2) which we expect to be important determinants of distribution and abundance for most breeding birds. In particular, the elevational gradient in Switzerland is severe, and this substantially affects climate/weather and vegetation.

To illustrate some basic concepts and motivate subsequent developments, we focus presently on modeling the effect of these landscape covariates on occurrence to derive a distribution map. However, we note that other covariates that describe deviations from a rigorously standardized protocol are also of interest to model the effects due to non-standardization. Such nuisance effects have been noted to be important in a number of large-scale wildlife surveys, including the North American Breeding Bird Survey (Sauer et al., 1994), Christmas Bird Count (Link and Sauer, 1999), the North American Amphibian Monitoring Program (NAAMP; Weir et al.,

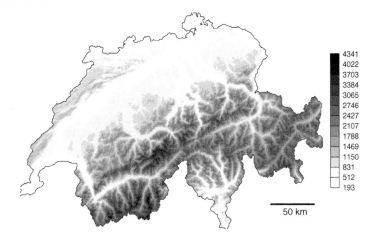

Figure 3.1. Elevation, m above sea level. Due to the extreme elevational gradient and its influence on climate/weather and vegetation, we expect that it should be an important factor leading to variation in abundance and occurrence of most bird species in Switzerland.

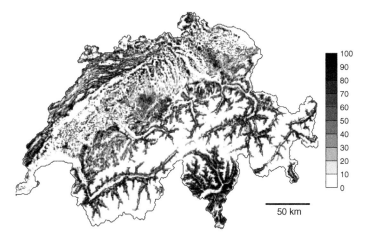

Figure 3.2. Forest cover, percent of quadrat that is forested. Forest should be an important indicator of habitat suitability for many bird species.

2005), and Swiss butterfly (Kéry and Plattner, 2007) and bird (Kéry and Schmid, 2004, 2006) surveys.

Some consideration needs to be given to how the replicate observations should be integrated into the analysis. Presently, they provide more information than is necessary to carry out a simple logistic regression. We will begin with an analysis based only on the data from the first survey period, since the species was found to be more detectable early in the breeding season (Kéry et al., 2005), and to carry out an ordinary logistic regression for a species of concern, we only require a single sample. In general, the manner in which one deals with such replicates depends largely on the formalization of a model for the underlying process of interest, and for the observation mechanism (i.e., detectability). We are intentionally avoiding these issues for the time being in order to focus on the basic logistic regression problem at hand.

We consider models containing forest cover and linear or quadratic elevation effects. Most species should respond to both of these factors, and the response to elevation is likely to be quadratic simply because of the severe environmental gradient that results from increasing elevation. Many species may have some intermediate 'optimum' elevation of occurrence that should be evident in the data. It is simple to carry out a logistic regression in all popular software packages, and **R** is no exception, having a fairly versatile routine (called `glm`) for fitting generalized linear models, of which logistic regression is a particular case. The `glm` function represents a reasonably user-friendly method for analysis, but one that obscures the technical details of how such models are analyzed. Thus, while the use of `glm` (or similar functionality) is sufficient for many purposes, we will do this the hard way to foster learning, and also because the basic construct becomes essential for some important generalizations to be described subsequently. As we have claimed previously, if you do the easy problems the hard way then, ultimately, the hard problems will be easy.

The basic strategy is to write an **R** function (or script) that evaluates the log-likelihood function, and then pass this to one of the available optimizers (e.g., `nlm`). The code to evaluate the log-likelihood and the call to `nlm` are given in Panel 3.1. The key elements of this function are the lines

```
probs<-expit(b0 + b1*elev + b2*(elev^2) + b3*forest)
```

which computes the binomial probabilities according to Eq. (3.1.1), as a function of covariates elevation (a quadratic) and forest cover. The other main calculation is the contribution of each observation to the binomial log-likelihood. This is done for the vector of observations simultaneously by the instruction:

```
tmp<-log(dbinom(z,1,probs))
```

The log-likelihood function allows for the possibility of missing values by defaulting the log-likelihood to 0 wherever a missing value occurs. Thus, missing values contribute nothing to the objective function (the negative of the summed log-likelihood). The log-likelihood is minimized and the output saved to an object **out** by executing the following **R** instruction:

```
out <- nlm(lik,c(0,0,0,0),hessian=TRUE)
```
The second argument to this function call (the vector of 4 zeros) are the starting values. In some cases, especially for very complex models or sparse data, numerical optimization routines can be sensitive to starting values. The object `out` is a list having several elements which are described subsequently (and see Chapter 2, especially Section 2.3.2). In order to execute this function, the various objects referenced within the function must be loaded into the **R** workspace. These include M, `elev`, `forest` and the data vector `z`. These are available in the Web Supplement, and we provide an **R** script for carrying out these analyses.

A slightly more complicated version of the function (Panel 3.2) allows one to easily fit a suite of models by repeated function calls. Note that it constructs the log-likelihood for the maximal model, but defaults all coefficients to zero unless their names are passed to the function using the `vars =` option. We can execute this by issuing a command such as:

```
out<- nlm(lik,c(0,0,0,0),
          vars=c("const","elev1","elev2","forest"),hessian=TRUE)
```

This fits a model with an intercept (which is named `const` in the **R** log-likelihood definition), a quadratic response to elevation and a linear response to forest cover.

```
lik<-function(parms){
    b0<-parms[1]
    b1<-parms[2]
    b2<-parms[3]
    b3<-parms[4]
    ones<-rep(1,M)
    ### Compute binomial success probabilities
    probs<-expit(b0*ones+b1*elev+b2*(elev^2)+b3*forest)
    lik<-rep(0,length(z))
    ### evaluate log of binomial pmf
    tmp<-log(dbinom(z,1,probs))
    ### substitute 0 for missing values
    lik[!is.na(z)]  <-  tmp[!is.na(z)]
    lik<-  -1*sum(lik)
    return(lik)
}
```

Panel 3.1. Construction of the likelihood for a logistic regression model in **R** for the willow tit data. Binomial models of this type can also be fitted using the **R** function `glm`.

The object `out` has a number of elements which can be manipulated using standard **R** conventions. In the present example, several of the elements are:

```
$minimum
[1] 100.8872


$estimate
[1] -0.5422133  1.8473435 -1.0608642  0.6469108


$gradient
[1] 2.349054e-05 3.146269e-05 6.520951e-05 4.888534e-05


  . .

  . .
```

Main interest is in the minimized negative log-likelihood (100.8872) and the 4 parameter estimates (in order of the `vars` specification). The numerical gradient should be close to 0 if the algorithm has converged (it has). Although some output is provided by default, the numerical hessian which is useful to obtain estimates of precision (see Section 2.3.2) is provided if we specify `hessian=TRUE` in the call to `nlm`.

The fitted response function (coefficient estimates rounded) is:

$$\text{logit}(\pi_i) = -0.542 + 1.847 \,\text{elev}_i - 1.061 \,\text{elev}_i^2 + 0.647 \,\text{forest}_i.$$

```
lik<-function(parms,vars){
    tmp<-c(0,0,0,0)
    names(tmp)<-c("const","elev1","elev2","forest")
    tmp[vars]<-parms
    ones<-rep(1,M)
    probs<-expit(tmp[1]*ones+tmp[2]*elev+ tmp[3]*(elev^2)+tmp[4]*forest)
    lik<-rep(0,length(z))
    tmp<-log(dbinom(z,1,probs))
    lik[!is.na(z)]  <-  tmp[!is.na(z)]
    lik<-  -1*sum(lik)
    return(lik)
}
```

Panel 3.2. A slightly generalized **R** formulation of the logistic regression likelihood to facilitate fitting multiple models. The likelihood is specified in terms of the most complex model of interest, and coefficients are set to 0 by default. Covariates are included by specifying them with the `vars = ` option.

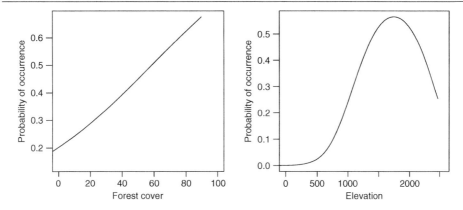

Figure 3.3. Fitted response in occurrence probability for the willow tit to forest cover (percent of quadrat covered in forest) and elevation (meters above sea level).

One implementation of this in **R** is

```
ests<-out$estimate
logitp <- ests[1] + ests[2]*elev + ests[3]*elev^2 + ests[4]*forest
```

where the new object `logitp` is a vector of length M.

There is a positive response in $\Pr(z = 1)$ to forest cover. i.e., a unit increase in forest cover (that would be a standard deviation – which is about 0.27) yields an increase of 0.647 in $\mathrm{logit}(\pi)$. Because of the scaling (logit scale) and standardization of covariates, it is most convenient to view the response to forest and elevation graphically, on the probability scale, and for unstandardized covariates. The response of occurrence probability to forest cover and elevation is depicted in Figure 3.3. The response function to elevation is concave, and the optimum probability of occurrence (taking the first derivative of the quadratic response function, setting that to 0, and solving) is calculated according to

$$\mathtt{elev}_{\mathrm{opt}} = -(1.847)/(2*-1.061) = 0.870.$$

As before, this is in units of standard deviations from the mean. The mean elevation was 1182.57 m and the standard deviation was 646.33 m. Thus, the maximum occurrence probability occurs at an elevation of $1182.57 + 0.870 \times 646.33 = 1744.87$ m. The effect of elevation on π is shown in the right panel of Figure 3.3. We also used the estimated model parameters to generate estimates of π for all 41365 1 km quadrats in Switzerland. These estimates are mapped in Figure 3.4.

We might typically consider whether one, both, or either of elevation and forest cover are important or significant. If we made a distinction between biological and

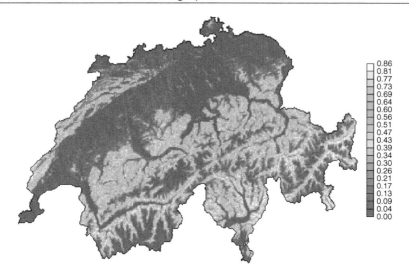

Figure 3.4. Estimated probabilities of occurrence of the willow tit for every 1 km quadrat in Switzerland.

statistical significance, then we might formally evaluate the latter using hypothesis testing or model selection ideas, such as Akaike's Information Criterion (AIC) (Burnham and Anderson, 2002, also see Chapter 2). Results obtained by fitting a number of different models are summarized in Table 3.2. We see that the model previously described (the model containing all effects) is favored by AIC.

3.1.2 Observation Covariates

For the Swiss BBS data, we noted (Section 3.1.1) that the survey is based on 3 samples during the breeding season and that a number of covariates should affect the response. However, some of these covariates might not affect whether a species occurs on a particular quadrat but, rather, may influence variation in *observed* presence – our ability to detect the species given that it is there. These covariates include (at least) sample duration and also perhaps the date of the survey (which might influence bird behavior). In practice, we also might consider certain environmental covariates such as temperature, rainfall, and wind – factors that influence the behavior of many species and hence their detectability. Such covariates influence the efficacy of sampling to some extent, but *not* the underlying occurrence process. That is, the presence of willow tit at a site should not depend on such things as sample effort, duration, sampling intensity, or environmental conditions.

Table 3.2. Results of fitting several logistic regression models to the willow tit data from the Swiss Breeding Bird Survey. The estimated effects under each model are given in columns labeled by the effect name. 'np' is the number of parameters in the model.

model rank	intercept	Elev	Elev^2	Forest	np	AIC
1	−0.542	1.847	−1.061	0.647	4	108.89
2	−0.091	1.851	−1.469	–	3	111.68
3	−1.328	1.186	–	1.025	3	116.05
4	−1.046	0.763	–	–	2	131.88
5	−1.019	–	–	0.660	2	135.15
6	−0.931	–	–	–	1	143.13

How do we accommodate these *sampling covariates* into the logistic regression model? We could just expand the logistic model by plugging such covariates into the model for logit(π). For example, suppose $\texttt{effort}_i$ is some measure of effort expended to obtain the observation at sample unit i, then we might consider the model:

$$\text{logit}(\pi_i) = \beta_0 + \beta_1 \,\texttt{elev}_i + \beta_2 \,\texttt{elev}_i^2 + \beta_3 \,\texttt{forest}_i + \beta_4 \,\texttt{effort}_i.$$

The problem with this approach is that interpretation of the expanded model is not straightforward. Because some factors are *not* expected to influence *occurrence* of species. Including these factors in the same model impairs our ability to interpret estimates and, especially, predictions of the occurrence or occurrence probability made under the model. For example, if we are required to make a prediction of z at some unsampled quadrat, what value of $\texttt{effort}$ would we use to make that prediction?

In Section 3.3 we develop a more flexible framework for handling these different types of covariates that is, one that distinguishes between sources of variation in both occurrence and detectability, and one that yields a more coherent framework for inference about the state variable of interest (occurrence). Before addressing that problem, we briefly consider Bayesian analysis of this ordinary logistic regression model to develop some tools that will be useful in later chapters of this book.

3.2 BAYESIAN LOGISTIC REGRESSION

Fitting a logistic regression model by maximum likelihood is easy, so why would anyone ever want to do a Bayesian analysis? While we probably would never have one do a Bayesian analysis of a simple logistic regression model where the inference was focused on estimating the coefficients of the regression model, a Bayesian analysis has advantages in some cases.

Consider the problem of estimating specific values or collections of values of the dependent variable, such as, the area occupied or the extent of a species range. In the context of the Swiss BBS data, we might hope to quantify the number of quadrats that are occupied by willow tits. We could do this by sampling every quadrat, say $i = 1, 2, \ldots, G$, and then computing the total (area occupied)

$$A_{\text{occ}} = \sum_{i=1}^{G} z_i.$$

Quantities such as this have to do with the particular realized values of z, not their expectations over some hypothetically infinite population. For estimating functions of random variables (random effects), we feel that a Bayesian formulation is usually more coherent, general, and easy to implement.

For the Swiss BBS, A_{occ} is the number of square kilometers containing at least 1 willow tit territory. Since we can't hope to sample all quads in most cases, we would instead use an estimator based on the logistic regression fit to a sample of quadrats (i.e., the situation just discussed in the previous Section), having the form:

$$\hat{A}_{\text{occ}} = \sum_{i}^{G} \widehat{\text{E}(z_i)} = \sum_{i}^{G} \hat{\pi}_i,$$

where $\hat{\pi}_i$ is the estimated or fitted probability of occurrence for quadrat i. For the willow tit data, the estimates were given in Table 3.2, and thus we can evaluate

$$\hat{\pi}_i = \text{expit}(\hat{\beta}_0 + \hat{\beta}_1\, \texttt{elev}_i + \hat{\beta}_2\, \texttt{elev}_i^2 + \hat{\beta}_3\, \texttt{forest}_i)$$

for all G quadrats. For the Swiss willow tit data, this yields

$$\hat{A}_{\text{occ}} = 9932.58 \text{ km}^2.$$

It is not immediately clear how to obtain a characterization of uncertainty for this estimator, although we imagine that some have attempted to conjure-up an approximate variance using the delta method (Williams et al., 2002, Appendix F), the general utility of which is unknown.

3.2.1 Analysis by Markov Chain Monte Carlo

Bayesian analysis of logistic regression models is straightforward using conventional methods of Markov chain Monte Carlo (MCMC). To proceed, we must address two issues. First, we need to specify prior distributions for all of the parameters.

Customary priors for logistic regression parameters in the mean function are flat normal priors. For example, for β_0 we might assume a normal prior with mean 0 and standard deviation 100, and assume that it is independent of all other parameters. This is a very diffuse prior, being relatively flat in the vicinity of $\beta_0 = 0$. Provided that covariates are standardized, such a prior is often colloquially referred to as non-informative. After describing prior distributions for the model parameters, we can focus on devising an MCMC algorithm, which we will use for obtaining a sample of the unknown quantities (parameters and state variables) from the posterior distribution. In practice, we iteratively sample the unknown quantities from their full-conditional posterior distributions (see Section 2.4.4). The full-conditional for β_0 is the product

$$[\beta_0|\cdot] \propto \left\{ \prod_{i=1}^{M} \pi_i(\boldsymbol{\beta})^{z_i} (1 - \pi_i(\boldsymbol{\beta}))^{1-z_i} \right\} \mathrm{N}(\beta_0|0, 100^2),$$

where $\boldsymbol{\beta} = (\beta_0, \beta_1, \beta_2, \beta_3)$. Standard methods (see Chapter 2) may be used to simulate β_0 from this distribution. We obtain a posterior sample by iteratively simulating from the complete set of full-conditional distributions, for all of the model parameters.

But what if we are interested in some new value of z that corresponds to an unsampled quadrat? What is its full-conditional? Under this simple logistic regression model, when we condition on the parameters of the logistic-regression model, then z for any unsampled quadrat, say z_j, is independent of all other values of z. Thus,

$$[z_j|\boldsymbol{\beta}] = \mathrm{Bern}(z|\pi_j) \tag{3.2.1}$$

where

$$\mathrm{logit}(\pi_j) = \beta_0 + \beta_1 \, \mathtt{elev}_j + \beta_2 \, \mathtt{elev}_j^2 + \beta_3 \, \mathtt{forest}_j.$$

Posterior samples of *predictions* of z_j obtained in this way constitute a sample from the *posterior predictive distribution* of z_j. Note that while Eq. (3.2.1) is conditional on $\boldsymbol{\beta}$, the fact that the parameters $\boldsymbol{\beta}$ are being sampled from the posterior distribution is effectively averaging posterior uncertainty in $\boldsymbol{\beta}$ out of Eq. (3.2.1). Thus, the MCMC algorithm provides a Monte Carlo evaluation of the following integral:

$$[z_j|\mathbf{y}] \propto \int [z_j|\boldsymbol{\beta}][\boldsymbol{\beta}|\mathbf{y}] \, \mathrm{d}\boldsymbol{\beta}.$$

This formal accounting for the fact that $\boldsymbol{\beta}$ is unknown, by averaging over uncertainty in $\boldsymbol{\beta}$, is often touted as a great virtue of a Bayesian analysis in such prediction problems. It is the primary distinction between Bayesian and classical approaches to prediction of random effects (see Section 2.6).

3.2.2 Logistic Regression in WinBUGS

Bayesian analysis of logistic regression models in **WinBUGS** requires little more than a pseudo-code rendering of the model. The model under consideration for the willow tit data is shown in Panel 3.3. The data and **R** instructions for fitting this logistic regression model in **WinBUGS** are provided in the Web Supplement. To undertake the larger prediction problem (i.e., on the grid G), we need to include some additional detail in the **WinBUGS** model specification describing the probability structure of the quantities for which predictions are desired. The relevant extension of the model for obtaining the predictions on the 1 km^2 grid that covers Switzerland is provided in the Web Supplement. Note that this can be a tedious calculation in **WinBUGS**. It is somewhat more efficient to run the code shown in Panel 3.3 and then use the resulting MCMC output to compute the predictions in **R**.

Returning to the willow tit data, we fit the model using this **WinBUGS** model specification. This yielded the posterior summaries shown in Table 3.3. Note that the posterior means are somewhat similar to the MLEs reported in Table 3.2. In general, they need not be since the MLE is the *mode* of the *multivariate* posterior distribution, given our choice of priors. The difference will generally depend on how asymmetrical the posterior is and the prior distributions. The estimate of A_{occ}, (posterior mean) was 9969.0 and the 95 percent credible interval was $(8223; 11900)$. This is an accurate assessment of uncertainty (if you believe in the model). There are no asymptotic approximations required.

```
model {
   for(i in 1:M){
      z[i] ~ dbin(mu[i],1)
      logit(mu[i]) <- b0 + b1*elev[i] + b2*elev2[i] + b3*forest[i]
    }
   b0 ~ dnorm(0,.001)
   b1 ~ dnorm(0,.001)
   b2 ~ dnorm(0,.001)
   b3 ~ dnorm(0,.001)
}
```

Panel 3.3. WinBUGS model specification for a simple logistic regression. Inputs are the objects M, elev, elev2, forest, and the observations of presence/absence z.

Table 3.3. Bayesian estimates of model parameters for the willow tit data. Estimates are based on 3 chains of length 10000 after 1000 burn-in and thinned by 2 for a net 15000 posterior samples.

parameter	mean	sd	2.5%	median	97.5%
A_{occ}	9969.000	940.000	8223.000	9943.000	11 900.000
b0	−0.565	0.280	−1.123	−0.564	−0.019
b1	1.936	0.331	1.335	1.918	2.636
b2	−1.117	0.295	−1.733	−1.103	−0.572
b3	0.661	0.217	0.242	0.658	1.094

3.3 MODELS OF OCCUPANCY ALLOWING FOR IMPERFECT OBSERVATION

In most practical settings, observed zeros are ambiguous due to imperfect detection of species, and there are many inference problems for which that becomes relevant. We addressed two problems in the previous sections. The first problem was estimating the area of occurrence of a species in which the apparent extent of occurrence will typically be under-stated because a species might be present but go undetected. Obviously this could be more pervasive for low-density or secretive species but the possibility might also be a consideration for more abundant species, depending on how effective the sampling method is. We also introduced a situation in which we wish to model the influence of factors that induce variation in the data but are not related to occurrence (e.g., environmental conditions at the time of sampling, or effort). Some attention to detectability, as being distinct from occurrence, is necessary when such ecological quantities are to be estimated from apparent presence/absence.

3.3.1 Sampling Error for Binary Observations

To formalize the distinction between variation in detectability and that in occurrence, we first have to introduce a distinction between *observations* and the underlying *state variable* that is only observed imperfectly. We will now denote the observations by y and the underlying state variable by z. To accommodate imperfect sampling, we recognize that y is equal to z only sometimes and, at other times, we may falsely observe $y = 0$. Thus, observations of non-occurrence arise by one of two possible mechanisms. First are *sampling zeros* – the species does occur, but it was not observed during sampling. Second are 'structural' or fixed zeros. These are deterministic zeros that arise because the species was not available to be detected by the sampling activity.

In practice, we have no way of knowing whether a particular observation of $y = 0$ is of one type or the other. Resolving these two distinct types of zeros requires that we make a formal distinction between the true occupancy state, z, and the *observed* occupancy state, y. This requires an increase in model complexity (more parameters). Specifically, we distinguish between *detection probability*, the probability of observing the species *given that it is present*, i.e.,

$$p = \Pr(y_i = 1 | z_i = 1)$$

and *occurrence probability*, which is

$$\psi = \Pr(z_i = 1).$$

In the willow tit example involving logistic regression we intentionally did not distinguish between these two parameters (nor between y and z even), as is typically (almost universally) the case in logistic regression applications in biogeography and elsewhere (e.g., Guisan and Zimmermann, 2000; Hames et al., 2002; Reunanen et al., 2002; Engler et al., 2004; Guisan and Thuiller, 2005).

Having made a formal distinction between observations y and the state variable z, and the notion of observation error in the form of non-detection, we may formalize a hierarchical model for this simple system. In particular, we can describe the relationship between y and z, in terms of p and ψ, by the compound Bernoulli model in which, for the observation model, we have:

$$y_i | z_i \sim \mathrm{Bern}(z_i p)$$

and, for the process or state model,

$$z_i \sim \mathrm{Bern}(\psi),$$

where we might allow ψ to vary in response to spatial covariates as before. Here, z_i are generally unobservable, latent variables (or random effects). In the observation model, we have expressed the Bernoulli success probability as the product $z_i p$. This implies that if the species is not present at a site i, so that $z_i = 0$, then y_i is a fixed 0. Otherwise, y_i is a Bernoulli trial with parameter p. This simple hierarchical model, which comprises two Bernoulli components – essentially, two logistic regression models – yields a vast array of interesting and useful extensions that we will develop in subsequent chapters of this book.

As we have formulated the problem here, it turns out that this is a degenerate model in the sense that parameters p and ψ are not uniquely identifiable (from a single observation y_i). This is because the marginal pmf of y_i, unconditional on z, is a Bernoulli with parameter π where

$$\pi = \Pr(y\!=\!1|z\!=\!1)\Pr(z\!=\!1) + \Pr(y\!=\!1|z\!=\!0)\Pr(z\!=\!0)$$
$$= p\psi$$

i.e., the product of detection and occurrence probabilities. Note that the second term evaluates to 0 under the assertion that false-positives are not possible, but this need not be the case; see Section 3.8. We shortly will consider a remedy to this problem, by expanding the design – the manner in which data are collected – so that information about both p and ψ is obtained.

3.3.2 Importance of Detection Bias

We are often confronted with questions about the relevance of detectability and whether it is important to any particular inference problem. This is an active debate in many areas of animal ecology where models of detectability (of individuals, or species) arise in the context of statistical modeling. Our view is that when inference about occurrence is the objective of a study, it is imperative that imperfect detectability of species be considered. In most problems it is the central component of the observation model and it seems difficult to argue successfully that ignoring it should somehow be innocuous because one must under-state occurrence probability and distribution unless sampling is exhaustive.

Even if the simple objective is to estimate covariate effects, then imperfect detection leads to biased estimates of covariate effects (Gu and Swihart, 2004; Mazerolle et al., 2005). This can be demonstrated empirically in a few lines of **R** code. In Panel 3.4, we simulate a covariate, x, for 10000 hypothetical sites and compute occurrence probability under the model $\psi = \text{expit}(-1 + 1 * x)$. Further, we suppose that the probability of detecting the species at an occupied site is 0.65. The data are realized, and then the model that ignores imperfect detectability is fitted using the function `glm`. We see that the estimated intercept is substantially under-estimated as we expect. But so too is the estimate of the covariate effect. In this case of a single realization, we have $\hat{\beta} = 0.80$ compared to its true value of $\beta = 1.0$. Due to the sample size, these differences are approximately equal to the actual bias due to model misspecification. Thus, when imperfect observation of occurrence state is unaccounted for, we will tend to under-state the distribution, extent of occurrence, and effects of covariates. We note that this bias is likely to be heterogeneous spatially (and temporally), e.g., it may be more substantial around the edge of a species' range. This is because, typically, the probability of detecting a species in most practical situations is related to species abundance (see Chapter 4). Thus, there is an inherent confounding of detectability with ecological processes that might be of interest. This consideration appears to be largely ignored in the biogeographical literature and elsewhere.

Given the influence of detection bias on models of occupancy, how do we accommodate this phenomenon in modeling and estimation from presence/absence data? This requires either some additional information obtained through alternative sampling protocols (e.g., those which produce more replicates), some additional model structure (e.g., Section 4.6) or a combination of both.

3.3.3 The Repeated Measures Design

We have introduced a new parameter, probability of detection, or p. Unlike the the parameter ψ that governs the process that is usually of biological interest, probability of detection characterizes our ability to detect a species and it is usually a nuisance parameter (i.e., not directly relevant to the science of the problem). We also noted that it is not always the case that these two parameters are uniquely identifiable. For example, if only a single replicate is obtained then p and ψ cannot be estimated separately when only a single observation at each site is available. To

```
R>  x<-rnorm(10000)
R>  psi<-expit(-1+1*x)
R>  z<-rbinom(10000,1,psi)
R>  y<-rbinom(10000,1,.65*z)
R>  glm(y~x,family=binomial())

    Call:  glm(formula = y ~ x, family = binomial())

    Coefficients:
    (Intercept)              x
        -1.5611        0.8003

    Degrees of Freedom: 9999 Total (i.e. Null);   9998 Residual
    Null Deviance:       9966
    Residual Deviance: 9088            AIC: 9092
```

Panel 3.4. R instructions to simulate occurrence data subject to detection bias and fit the model which ignores detection bias.

obtain data that are informative about both parameters, we need to make certain strict model assumptions (see Section 4.5), or make a modification to the sampling protocol. Here, we consider the modification in which replicate samples of at least some spatial sample units are taken. That is, we repeat the survey $J > 1$ times at each site, or at a sample of sites. These replicates yield information on the detection process.

Formally, the data form a two-dimensional array, $\{y_{ij}\}_{M \times J}$. A hypothetical data set is displayed in Table 3.4. To understand why this expanded data structure yields additional information, note that each row constitutes a multivariate observation for a particular site. In fact, the replicate observations are analogous to repeated measures such as arise in many classical areas of statistics (animal science, biostatistical applications, etc.). This repeated measures structure induces a certain dependence structure, such that replicate observations (at the same site) are correlated with one another. This is classically referred to as the intra-class correlation. In particular, using the relationship between marginal and conditional moments (e.g., see Section 1.3.2), it can be shown that the correlation between successive measures of presence/absence under specific model assumptions (described subsequently) is given by:

$$\mathrm{Corr}(y_1, y_2) = \frac{p(1 - \psi)}{1 - p\psi},$$

where p and ψ are the probability of detection and occurrence, respectively. Note that the correlation is monotone, increasing to 1 as p increases. Thus, correlation, repeatability (of samples), and detectability are largely synonymous in a precise, mathematical way under a model in which imperfect observation is described by false-negative errors with probability of detection p. The extra information under the repeated measures design derives from this correlation, which can be estimated from the replicate observations.

Statisticians would typically approach the modeling of such data using various regression procedures, such as a mixed logit type of model with individual effects (i.e., a generalized linear mixed model). This is the basic approach that we referred to as an implicit process model in Section 1.4.3. However, we prefer the explicit process model formulation, in which the hierarchical model is parameterized in terms of biologically meaningful processes such as detection and occurrence. The correlation might be modeled adequately under either approach, but it actually means something ecological under the explicit process model formulation.

3.3.3.1 Relationship to capture–recapture

We note that these data resemble data that would arise from classical capture–recapture studies of marked individual animals. That is, we have encounter or detection histories for a number of sample units, which happen to be sites instead of individual animals (i.e., rows of Table 3.4 could be thought of as corresponding to an individual encounter history over time). One important distinction is that, with *occurrence data* we observe the *all-zero* encounter histories, which we are not able to do in capture–recapture type problems. Thus, when confronted with these all-zeros, the statistical modeling/inference problem is essentially to partition the zeros into structural and sampling zeros. Nichols and Karanth (2002) recognized the duality between the problem of estimating occupancy and estimating the size of a closed population. When confronted with data of the form Table 3.4, they discarded the all-zero encounter histories and applied methods of closed population size estimation (see Chapter 5) to estimate the effective population size – which, in the context, is equivalent to the number of occupied sites. Then they divided this estimate by the number of sites that were sampled to obtain an estimate of ψ. We discuss this issue more in Section 3.7.

3.3.4 The Closure Assumption and Temporal Scale

To estimate both probability of detection and occurrence probability, we need to assert something analogous to the classical closure assumption in capture–recapture – that is, occupancy status of sites does not change. Biological context and inference objectives are crucial to the assertion of closure. For example, in monitoring most biological systems, we probably would not use replicates that are separated by 1 year because local populations are experiencing mortality and recruitment and therefore habitat patches are experiencing extinction and colonization at that time scale. For most vertebrates, we might be satisfied with sampling intervals of a day or even a few weeks, such that sampling covers the breeding season in the case of birds. While demographic closure is rarely strictly true, we hope that departures from it are negligible in practice. In some cases, violations of closure do not have detrimental effects. For example, in modeling anuran occupancy during the breeding season, suppose that replicate samples are spaced over 3 months. Some sample periods might be too early or too late for certain species of anurans to be detected, even if a particular wetland is occupied by breeding anurans during the breeding season. Are those early and late surveys non-detection, or non-occurrence? The scope of inference is crucial to making this assessment. If we view closure too narrowly, so that all ponds have to be physically occupied during the sampling, then we might under-state the relevant biological summary in this case. It might be that 80 out of 100 sampled wetlands were ever occupied during the breeding season but, when

Table 3.4. Encounter history data obtained under a design having replicate samples at each site – the repeated measures design.

site	replicate 1	replicate 2	replicate 3
1	0	0	0
2	0	0	0
3	0	1	0
4	1	0	1
5	0	1	1
6	0	0	0
7	0	1	1
8	1	1	1
9	0	0	0
10	1	1	1

restricted to the limited period of strict closure, perhaps only 20 out of 100 are. Clearly the 20 is less biologically relevant than 80 if the focus is on the breeding season. The solution is to let objectives and biology determine the relevant period over which $z = 1$ is relevant and meaningful. In the present case, sampling anurans, the event $z = 1$ is the event that a wetland is occupied *at all* during the breeding season. Thus an absence $y = 0$ that is too early in the season is a legitimate *sampling error* (i.e., it is a result of incorrect observation of the relevant state – not a correct observation of the process – that the pond was not used for breeding). Thus, we obtain replicate samples over the breeding season and we 'average over' this within-season variation in occurrence. There is a formal confounding of parameters in this case that is analogous to the effect of temporary emigration in certain capture–recapture models (Kendall, 1999).

3.3.5 Power of the Survey Method

Detection probability can be thought of as being related to the 'power' of the survey method. Consider the statistical hypothesis (which might be relevant to some assessment problems; Kéry (2002)) $H_0 : z = 0$ (species doesn't occur), against the alternative hypothesis $H_a : z = 1$ (species does occur). In this context, power is the probability of rejecting the null hypothesis (non-occurrence) given that the alternative is true (occurrence). This is precisely the probability of detection given presence (i.e., p). In a survey based on J replicates where the observation y is the number of times that detection occurs, then, under binomial sampling, we have

$$\text{Power} = \Pr\left(\text{reject } H_0 \text{ given } H_a\right)$$
$$\Pr(\text{detect}|\text{present}) = \Pr(y > 0|z = 1)$$
$$= 1 - (1 - p)^J.$$

3.4 HIERARCHICAL FORMULATION OF A MODEL FOR OCCUPANCY

The model for occupancy allowing for imperfect detection has a natural formulation as a hierarchical model, containing explicit models for both the observations and the ecological process that is the focus of inference. We describe this formulation explicitly here. In Section 3.5, we describe the classical formulation (a zero-inflated binomial model) that is prevalent in the literature.

To introduce this class of models, we consider the simplest interesting case in which the occupancy state variable z has 2 states $z \in \{0, 1\}$ (as such, we will sometimes refer to this model as the two-state occupancy model). Suppose further that M sites are sampled J times, and that p is constant across replicate surveys. The observations are y_{ij} for $j = 1, 2, \ldots, J$, and for $i = 1, 2, \ldots, M$ sites. To describe the model, note that for an *occupied site*, y_{ij} are *iid* Bernoulli outcomes having parameter p. Thus, the site totals $y_i = \sum_{j=1}^{J} y_{ij}$ have a binomial distribution with parameter p and index, or sample size, J. Conversely, for sites that are *unoccupied*, y_{ij} must be zero with probability 1 and so also must be the site totals y_i. Our definition of y_i is something of an abuse of notation, being inconsistent with the usual 'dot notation' which would have us write $y_{i.} = \sum_{j=1}^{J} y_{ij}$. But we find that the '.' is difficult to read and sometimes gets lost.

These considerations fully specify the hierarchical model that is composed of two components. The observation model is described conditional on the state variable z, according to:

$$y_i \sim \text{Bin}(J, p\, z_i)$$

i.e., a binomial distribution with parameter $p \times z_i$. Thus, if the site is occupied ($z_i = 1$), the observations are binomial with parameter p. Conversely, if the site is unoccupied ($z_i = 0$), then observations are binomial with probability 0 (i.e., the observations must be zero). The process model is simply

$$z_i \sim \text{Bern}(\psi_i)$$

with, for some covariate x_i that is thought to influence occupancy,

$$\text{logit}(\psi_i) = \beta_0 + \beta_1 x_i.$$

Note that this process component of the model is precisely the logistic regression model that we considered previously in Section 3.1. The two-part observation model could also be expressed according to:

$$\begin{aligned}
y_i &\sim \text{Bin}(J, p) \quad && \text{if } z_i = 1 \\
y_i &= 0 && \text{if } z_i = 0.
\end{aligned}$$

3.4.1 Bayesian Analysis of Hierarchical Representation

Because of the simple conditional structure of the model, a Bayesian analysis of the hierarchical specification is straightforward. We need first to prescribe prior distributions for the model parameters. In this case, it is customary (and natural) to suppose that $p \sim \mathrm{U}(0,1)$ and $\psi \sim \mathrm{U}(0,1)$, since both parameters are probabilities. Having identified the unknown quantities of the model, and specified prior distributions, we need to characterize the conditional posterior distributions of each unknown. It can be shown that the conditional posterior distribution of p given the remaining parameters and the data is $\mathrm{Be}(a_p, b_p)$ with $a_p = 1 + \sum_{i=1}^{M} y_i z_i$ and $b_p = 1 + J \sum_{i=1}^{M} z_i - \sum_{i=1}^{M} y_i z_i$. The conditional posterior distribution of ψ is $\mathrm{Be}(a_\psi, b_\psi)$ with $a_\psi = 1 + \sum_i z_i$ and $b_\psi = 1 + M - \sum_i z_i$. Finally, we need the full-conditional for the state variables z. Note that for each sample location where $y_i > 0$, it is necessarily the case that $z_i = 1$, and these values of z can be fixed. However, wherever $y_i = 0$, we need to obtain a realization of z_i from the conditional distribution $z_i | y_i = 0$. For this, we draw a Bernoulli outcome having probability

$$\Pr(z_i = 1 | y_i = 0) = \frac{(1-p)^J \psi}{(1-p)^J \psi + (1-\psi)}. \qquad (3.4.1)$$

Implementation of an MCMC algorithm for this problem is thus very simple, requiring repeated sampling from beta and Bernoulli distributions. A Bayesian analysis of this model can just as easily be carried out in **WinBUGS**. The **WinBUGS** model specification is shown in Panel 3.5.

3.4.2 Example: Analysis of the Willow Tit Data

We return now to the willow tit data to which we previously fitted simple logistic regression models containing effects of elevation and forest cover. Inclusion of these effects in a model for occurrence probability can be accomplished directly, as shown in Panel 3.6. The estimates under this model are given in Table 3.5. We computed the posterior distribution of the area occupied by recombining the posterior samples of the model parameters in **R**, which is somewhat more efficient than sampling each z_i for every potential grid cell in **WinBUGS**. This yields an estimate (posterior mean) of $12\,080$ km^2 occupied by the species, as compared to 9969 from the previous analysis, which was based on apparent presence/absence from a single sample. Note that the power to detect this species in 3 visits is $\Pr(\text{detection}|\text{occupied}) = 1 - (1 - 0.79)^3 \approx 0.99$. As such, there may be little advantage to formally accounting for detection probability (given 3 visits) as we obtain almost a perfect observation of presence/absence. However, there does not

Table 3.5. Bayesian estimates of model parameters for the willow tit data, under the ordinary logistic regression and the two-state occupancy model allowing for imperfect detection.

parameter	Logistic reg		Imperfect detection	
	mean	sd	mean	sd
A_{occ}	9969.000	940.000	12 080.000	978.000
intercept	−0.561	0.281	−0.208	0.282
elev	1.941	0.329	2.081	0.325
elev^2	−1.122	0.296	−1.150	0.284
forest	0.659	0.220	0.870	0.240
p	−	−	0.787	0.029

```
model {
      psi ~ dunif(0,1)   # prior distributions
      p   ~ dunif(0,1)
      for(i in 1:M){
          z[i] ~ dbin(psi,1)       # process model
          mu[i]<-z[i]*p            #
          y[i] ~ dbin(mu[i],J)     # observation model
      }
}
```

Panel 3.5. WinBUGS model specification for the simple two-state occupancy model with constant p and ψ. The data, passed from **R**, are the number of sites M, the number of replicates J, and the vector of detection frequencies y.

appear to be a flexible alternative to modeling such data when observations are subject to error.

Based on the present analysis, we conclude that the effective detection probability (per sample) for the simple logistic regression model in Section 3.1.1 was about 0.79. This influences the estimated covariate effects in the expected way. For example, we see that the effect of forest cover is 0.659 under the simple logistic model, and 0.870 under the model which adjusts for imperfect detection. The response of occurrence probability to both forest cover and elevation are displayed graphically[3] in Figure 3.5. We see that the effect is not constant across either forest or elevation. There is a more pronounced difference in estimated occurrence where conditions favor occurrence.

---

[3] The figure uses the estimates obtained from the Bayesian analysis of the simple logistic model, which differs only slightly from the MLEs.

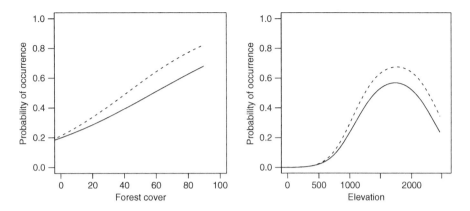

Figure 3.5. Fitted response in occurrence probability for the willow tit to forest cover (percent of quadrat covered in forest) and elevation (meters above sea level). The solid line is based on estimates under the ordinary logistic regression model. The dashed line is under the occupancy model that allows for imperfect detection.

3.4.3 Bayesian Model Selection: Calculation of Posterior Model Weights

Model selection using classical likelihood methods is often based on Akaike's Information Criterion (AIC). The use of AIC has become widespread owing to the popularity of the book by Burnham and Anderson (2002), and some harbor strong feelings and beliefs regarding the use of AIC to the exclusion of alternative methods of model selection. This has recently been called into question by a number of authors, for various reasons (Stephens et al., 2005; Guthery et al., 2005; Link and Barker, 2006), and resulted in some enthusiastic debate around coffee pots, water coolers, at the pub, and in the literature (Lukacs et al., 2007; Stephens et al., 2007). There is a similar model selection criterion, referred to as the deviance information criterion (DIC) (Spiegelhalter et al., 2002; van der Linde, 2005, also, see Chapter 2), which **WinBUGS** computes for certain models[4].

Perhaps one of the reasons that model selection in Bayesian analyses has lagged behind the adoption of such methods in non-Bayesian applications is because there is no automatic way to do it[5]. Bayes factors can be difficult to compute, and there is extreme sensitivity to prior distributions (Kass and Raftery, 1995; Kadane and

---

[4]The DIC FAQ is available at http://www.mrc-bsu.cam.ac.uk/bugs/winbugs/dicpage.shtml, current as of May 13, 2007.

[5]A virtue of classical statistical methods cited by Efron (1986) for which he got severely pummeled by Lindley (1986).

Lazar, 2004; Link and Barker, 2006). The computation problem might now be less of a practical issue as software becomes more available. But, as of yet, we don't know a way around the prior sensitivity issue.

```
model {
      for(i in 1:M){
         z[i] ~ dbin(psi[i],1)
         logit(psi[i]) <- b0 + b1*elev[i] + b2*elev2[i] + b3*forest[i]
         tmp[i]<-z[i]*p
         y[i] ~ dbin(tmp[i],J[i])
       }

      p ~ dunif(0,1)
      b0 ~ dnorm(0,.001)
      b1 ~ dnorm(0,.001)
      b2 ~ dnorm(0,.001)
      b3 ~ dnorm(0,.001)
}
```

Panel 3.6. **WinBUGS** model specification for occupancy model with covariates on occurrence probability, ψ, but not detection probability p.

Nevertheless, given a particular choice of prior distributions for model parameters, one can compute posterior model weights by explicit specification of a prior distribution on the *model set* itself. This can be done (e.g., in **WinBUGS**), by specifying a set of latent indicator variables, one for each model effect, say w_k for effect k, and imposing a Bernoulli prior on each w_k, say having parameter π_k. This notion was suggested by Kuo and Mallick (1998) (see also Congdon, 2005, Section, 3.2). Specification of the π_k dictates the prior probability for each model. As an example, using the willow tit data with covariates elev, elev^2, and forest, we introduce the set of latent *indicator variables*

$$w_k = \begin{cases} 1 & \text{if covariate } k \text{ coef} \neq 0 \\ 0 & \text{if not} \end{cases}$$

having prior distributions:

$$w_k \sim \text{Bern}(0.5)$$

which we assume here to be mutually independent. The model is expanded by specifying the linear predictor as

$$\text{logit}(\psi_i) = \beta_0 + w_1\beta_1 \, \text{elev}_i + w_1 w_2 \beta_2 \, \text{elev}_i^2 + w_3\beta_3 \, \text{forest}_i.$$

The null model, having no covariates, occurs if $w_1 = 0$ and $w_3 = 0$, which has prior probability 0.25. The model with linear and quadratic elevation effects, but not forest, occurs with probability 0.125, etc. Note that the quadratic elevation term cannot enter the model unless the linear term is also in the model (since the quadratic term interacts with the product $w_1 w_2$). Whether or not this is a sensible prior on the model set is debatable, as it should be. The **R** and **WinBUGS** model specification to obtain the results reported here are available in the Web Supplement.

As we noted above, prior distributions, both on the model set and on the parameters of the models, can have an acute influence on calculation of posterior model probabilities (Link and Barker, 2006). Currently, there does not seem to be a general or automatic solution to this issue, but it should be admitted in any analysis and priors should be clearly stated. Adherents to AIC seem to avoid this issue primarily by considering only a single prior distribution, which they have termed (Burnham and Anderson, 2004, p. 282) the 'savvy prior.' To illustrate this sensitivity to prior, we did an analysis of the willow tit using a number of priors on the parameters. To make this illustration somewhat more interesting, we added three additional covariates, a quadratic effect of forest cover, a covariate that was simulated $N(0, 1)$ noise, and also `length` (route length). Results of the Bayesian model selection exercise are reported in Table 3.6.

The model containing quadratic elevation and forest effects is the best model when $\tau = 0.1$, followed by the model with a linear elevation and quadratic forest effects. The third model includes the quadratic elevation, and forest effects and route length. When the prior is made more diffuse, the order switches and the model without `elev`^{2} gets more weight followed by that with quadratic forest and elevation effects, but no route length. For $\tau = 0.001$, the posterior model weight concentrates on the model with quadratic forest and linear elevation effects. We see that there is demonstrable sensitivity to the prior distributions on the regression parameters. In addition, there seems to be an important quadratic effect of forest cover, a possibility that we have not previously considered (e.g., Kéry et al. (2005); Royle et al. (2005)). The estimate (not reported) indicated a concave response which has a good biological explanation[6].

3.5 OCCUPANCY MODEL AS A ZERO-INFLATED BINOMIAL

In Section 3.4, we described a hierarchical formulation of models for occurrence. The hierarchical formulation in which the latent (occupancy) state variables are retained

---

[6]The willow tit likes trees....but not too many.

in the model yields naturally to a Bayesian analysis since the component models have simple conditional specifications. This hierarchical formulation is distinct from the typical approach (e.g., described in MacKenzie et al. (2006) and related literature), which is based on the marginal distribution of the observations, i.e., having removed the latent process z_i from the likelihood by *marginalization* (or integration) (Section 2.6). This is the basic idea underlying integrated likelihood, the standard classical/frequentist procedure for the analysis of models with random effects or latent variables (Laird and Ware, 1982; Searle et al., 1992). When the latent variable is discrete, as it is in this case, then integrated likelihood is easy to calculate, being equivalent to averaging the conditional (on z) likelihood over only two possible states. This can be seen as an application of the law of total probability (Chapter 2). In the present case,

$$\Pr(y) = \text{Bin}(y|J,p)\Pr(z=1) + \text{Bin}(y|J,p)\Pr(z=0).$$

Under the assumption that false positives are not possible, the second term on the right-hand side evaluates to a point-mass at $y = 0$. Since $\psi = \Pr(z=1)$ and $1 - \psi = \Pr(z=0)$, we have

$$\Pr(y|p,\psi) = \text{Bin}(y|J,p)\psi + I(y=0)(1-\psi), \qquad (3.5.1)$$

where $I(arg)$ is the indicator function, $I(arg) = 1$ if arg is satisfied. The *marginal* distribution (i.e., not conditional on occupancy status) of y is a mixture of a binomial and a point-mass at zero, a distribution that is commonly referred to as a *zero-inflated binomial* (ZIB). That is, the observed counts are binomial counts with extra zeros mixed in. The likelihood is the product over sites of terms having the basic form as in Eq. (3.5.1):

$$L(\psi, p|y_1, y_2, \ldots, y_M) = \prod_{i=1}^{M} \left\{ \text{Bin}(y_i|J,p)\psi + I(y=0)(1-\psi) \right\}.$$

The ZIB model can be readily implemented in **R**, using the likelihood construction given in Panel 3.7. This function is implemented using the Swiss bird data in the Web Supplement.

3.6 ENCOUNTER HISTORY FORMULATION

We have so far considered models having covariates that influence occurrence (although in the simple logistic regression the covariates influence the *apparent* occurrence probability). In practice, there is often considerable variation in conditions across surveys that occur at different times or under different conditions,

Table 3.6. Bayesian model selection results for the willow tit data. The model is described in each row by a binary sequence indicating whether each effect is in the model (1) or not (0). The right-most 3 columns are the posterior model probabilities under 3 prior regimes for the regression coefficients: The priors for all regression coefficients were $\beta \sim N(0, \tau^{-1})$ for $\tau \in \{0.1, 0.01, 0.001\}$.

Model description						Posterior model probabilities		
el	el2	fo	fo2	no	len	$\tau = 0.1$	$\tau = 0.01$	$\tau = 0.001$
1	1	1	1	0	0	0.447	0.312	0.136
1	0	1	1	0	0	0.284	0.589	0.832
1	1	1	1	0	1	0.132	0.028	0.003
1	0	1	1	0	1	0.065	0.048	0.019
1	1	1	1	1	0	0.032	0.011	0.002
1	0	1	1	1	0	0.018	0.009	0.007
1	1	1	1	1	1	0.008	0.001	0.000
1	0	1	1	1	1	0.007	0.001	0.000
1	1	1	0	0	0	0.005	0.001	0.000
1	1	1	0	0	1	0.001	0.000	0.000
1	1	1	0	1	0	0.001	0.000	0.000
1	1	1	0	1	1	0.000	0.000	0.000

things that influence detection probability but not occurrence probability. In addition to weather or environment, we might have observer differences, variation in sampling intensity or effort (per unit area), or even protocol changes which may

```
lik<-function(parms,vars){
    tmp<-c(0,0,0,0,0)
    names(tmp)<-c("pconst","psiconst","elev1", "elev2","forest")
    tmp[vars]<-parms
    ones<-rep(1,M)
    p<-expit(tmp[1])
    b0<-tmp[2]
    b1<-tmp[3]
    b2<-tmp[4]
    b3<-tmp[5]
    psi<-expit(b0*ones + b1*elev + b2*(elev^2) + b3*forest)
    -1*sum(log(dbinom(y,J,p)*psi  +  ifelse(y==0,1,0)*(1-psi)))
}
```

Panel 3.7. Implementation of the zero-inflated binomial likelihood for the 2-state occupancy model in **R**. As described here, the model being fitted has constant ψ and p.

or may not be precisely described. Such covariates may vary by site and survey occasion and should influence p accordingly.

For the Swiss BBS, several covariates are available. In particular, sampling occurs over the breeding season (a period of about 3 months), and it stands to reason that the detectability of most bird species will vary seasonally, as a result of breeding/nesting behavior. Thus, we consider $\mathtt{day}_{ij}$, the integer day (from April 1) on which sampling occurred for replicate j of quadrat i. Secondly, we consider a measure of sampling intensity, $\mathtt{intensity}_{ij} = \mathtt{duration}_{ij}/\mathtt{length}_i$ where $\mathtt{length}_i$ is the length (km) of route i and $\mathtt{duration}_{ij}$ is the period of time (minutes) that it took to complete replicate j at quadrat i. Finally, we should consider route length itself as a covariate (i.e., not interacting with duration). It seems reasonable to view this as a covariate relating to effort, but our view, following Kéry et al. (2005) (see also Royle and Dorazio (2006), and Royle et al. (2007b)), is that it influences occupancy directly by serving to increase the effective sample area. That is, the area exposed to sampling should increase monotonically with increasing route length. Area accumulation should affect the process component of the model when the state variable is related to abundance or occupancy. Some of the covariate data are given in Table 3.7. We suppose that the effect of Day could possibly be quadratic. As noted previously, we have standardized these covariates to have mean 0 and variance 1.

Whereas previously we had aggregated data into site-totals, y_i, which we noted was reasonable provided that samples of the same site were *iid* Bernoulli trials, the analysis of models having such effects requires a more general formulation of the model, a formulation based on site-specific encounter histories. In the more general case, the observations may have distinct detection probabilities in each replicate sample.

3.6.1 Likelihood Analysis

In the presence of observation-level covariates, we require a formulation of the likelihood of the site-specific encounter history $\mathbf{y}_i = \{y_{ij}; j = 1, 2, \ldots, J\}$, a sequence of 0s and 1s. There are 2^J possible encounter histories – all of which are potentially observable.

With observation-level covariates, under the assumption that replicate samples yield independent observations, the individual encounter history probabilities are of the form:

$$\Pr(\mathbf{y}_i | z_i = 1) = \prod_j p_{ij}^{y_{ij}} (1 - p_{ij})^{1 - y_{ij}}$$

with

$$\mathrm{logit}(p_{ij}) = \alpha_0 + \alpha_1 x_{ij}$$

Table 3.7. A subset of the Swiss data, showing the covariate data for a few sample quadrats. The 'NA' entries correspond to missing data in **R** data files. The variables day1–day3 are the integer day-of-year on which each sample was taken; dur1–dur3 are the duration of each sample in minutes, and length is the route length (km) walked through the quadrat.

day1	day2	day3	dur1	dur2	dur3	length
29	58	73	240	240	240	6.2
13	39	62	160	155	140	5.1
30	47	74	120	105	85	4.3
23	44	71	180	170	145	5.4
28	56	73	210	225	235	3.6
17	56	73	150	180	195	6.1
16	37	76	115	105	95	5.1
24	47	74	155	140	135	3.7
25	46	70	165	165	180	3.8
21	38	50	220	230	255	7.7
30	60	74	135	150	135	3.1
39	64	90	NA	NA	NA	7.9
28	49	81	240	230	250	4.4
41	69	71	145	180	165	6.0
18	50	72	175	270	210	6.2
35	58	75	260	270	290	6.1
28	39	61	176	145	150	4.5
21	40	86	180	160	165	6.2
20	42	63	195	240	270	6.9
35	50	75	150	130	140	4.6

for some covariate x_{ij} that varies by replicate and spatial sample unit i. To account for the sampling of (potentially) unoccupied sites, we have to zero-inflate these encounter history probabilities. Because these cell probabilities constitute those of a multinomial distribution for the 2^J potential encounter histories, the resulting zero-inflated distribution is a zero-inflated multinomial. The zero-inflated multinomial is the product of components having the general form:

$$\Pr(\mathbf{y}) = \prod_i \left\{ \Pr(\mathbf{y}_i | p_{i1}, p_{i2}, \dots, p_{iJ})\psi_i + I(\mathbf{y}_i = \mathbf{0})(1 - \psi_i) \right\}.$$

This can be implemented directly in **R**, (see Panel 3.8). The key feature of the **R** formulation of the likelihood is that, in the loop over M (the number of sites), each observation consists of the encounter history (a vector of length J), which is extracted as a row from the $M \times J$ data matrix, ymat. The corresponding detection probabilities vary by site and sample, whereas the occurrence probability ψ remains a function of site-specific covariates as we've seen in several previous analyses. The detection covariates are also $M \times J$ matrices. For example, the object date has elements date$[i, j]$, the date on which sample j was obtained for quadrat i. A number of models were fit to the willow tit data and the results are

Table 3.8. Models of occupancy fitted to the willow tit data from the Swiss BBS. These models contain covariates on both occurrence probability (ψ) and also detection probability (p). The models were fitted using maximum likelihood methods applied to the zero-inflated multinomial likelihood.

	Detection parameters				Occurrence parameters				
α_0	intensity	date	date^2	β_0	elev	elev^2	forest	length	AIC
1.329	–	–	–	−0.207	1.985	−1.091	0.846	–	433.468
1.330	–	–	–	−0.171	2.032	−1.127	0.823	0.141	434.945
1.252	0.196	–	–	−0.161	2.036	−1.130	0.823	0.153	436.086
1.241	0.146	0.153	–	−0.163	2.029	−1.131	0.825	0.15	437.376
1.307	0.155	0.170	−0.074	−0.163	2.028	−1.130	0.824	0.15	439.145
1.316	–	–	–	−1.050	1.348	–	1.219	–	452.311
1.223	0.237	–	–	−0.656	–	–	–	–	529.876
1.200	0.190	0.177	–	−0.656	–	–	–	–	530.929
1.280	0.206	0.195	−0.094	−0.654	–	–	–	–	532.553
1.292	0.188	0.192	−0.095	−0.661	–	–	–	−0.164	533.243

given in Table 3.8. The best model by AIC favors a quadratic elevation effect, forest cover, but no other covariates. It is somewhat surprising that none of the observation covariates appear to be relevant. We address this issue in subsequent analyses of these data (e.g., Sections 4.3.1 and 8.3.1.4).

3.6.2 Hierarchical Formulation

The hierarchical formulation of such models is straightforward. We have to assume that each y_{ij} has a distinct probability of detection. The observation model is of the form

$$y_{ij} \sim \text{Bern}(p_{ij} z_i)$$

with

$$\text{logit}(p_{ij}) = \alpha_0 + \alpha_1 \, \text{day}_{ij} + \alpha_2 \, \text{day}_{ij}^2 + \alpha_3 \, \texttt{intensity}_{ij}.$$

The occupancy process model is unchanged

$$z_i \sim \text{Bern}(\psi_i)$$

with

$$\text{logit}(\psi_i) = \beta_0 + \beta_1 \, \texttt{elev}_i + \beta_2 \, \texttt{elev}_i^2 + \beta_3 \, \texttt{forest}_i + \beta_4 \, \texttt{length}_i$$

Bayesian analysis of the model in **WinBUGS** is also straightforward. We provide the implementation from **R** in the Web Supplement.

3.7 FINITE-SAMPLE INFERENCE

We believe that few appreciate the distinction between the estimation of population parameters (e.g., the probability that a site is occupied), and estimation of finite-sample manifestations of those quantities such as the number of occupied sites in the sample. The probability of occurrence parameter ψ is a population average, and associated estimates apply to a theoretically infinite population of sites from which the sample of size M was drawn. However, given a particular collection or sample of M sites, the actual proportion of those sites occupied is a function of the latent state variables:

$$\psi^{(fs)} = \frac{1}{M} \sum_i z_i, \qquad (3.7.1)$$

where 'fs' is for finite sample. We might naturally refer to $\psi^{(fs)}$ as the occurrence or occupancy *rate*, whereas ψ is the occurrence *probability*.

```
lik<-function(parms,vars){
    tmp<-c(0,0,0,0,0,0,0,0)
    names(tmp)<-c("pconst","psiconst","elev1","elev2","forest",
                "intensity","date1","date2")
    tmp[vars]<-parms
    ones<-rep(1,M)
    pmat<-expit(tmp[1]*ones + tmp[6]*intensity + tmp[7]*date + tmp[8]*date^2)
    psi<-expit(tmp[2]*ones + tmp[3]*elev + tmp[4]*elev^2 + tmp[5]*forest)
    loglik<-rep(NA,M)
    for(i in 1:M){
        yvec<-ymat[i,]
        navec<-is.na(yvec)
        nd<-sum(yvec[!navec])
        pvec<-pmat[i,]
        cp<- (pvec^yvec)*((1-pvec)^(1-yvec))
        cp[navec]<-1
        loglik[i]<-log(prod(cp)*psi[i] + ifelse(nd==0,1,0)*(1-psi[i]))
    }
    sum(-1*loglik)
}
```

Panel 3.8. R construction of the likelihood for the occupancy modeling having covariates that influence both detection probability and occurrence probability. M is the number of sites. The data objects are the $M \times J$ matrices intensity, and date, and the $M \times 1$ vectors elev, and forest. The observations are contained in the $M \times J$ matrix ymat.

The distinction between ψ and $\psi^{(fs)}$ has to do primarily with the scope of inference. If one is interested in the particular sites for which data were collected (or, in fact, the occupancy states of any specific collection of sites), then the quantity of interest is the finite-sample *proportion* of occupied sites, $\psi^{(fs)}$ (so-called 'proportion of area occupied', or PAO). Conversely, if one is interested in a (much) larger collection of sites from which the sampled sites are representative (e.g., randomly selected), then ψ is the relevant parameter. While the distinction may seem subtle, there can be important practical differences. In particular, while point estimates of the two quantities will be similar, their uncertainty will typically be very different, by virtue of having observed $y = 1$ for some of the M sites. There is no uncertainty about their corresponding occupancy states (i.e., $z_i = 1$ wherever $y_i = 1$) and they contribute nothing to the total variance of estimating $\psi^{(fs)}$.

Consider this issue in a specific context. Suppose a wildlife refuge has exactly 100 distinct wetlands on it, all of which are known and sampleable. And suppose the manager is tasked with some basic monitoring and assessment of those wetlands, so that she has to report, on an annual basis, how many of them were used by some rare species of frog that only occurs on the refuge. The refuge statistician convinces the refuge manager that the proper analytical framework for carrying out this assessment is to let z_i denote the occupancy status of wetland $i = 1, 2, \ldots, 100$ on the refuge, assume that $z_i \sim \text{Bern}(\psi)$, and then obtain replicate samples on a number of basins. It is decided that 50 of those wetlands will be sampled (randomly), two times each. Clearly, the refuge manager is not interested in making an inference about the parameter ψ, which is the mean of a hypothetically infinite population of wetlands from which those on the refuge represent a random (or at least representative) sample. This hypothetical population is vague and may not even exist. Thus, the refuge manager and statistician determine that they would like to obtain an estimate of $\sum_{i=1}^{100} z_i$ and a characterization of its uncertainty.

Of the 50 sampled wetlands, sampling revealed that 30 were occupied and 20 were unoccupied. The estimation problem is composed of two parts. First, we need to estimate the fraction of the 20 apparently unoccupied wetlands that really were occupied, a quantity which we will call n_0. Secondly, we need to obtain a *prediction* of how many of the remaining 50 (those that went unsampled) are occupied, say x_0. Then, an estimator of N, the number of occupied sites, including these 50 unsampled sites, is:

$$\hat{N} = 30 + \hat{n}_0 + \hat{x}_0.$$

Clearly the uncertainty affects this estimator only through the last two components of this expression – i.e., the two unobservable components.

Let's first consider inference about the number occupied out of the 20 apparently unoccupied sites. A wrong answer for estimating the number occupied out of these

20 is to assert that $n_0 \sim \mathrm{Bin}(20, \psi)$ and to use $\hat{\psi}$ in the estimator $\hat{n}_0 = 20\hat{\psi}$ for which we might claim the variance to be $400 \times \mathrm{Var}(\hat{\psi})$. This is wrong for two reasons. First, that quantity is an estimator of $E[n_0]$, not n_0, and it does not include the binomial variance of n_0. Thus, confidence intervals will be too narrow and will not include the true value n_0 with the correct frequency. Secondly, these 20 sites are not like 20 random sites on the landscape about which we know nothing. We know that $\mathbf{y}_i = (0,0)$ for these sites. Thus, what we really need for inference is the probability that such a site is occupied *given* that it was observed to be unoccupied in two samples. This probability is, by Bayes' rule:

$$\pi = \mathrm{Pr}(z = 1 | y = 0) = \frac{(1-p)^2 \psi}{(1-p)^2 \psi + (1-\psi)}$$

which is the same as Eq. (3.4.1) where we used it to simulate realizations from the posterior distribution. Thus, the variance of n_0 given p and ψ, is

$$\mathrm{Var}(n_0 | p, \psi) = 20\pi(1 - \pi).$$

Note that this variance diminishes as $p \to 1$. We don't know p and ψ in this case, and so we require an estimate of π which we could plug into this variance expression. This will introduce additional variance that we need to account for. We can apply the formula relating marginal to conditional variance, yielding

$$\mathrm{Var}(n_0) = 20\pi(1 - \pi) + 20^2 \mathrm{Var}(\hat{\pi}).$$

So the uncertainty due to estimating n_0 has a component due to how far p is from 1 (the first part), and how well (or poorly) we estimated π (the second part), which is a function of p and ψ.

Now, to this we have to add an estimate of how many of the 50 wetlands that weren't sampled are occupied, and we need to characterize our uncertainty about that prediction. According to our process model, we have $x_0 \sim \mathrm{Bin}(50, \psi)$. But we need to use an estimate of ψ and there should be some additional uncertainty induced as a result. The variance of x_0 is

$$\mathrm{Var}(x_0) = 50\psi(1 - \psi) + 50^2 \mathrm{Var}(\hat{\psi}).$$

Note that this will tend to the nominal binomial variance as $p \to 1$ (in which case the second term on the right-hand side diminishes to zero).

We can do all of this accounting in an ad hoc way, as we have just done, but this approach seems more difficult to extend for example, to models that include covariates that influence detection and occurrence. Alternatively, we can adopt a Bayesian framework for drawing posterior samples of each unknown z_i of interest.

Posterior samples of sample quantities or predictions can easily be computed as functions of the posterior draws of the latent z variables. This is one of the main benefits of the hierarchical formulation of the two-state occupancy model. That is, it permits the construction of these finite-sample estimators, as functions of the latent z variables. That some of the z's are unobserved does not pose any difficulty in estimating functions of them, such as $\psi^{(fs)}$, using common methods of Bayesian analysis based on MCMC. In Section 3.4.1 we provided an MCMC algorithm for a simple two-stage occupancy model for which functions of latent state variables can be computed. To make predictions at unsampled sites, one only has to include one additional step in the MCMC algorithm, that being a random draw of some new z from a Bern(ψ) distribution. Alternatively, estimation of functions of z can be achieved using **WinBUGS** (e.g., Panel 3.6).

3.7.1 Example using Swiss BBS

Royle and Kéry (2007) illustrated the distinction between population quantities and their finite-sample manifestations, and we provide a portion of their results here (we revisit the problem in Chapter 9). They considered data from the Swiss BBS over 4 years (2001 to 2004) for the European Crossbill (*Loxia curvirostra*), a species expected to be very dynamic in a metapopulation sense. The crossbill is a medium-sized (34g) pine-seed-eating finch widespread in Switzerland. Its abundance and occurrence depend greatly on the cone set of conifers, and in mast years, crossbills appear irruptively in many regions where otherwise they do not occur or are scarce. Royle and Kéry (2007) fitted a temporally dynamic model (described in Chapter 9) and provided annual estimates ψ_t for each year, in addition to the proportion of occupied quadrats from among the 267 *sampled* quadrats. Interest in this summary of the data could be motivated by concern over the existence of a relevant sample frame, which might be an important problem if the quadrats were selected arbitrarily instead of systematically or randomly. The parameter estimates, including both population and finite sample occupancy estimates, are given in Table 3.9. We see that while the estimates of ψ_t and $\psi_t^{(fs)}$ are equivalent, the sample quantities are estimated much more precisely for the reason described previously.

3.7.2 Likelihood Estimation

It is possible to formalize estimation and inference for the number of occupied sites using conventional likelihood methods. That is, we can expand the zero-inflated binomial likelihood to include a parameter for the number of occupied sites, N

Table 3.9. Estimates of occupancy model parameters for Swiss BBS data on the European Crossbill (*Loxia curvirostra*) from 2001–2004. q_x is the $100 \times x$th percentile of the posterior distribution. Taken from Royle and Kéry (2007). The parameters p_t, ψ_t and $\psi_t^{(fs)}$ are annual detection probability, occurrence probability and proportion of occupied quadrats of the 267 sampled quadrats.

parameter	mean	sd	$q_{0.025}$	$q_{0.500}$	$q_{0.975}$
p_1	0.584	0.044	0.493	0.584	0.666
p_2	0.493	0.037	0.422	0.493	0.564
p_3	0.566	0.033	0.504	0.566	0.629
p_4	0.574	0.037	0.499	0.574	0.643
ψ_1	0.242	0.029	0.190	0.241	0.300
ψ_2	0.391	0.035	0.323	0.390	0.461
ψ_3	0.450	0.034	0.386	0.450	0.517
ψ_4	0.346	0.032	0.286	0.345	0.409
$\psi_1^{(fs)}$	0.240	0.0124	0.222	0.237	0.271
$\psi_2^{(fs)}$	0.389	0.0210	0.353	0.387	0.436
$\psi_3^{(fs)}$	0.449	0.0149	0.425	0.447	0.481
$\psi_4^{(fs)}$	0.345	0.0148	0.320	0.342	0.380

and estimate that parameter directly. This is motivated by the similarity between replicate presence/absence data and normal encounter history of individuals that arises from conventional capture–recapture studies (which we mentioned in Section 3.3.3). In Royle et al. (2007a), we noted that the zero-inflated binomial likelihood could be modified by conditioning the observed encounter histories on N. That is, sites can be classified into three types: those that were occupied and where detection occurred, occupied and not detected, and unoccupied. This yields a trinomial distribution for the three frequencies n (the number of observed occupied sites), $N - n$ and $M - N$. The trinomial distribution can be rearranged to yield the following joint likelihood of ψ, p and N:

$$L(N, p, \psi | \mathbf{y}, M, n) = \left[\frac{N!}{(N-n)!} \ p^{\sum_{i=1}^n y_i} \left(1 - p\right)^{J \cdot N - \sum_{i=1}^n y_i} \right]$$
$$\times \left[\frac{M!}{N!(M-N)!} \ \psi^N (1 - \psi)^{M-N} \right], \qquad (3.7.2)$$

where y_i is the total number of detections from J replicate samples, as before. This likelihood is the product of two components – the likelihood of the site-specific detection frequencies and the observed number of occupied sites, n, conditional on N (in the first set of square brackets), and the conditional (binomial) likelihood of N given M (in the second set of square brackets). Note that the first term is the likelihood for N and p that arises in closed population sampling, say $L_0(N, p|\mathbf{y})$ (see Chapter 5 and Williams et al., 2002, p. 299. Eq. (3.7.2) can be maximized to obtain the MLE of (p, N, ψ) directly. If one is interested only in estimating N, ψ may be removed from the joint likelihood by integration, say, over a $U(0, 1)$ prior distribution. This yields the truncated likelihood

$$L_T(N, p|\mathbf{y}, M) = L_0(N, p|\mathbf{y}) \; \frac{I(N \leq M)}{M + 1}. \tag{3.7.3}$$

Note that if M is sufficiently large, we could ignore the second component of Eq. (3.7.3), obtain $\hat{N}$ by maximizing L_0, and equate $\hat{\psi} = \hat{N}/M$, which is the solution for estimating ψ suggested by Nichols and Karanth (2002), as we noted in Section 3.3.3.

The factorization (Eq. (3.7.2) and hence Eq. (3.7.3)) establishes an approximate duality between estimating the number of occupied sites, wherein the all-zero capture histories are observed, and estimating the population size, where they are not. Specifically, the likelihood for the standard closed population capture–recapture model is one component of the factorization given by Eq. (3.7.2). As such, the occupancy model is an extended version of the closed population model, having a prior distribution on the number of occupied sites N that is discrete uniform on the integers $0, 1, 2, \ldots, M$, and an additional parameter ψ which is the probability that an element on the list of size M is an occupied site.

This basic relationship motivates certain Bayesian analyses of multinomial models that are encountered later in this book. That is, we can apply this duality in the opposite direction – given capture–recapture data (on animals) wherein the zero detection frequencies are *not* observed, we can 'augment' the data with an arbitrarily large (relative to n) number of all-zero capture histories and then estimate the proportion of those zeros that were exposed to sampling (i.e., estimate site occupancy). This provides an efficient method of analysis for many models including models described in Chapters 6 and 7.

3.8 OTHER TOPICS

Modeling and inference for occupancy and occurrence probability is a vast topic which we have only given a brief treatment of here. As a chapter in a book, it is impossible to cover every useful or interesting aspect of the topic. There are a number of topics related to modeling occupancy or occurrence that fit nicely into the framework of hierarchical models. We mention three of them here.

3.8.1 False-Positive Errors

The formulation of the two-state occupancy model described in this chapter (e.g., MacKenzie et al., 2002; Tyre et al., 2003, etc.) is predicated on the assertion that false *positives* are impossible, i.e., that the species will not be detected where it does not occur. While this might appear to be a non sequitur, misidentification of species can occur in field settings, and this can lead to false positives. If false positives do occur, it is absolutely critical that they be accommodated in the model. Otherwise, under a simple binomial sampling scheme, apparent occupancy will tend to 1.0 as the number of visits to sites increases (Royle and Link, 2006). Indeed, even for very low rates of misclassification, the bias in apparent occupancy rates, or those estimated under a model that does not permit false positives, will be extreme. Attention should also be paid to false positives because surveys for many taxa (in particular of birds and anurans) involve the simultaneous sampling of large numbers of species by volunteer observers with variable skill levels. This circumstance is ideal for the introduction of false positives by misidentification of species. It seems plausible that false positives will occur in many survey situations where they have previously been disregarded.

Royle and Link (2006) described a generalization of the MacKenzie et al. (2002) model to the situation where both false negative and false positive observations are possible. The model can be represented as a finite-mixture of binomial random variables in which p_1 is the probability of detection at sites that are truly occupied, and p_2 is the probability of (false) detection at unoccupied sites. The main conceptual limitation of this model is that there is no contextual information about p_1 and p_2. That is, there is no basis for knowing whether it is p_1 that goes with occupied sites or *vice versa*. Thus, one must impose constraints on the model parameters in order to interpret the parameters sensibly. Royle and Link (2006) argued that it should be reasonable to assume that the larger of p_1 and p_2 corresponds to occupied sites (unless statisticians are doing the sampling). The other conceptual problem is that it is not possible to distinguish this model from a model in which $\psi = 1$ and p is itself heterogeneous, according to a finite mixture of the Norris and Pollock (1996) type. That is, heterogeneity in p can appear to be misclassification of the occupancy state in the form of both false positives and negatives. This is unfortunate because it suggests that one cannot allow for both phenomena in the absence of auxiliary data.

3.8.2 Multi-State Occupancy Models

In this chapter we considered models of a simple two-state (binary) occupancy variable defined as presence or absence of a species. However, the basic conceptual and technical framework extends readily to the case where there may be several or many distinct occupancy states, which may correspond to distinct demographic states of the species under study. For example in bird atlas data it is typical to collect data on many different occurrence states relating to evidence of breeding. For example, suppose we categorized sites into classes 'not present,' 'present but not breeding,' 'present and breeding,' and 'successful breeding.' There may be constraints among those states (e.g., breeding is mutually exclusive of not present and present but not breeding), and we may correctly classify (i.e., *observe*) each state with state-dependent (detection) probabilities. Royle (2004b) and Royle and Link (2005) considered multi-state models for vocal anuran surveys where the state variable was defined in terms of a 'calling index'. This is an ordinal variable taking on values 0 to 3 (Weir and Mossman, 2005). A broader conceptual view of multi-state occupancy models is given by Nichols et al. (2007).

3.8.3 Spatial Models of Occupancy

We have assumed that the occupancy states $z_1, z_2, \ldots, z_M$ are independent of one another. In practice, they are likely to be dependent for a host of reasons, including variation in abundance within the range of a species, spatial dependence in landscape structure or, in some cases, that individuals are exposed to sampling at multiple spatial units (e.g., tigers exposed to a camera trap array). The standard framework for modeling spatial dependence in a binary state variable is the use of so-called autologistic models. A good description in the context of modeling species distribution can be found in Augustin et al. (1996). The framework was recently applied to modeling the distribution of carnivores from survey data subject to imperfect detection by Sargeant et al. (2005). We discuss these models in Chapter 9.

3.9 SUMMARY

In this chapter we began with a brief introduction to logistic regression. This is a major building block topic conceptually, practically, and technically. Logistic regression is fundamental to many problems of interest in biogeography, metapopulation biology, and landscape ecology. One important theme in our treatment of logistic regression is the problem of imperfect observation of the occupancy state variable (detection bias). It is difficult to attribute strict ecological interpretations to observational data and resulting models when faced with data subject to detection bias (Moilanen, 2002; MacKenzie et al., 2002; Gu and Swihart, 2004). Non-detection bias is important because it biases effects toward 0 so we generally underestimate the effect of factors that influence occupancy (and they become less significant), and it leads to an under-statement of range or distribution. This problem is considerably more profound than the simple statement that it induces bias, because the detectability of species is often fundamentally related to the process that is the object of inference. For example, we would typically expect that the detectability of a species increases with increasing abundance (see Chapter 4). Thus the bias induced by observation can be related to the process that we are studying. As such, some attention to detection bias, as an important component of the observation model, is critical in studies of distribution and occurrence.

Motivated by this problem of non-detection bias, we considered the important extension of the logistic regression framework to include observation error in the form of non-detection or false negative errors. It is the introduction of this observation component of the problem that makes the logistic regression model an interesting hierarchical model. Indeed, the two-state occupancy model described in MacKenzie et al. (2002, 2006) and elsewhere, which is now widely used, has a very natural reformulation as a hierarchical model. We developed the two-state occupancy modeling framework from this hierarchical perspective first because it seems most natural when you begin with a basic logistic regression model – which forms the 'process model' for the (true) occurrence state. We note that this motivation suggests the general method of formulating hierarchical models which can be enormously useful in practice. That is, if you could observe the occurrence process directly, what would the natural choice of a model model be? In the present case, if we could observe z then we would do a logistic regression, or some variant, and so that remains the process model as the framework is extended to include observation error.

While we believe that logistic regression is a building block topic, we have omitted a detailed development of many important topics in inference such as model fit and predictive evaluation, and many procedural extensions. For example, instead of using a linear-logistic model, substitute some flexible non-parametric form such as a generalized additive model (GAM), or a spline smoothers, etc. While we think that many of these extensions are important, they are beyond the scope of this chapter. We do return to the problem of modeling occupancy in several subsequent chapters. We consider models of occupancy dynamics (spatial and space–time systems) in Chapter 9 and multi-species systems in Chapter 12. In addition, we noted (Section 3.7.2) a formal (technical) duality between occupancy models and classical closed populations models. We exploit this in a number of other chapters to reformulate closed population models as occupancy models to further certain model extensions.

4

OCCUPANCY AND ABUNDANCE

In the segment of ecological science concerned with quantitative methods and sampling, occupancy has become a focus of considerable recent activity. In part this is because the design is simple and the sampling protocol is efficient – only requiring apparent presence/absence data. But occupancy is a crude summary of demographic state that produces an imprecise characterization of population dynamics except in limited situations. For example, consider a metapopulation system in which local populations are characterized mostly by high abundance. In such cases, the overall population can decline precipitously before this shows up as local extinctions, and hence declines in net occupancy. We can improve inferences about population change, even when focused on summaries of occupancy, when we explicitly consider the linkage between occupancy and abundance (Dorazio, 2007).

Indeed, one of the motivating interests in occupancy as a summary of population status is its use as a surrogate for abundance (MacKenzie and Nichols, 2004). That is, at least heuristically, it is widely regarded that ψ is related to N, at least in some kind of average sense, e.g., spatially or temporally. That is, high density corresponds to more occupied area (e.g., better habitat) and *vice versa*. And, common species tend to be both widely distributed and also more locally abundant. This is not merely a heuristic folk theorem, however, and considerable mathematical theory and empirical support exists (Brown, 1984; Lawton, 1993; Gaston et al., 1997). The broader theory relating abundance to occupancy is important because it supports the use of occupancy as an informative summary of population status. Conversely, there does not appear to be, at this time, a strong body of theory that motivates an interest in occupancy free of abundance. Occupancy is, fundamentally, the outcome of a process that governs how individuals are distributed in space. Therefore, it is necessarily a product of abundance or density and the parameters that govern the dynamics of such processes.

The underlying theory relating occupancy to abundance stems from basic probability considerations. Namely, if we let N be the local population size, which we regard as the outcome of some random variable having probability mass function $g(N|\theta)$, then precise mathematical relationships among percentiles, moments, and other characteristics of g follow directly. For example, $\text{Pr}_g(N > 0)$ (occupancy)

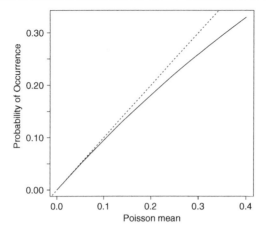

Figure 4.1. Occupancy and abundance under a Poisson model with λ near 0.

is related in a mathematically precise way to $E_g(N)$ (expected abundance), via parameter(s) θ. This was elaborated on in the series of papers by He and Gaston (2000a; 2000b; 2003), and formalized in an estimation context (in the presence of imperfect detection) by Royle and Nichols (2003).

The linkage between abundance and occurrence is very much scale dependent because occupancy and abundance are both fundamentally related to density. As patch size or spatial sample unit size decreases, mean abundance must decrease. That is, as the area, A, of a patch or sample unit decreases to zero then the mass of $g(N)$ will concentrate on 0 or 1, i.e.,

$$N(A) \in \{0,1\} \quad \text{as } A \to 0.$$

If this is approximately the case, then occupancy is nearly a sufficient characterization of abundance. Small spatial units might therefore be preferred if obtaining abundance information from occupancy is desired. This is exemplified under an assumption in which N has a Poisson distribution. In this case, $\psi = 1 - e^{-\lambda}$ and, as $\lambda \to 0$, then $\psi \approx \lambda$ (see Figure 4.1). It is useful to note the importance of A to this concept because A is often a controllable feature of data collection, at least partially so.

In this chapter, we develop a formalization of models of abundance and occupancy based on hierarchical models. Hierarchical models provide a natural framework for exposing and investigating the relationship between abundance and occupancy. Within the hierarchical modeling framework, a model describing variation in abundance serves as one level of the hierarchical model. A model for observations of presence/absence conditional on abundance forms a second component of the model. The observation mechanism and ecological process are linked formally by a model that describes how variation in abundance affects observed presence/absence of species. The conceptual framework allows us to devise models for occurrence by modeling the fundamental abundance process, thus inducing structure in occurrence in the form of a local abundance distribution $g(N|\theta)$ and derived quantities including $\Pr(N > 0)$. We also discuss some extensions to include alternative abundance distributions and observation models.

4.1 ABUNDANCE-INDUCED HETEROGENEITY IN DETECTION

In Chapter 3, we described a model for occupancy in the presence of imperfect detection for the case where p is constant or a function of known covariates. However, in many cases this assumption will prove to be overly restrictive. One interesting departure from this class of models is that in which p is thought to vary among sites not in response to measurable covariates, but due to variation in abundance among sites, something that might be expected in most sampling situations. That is, variation in abundance *induces* spatial heterogeneity in detection probability. In fact, there is a formal duality between variation in N and variation in p, and this was exploited by Royle and Nichols (2003) in order to estimate summaries of local abundance from simple observations of detection/non-detection. We begin with a review of that model (henceforth the 'RN model') followed by some extensions.

4.1.1 Royle–Nichols Model Formulation

We adopt the design and data structure that was described in Chapter 3. We suppose that $i = 1, 2, \ldots, M$ sites are sampled J times and the apparent presence/absence of a species is recorded during each visit. As in the models of occupancy described in Chapter 3, the data resulting from sampling are site-specific encounter histories, $\{y_{ij}; j = 1, 2, \ldots, J\}$. We require now a model for these binary observations of detection/non-detection.

Here, we suppose that the population size exposed to sampling varies among sites. Let N_i be the population size at site i (henceforth 'local population size'). In this case, we anticipate heterogeneity in detection probability among sites because sites with higher local population sizes should yield more net detections, and *vice versa*.

If individuals are detected independently of one another with probability r, then site-specific detection probability is

$$p_i = 1 - (1 - r)^{N_i}. \tag{4.1.1}$$

Here, p_i is the 'net' probability of detection, i.e., Pr(*at least* 1 individual is detected), and r is the individual-level detection probability. We emphasize that the particular function relating local abundance and detection probability specified by Eq. (4.1.1) arises under binomial sampling. This is one plausible model for the detection process, but this does not preclude other models relating detection to local abundance (e.g., see Section 4.4.1). While it is a favorite pastime of statisticians to quibble over model assumptions, we emphasize the conceptual motivation – the eminent sensibility of being more likely to detect a species at a site if more individuals are present, regardless of how one chooses to model the phenomenon. As a final remark, note that, despite the explicit appearance of N_i in Eq. (4.1.1), the N_i are not observed, but rather are unknown (latent) parameters of the model.

We consider the case here where individual detection probability, r, is constant across replicate samples of each local population. In this case, the detection frequencies, y_i for site i, are binomial counts based on a sample size of J and parameter p_i. We can be explicit about the dependence of p on N by denoting this as $p(N_i)$. This notation is not typical, e.g., in logistic regression, but we want to emphasize that N_i is unknown. The pmf of y_i is

$$f(y_i|N_i, r) = \frac{J!}{y_i!(J - y_i)!} p(N_i)^{y_i} (1 - p(N_i))^{J - y_i}. \tag{4.1.2}$$

While the model is parameterized directly in terms of the local abundance parameters $(N_1, N_2, \ldots, N_M)$, they are not, in general, uniquely identifiable. However, if we extend the model slightly to include additional structure on the N_i parameters, then we can make progress in terms of inference under this model. Subsequently, we consider the class of models in which the N_i parameters are themselves realizations of a random variable – i.e., random effects. This assumption will be denoted as $N_i \sim g(N|\theta)$. A natural assumption is to let $g(N)$ be Poisson,

$$N_i \sim \text{Po}(\lambda_i).$$

Note that our parameter-naming convention is to use θ to represent the parameter of $g(N)$ generically, but when a specific parameter of a prescribed distribution is in question we may label those uniquely. For the Poisson model, the null model might assume that $\lambda_i \equiv \lambda$, and we can extend the model to allow for variation due to measurable covariates:

$$\log(\lambda_i) = \beta_0 + \beta_1 x_i$$

for some x_i (or several).

It could be debated as to whether or not the Poisson assumption constitutes a natural choice. However, it is the standard null model for density in many settings, in which it arises under spatial aggregation of a Poisson point process (see Chapter 7). And, the Poisson model is probably the most *common* choice for a model when confronted with count data in ecology. That is, if one were fortunate enough to have observed $N_1, N_2, \ldots, N_M$, there is a good chance that the Poisson model would have been investigated, or even considered exclusively.

4.1.2 Derived Parameters: Occupancy and Flavors of p

Fundamentally, this model is an occupancy type model but one that has many occupancy states (indexed by N), and a parametric model that governs the distribution of occupancy states among sites. Two things are worth noting.

First, while the model is not parameterized directly in terms of a parameter ψ that corresponds to occupancy probability, we can derive such a parameter under the model. That is, $\psi = \Pr(N > 0) = 1 - g(0|\theta)$ and thus, having obtained estimates of θ, we can evaluate $1 - g(0|\hat{\theta})$ to obtain $\hat{\psi}$. By invariance of MLEs (Chapter 2), this corresponds to the MLE of ψ. In this sense the hierarchical models considered here are more general than the class of models considered in Chapter 3 – that is, we get occupancy as a by-product of the model describing heterogeneity in N and p among sites (Royle and Dorazio, 2006; Dorazio, 2007).

The second contrast between the model considered here and the simpler occupancy model (Chapter 3) is that the detection probability parameters have different interpretations. Here, p_i is the probability of detection conditional on N_i. We can have $N_i = 0$, which yields $p_i = 0$. In the models of Chapter 3, p was defined as the probability of detection *at an occupied site*. That is, *given* that $N > 0$. The two quantities are related to one another in the sense that the former is conditional on N whereas the latter omits the $N = 0$ category (which corresponds to $p = 0$) by conditioning. The p that is conditional on occupancy ($N > 0$) is usually called the *conditional detection probability*, and the p of the RN model as the *unconditional detection probability*. This could be a little confusing since both are conditional on abundance, but the conditional dependence is expressed differently. We can establish the precise relationship between these two variations of p by noting that, by the law of total probability,

$$
\begin{aligned}
p_{rn} &= \Pr(y > 0) \\
&= \Pr(y > 0 | N = 0)\Pr(N = 0) + \Pr(y > 0 | N > 0)\Pr(N > 0), \quad (4.1.3)
\end{aligned}
$$

where p_{rn} is the expected value of p for a site under the RN model. Note that the first term on the RHS evaluates to 0 by assumption. Therefore,

$$p_{rn} = \Pr(y > 0 | N > 0) \Pr(N > 0). \tag{4.1.4}$$

The quantity $\Pr(y > 0 | N > 0)$ is the usual 'conditional on occupancy' detection probability, i.e., of the MacKenzie et al. (2002) model, which we encountered repeatedly in Chapter 3, where we used the notation p. To avoid confusion here, we will temporarily (i.e., in this Chapter) denote this quantity as p_c. In Eq. (4.1.4), $\Pr(N > 0) = \psi$. Hence, $p_{rn} = p_c \psi$, which allows us to compare like quantities across models.

4.1.3 Induced Heterogeneity

It is clear from Eq. (4.1.1) that heterogeneity in abundance among sample units induces heterogeneity in detection probability. That is, sites with high local abundance will yield more net detections, etc. In particular, we have:

$$N_i = 0 \implies p_i = 0$$
$$N_i = 1 \implies p_i = r$$
$$N_i = 2 \implies p_i = 1 - (1 - r)^2$$
$$\vdots \qquad \vdots$$
$$N_i = \infty \implies p_i = 1.$$

Thus, as N_i gets large, p_i tends to 1.0, as we would expect under reasonable circumstances.

This formulation of the problem as a model of heterogeneity in detection probability establishes a direct conceptual linkage with a class of models known as 'Model M_h' in classical capture–recapture (Otis et al., 1978; Norris and Pollock, 1996; Coull and Agresti, 1999; Dorazio and Royle, 2003; Royle, 2006). Model M_h assumes an individual-specific detection probability, p_i, having probability density $h(p|\theta)$. Since p is inherently continuous, common choices of h in this context include the beta distribution and normal distribution for the logit-transformation of p (see Chapter 6). If we impose a model describing heterogeneity in N among sites, say $N \sim g(N|\theta)$, then heterogeneity in detection probability p derives directly from $g(N)$. That is, there is a formal duality between $g(N)$ and $h(p)$. The heuristic motivation of Royle and Nichols (2003) was that we should be able to estimate the heterogeneity distribution, $h(p)$, and then, in a sense, 'back-transform' to get $g(N|\theta)$. Thus, the RN model can be viewed simply as an alternative version of

Model M_h, i.e., a different pdf for p – one that has discrete support, and that is motivated by a plausible mechanism that generates heterogeneity in detection probability.

The appeal of the abundance-induced heterogeneity model is that, in some cases, it may be reasonable to view λ (or, more generally, $\mathrm{E}[N]$) as an estimate of abundance or, when sample units have known area, density. However, there is no reason that N_i must be interpreted as abundance *per se*. It could be viewed merely as some random effect that yields variation in p_i. That is, an alternative mixing distribution that accommodates heterogeneity in detection probability. Specifically, note that Eq. (4.1.1) implies that the complementary log-log link of p_i is linear in a discrete-valued random effect (Royle, 2006), i.e.,

$$\log(\log(1 - p_i)) = \log(\log(1 - r)) + \log(N_i). \qquad (4.1.5)$$

In this case, choice of pmf for N is equivalent to choice of pmf for the random effect $\log(N_i)$.

4.1.4 Analysis by Integrated Likelihood

Recognizing (in Eq. (4.1.2)) that p_i is itself a random variable because it depends on the unobserved N_i (via (4.1.1)), the model is analogous to classical random effects type models where N_i here is the random effect, having pmf $g(N|\theta)$. The classical treatment of random effects (e.g., Laird and Ware, 1982), is to remove them from the likelihood by integration to yield the marginal, or integrated likelihood, of the fixed-effect parameters.

Here, the random effect is discrete-valued, so we remove it from the likelihood by summation. That is, the marginal probability mass function of y can be expressed as (Royle and Nichols, 2003)

$$q(y|r, \theta) = \sum_{N=0}^{\infty} f(y|p(N)) \, g(N|\theta) \qquad (4.1.6)$$

or, if we segregate the $N = 0$ case to yield a form that is more compatible with the zero-inflated binomial form (Chapter 3):

$$q(y|r, \theta) = \left\{ \sum_{N=1}^{\infty} f(y|p(N)) \, g(N|\theta) \right\} + I(y = 0) \, g(0|\theta). \qquad (4.1.7)$$

For implementation, i.e., for a particular choice of $g(N|\theta)$, we substitute the pmf of N into Eq. (4.1.7). For the Poisson case,

$$q(y_i|\lambda, r) = \sum_{N_i=0}^{\infty} f(y_i|N_i, r) \frac{e^{-\lambda} \lambda^{N_i}}{N_i!} \qquad (4.1.8)$$

Table 4.1. Catbird detection frequencies based on $J = 11$ replicate samples of a BBS route.

# detections	0	1	2	3	4	5	6	7	8	9	10	11
# sites	31	5	5	5	2	0	1	1	0	0	0	0

which doesn't seem to simplify in any meaningful way (i.e. that yields a great improvement in computational efficiency), but can be computed easily in **R**, as we demonstrate in the next section. To evaluate the likelihood numerically, we have to choose the upper limit of summation in Eq. (4.1.8) as a compromise between computational efficiency and accuracy. Too low a value causes truncation of the likelihood, which can be especially problematic when skewed distributions for N are used (e.g., negative binomial).

4.1.5 Bird Point Counts

Here we consider data from the North American Breeding Bird Survey (BBS). The BBS is a roadside survey conducted using standard point-count methods at 50 locations ('stops') separated by approximately 0.5 miles, along a 25 mile route. Because of the small spatial scale of the observations, the relatively low density of many species, and difficulties in determining the number of unique individuals detected, the use of presence–absence data for the analysis of occupancy may have some appeal. The data used here are from a BBS route sampled in New Hampshire in 1991. In the operational survey, each route is surveyed only once. However, the data used here come from a study in which a number of routes were surveyed up to 14 times during approximately one month. This particular route was surveyed 11 times ($J = 11$). For general background on the BBS, see Robbins et al. (1986). For further description of the data used here, see Link and Barker (1994). These data have been analyzed previously by Royle and Nichols (2003) and Royle (2004c).

We provide an analysis of the data for the catbird (*Dumetella carolinensis*). The observed counts were reduced to presence/absence data, and the frequency of detections (out of $J = 11$) is shown in Table 4.1. For the record, there were 6 observations (out of 550) with a count of 2 individuals, and no count was greater than 2. Thus, apparent presence/absence is almost a sufficient summary of the data set, in the sense that very little information is lost.

The brute-force implementation of the likelihood in **R** is shown in Panel 4.1. This implementation is based on the assumption that N has a Poisson distribution with mean λ. While this implementation is not very efficient, it more clearly demonstrates the required calculations. The main thing to be aware of is that caution should be exercised in setting the upper limit of summation `Ninfty`. Too small of a value truncates the likelihood and will yield the appearance of convergence to the MLE. However, too large of a value induces some inefficiency. We have not discovered an implementation of this model in **WinBUGS** that works, although Bayesian analysis by standard methods of MCMC is straightforward.

Fitting the Poisson-abundance model in **R** yields the MLEs $\hat{r} = 0.179$ and $\hat{\lambda} = 0.546$, and the associated negative log-likelihood was 65.97. For comparison, the mean (over sites) of the maximum *count* per site (over all $J = 11$ replicate counts) was 0.46. We see that the model produces an upward adjustment to apparent abundance (per point count) of approximately 20 percent. Suppose we

```
RN.fn<-function(y=catbird.y,J=11,M=50,Ninfty=100){
    ####
    lik<-function(parms){
        r<-expit(parms[1])
        lambda<-exp(parms[2])
        pvec<-1-(1-r)^(0:Ninfty)
        gN<-dpois(0:Ninfty,lambda)
        gN<-gN/sum(gN)
        lik<-rep(NA,M)
        for(i in 1:M){
            lik[i]<-sum(dbinom(y[i],J,pvec)*gN)
        }
        -1*sum(log(lik))
    }
    ####
    tmp<-nlm(lik,c(0,0),hessian=TRUE)
    ests<-tmp$estimate
    aic<-tmp$minimum*2 + 2*length(ests)
    se<- sqrt(diag(solve(tmp$hessian)))
    list(ests=ests,se=se,aic=aic)
}
```

Panel 4.1. R specification of the likelihood of the Royle–Nichols (RN) model with Poisson abundance, and instructions for obtaining the MLEs for the catbird data. The likelihood and its optimization using the function nlm are contained in the function RN.fn which can be executed providing the arguments y, the vector of length M containing the number of detections at each site out of J trials and other arguments. A list is returned having elements that correspond to the estimates, their standard errors, and the AIC for the model.

had not recorded counts; then how might we obtain an estimate of abundance if we didn't think that imperfect detection was a concern? One possibility (justified in Section 4.5) is to equate the observed proportion of occupied sites to that expected under the model, i.e., an estimate of $\Pr(y > 0)$ to $1 - \exp(-\lambda)$, and then solve for λ. In the present example, the proportion of sites occupied was $19/50$ and solving for λ yields an apparent λ of 0.478, also lower than the fitted value obtained under the RN model.

It might be interesting to compare these results to those obtained by fitting the two-state occupancy model described in Chapter 3 – that is, the model that assumes that p_i is constant across sites (MacKenzie et al., 2002) and sites are characterized by two abundance states, either occupied ($N > 0$) or unoccupied ($N = 0$). The RN model can be characterized as a kind of 'multi-state' model, as we have noted previously. Working with the likelihood construction shown in Panel 4.1, we can modify this to fit the simpler model by setting only 2 possible states for N, $N \in \{0, 1\}$, as shown in Panel 4.2. While the implementation in Panel 4.2 is somewhat inefficient, it is a clear conceptual implementation and it preserves the form of the likelihood so that we may compare the negative log-likelihoods, or AIC, etc. Fitting the two-state model yields a negative log-likelihood of 67.26. Since they have the same number of parameters, we would conclude that the Poisson version of the RN model is favored.

Under the two-state model, $\hat{\lambda} = 0.515$, which is slightly lower than that obtained under the Poisson RN model. The difference is consistent with our expectation that heterogeneity in detectability should bias p high, and hence abundance low (see Royle and Nichols (2003) and Chapter 6 for justification). There is a concomitant effect on estimated probability of occurrence, ψ. Under the two-state model, we obtain $\hat{p} = 0.230$ – which is not directly comparable to the estimate of *individual* detection probability, r, obtained under the RN model (as discussed in Section 4.1.2). We note that the implied equivalent can be obtained by using $\hat{r}$ and $\hat{\lambda}$ under the RN model and solving Eq. (4.1.4). This calculation is carried out in **R** as follows:

```
# "net" or total probability of capture under RN model:
    prn<- sum((1-(1-0.179)^(0:100))*dpois(0:100,0.546) )
# occupancy under RN model
    psi<- 1-dpois(0,0.546)
# "conditional p" comparable to ordinary occupancy model
    pc<- prn/psi
```

The equivalent value of the 'conditional on occupancy' detection probability under the RN model is $p_c = 0.221$, just slightly less than the estimated 0.230 under the ordinary two-state occupancy model.

4.1.6 Application to Carnivore Survey Data

Carnivore surveys represent a natural area of application of occupancy type models because, typically, carnivores occur at relatively low densities and many species are territorial, and so spatial units have small exposed population sizes. Moreover, they are often sampled using non-invasive techniques such as hair snares, camera traps and scat sampling (Long et al., 2008) that yield only presence/absence data.

Here we consider forest carnivore survey data described in Zielinski (1995) and Zielinski et al. (2005). This survey focused on determining the distribution of mesocarnivores in forests of northern and central California using a common passive detection device known as a track plate (Zielinski, 1995, Chapter 4). A track plate is a carbon-blackened aluminum plate partially covered with white contact paper. It is typically enclosed in a baited wood box, and animal tracks are recorded when they visit the device. A sample unit consists of 6 such track plates, 500 m apart, in a pentagonal arrangement. The survey was based on 464 of these sample units located using the nationwide Forest Inventory and Analysis (USDA Forest Service) survey grid. Every other FIA grid was selected, resulting in from 7–10 km between the centers of adjacent sample units. The data used here are for fishers (*Martes pennanti*), taken from Royle et al. (2008) in the volume devoted to non-invasive carnivore sampling (Long et al., 2008). For the present purposes, we will define a detection to occur if any of the 6 detectors (track plates) registers a detection. If the subsamples are highly dependent (which might occur if they are sufficiently

```
lik<-function(parms,Ninfty=100){
    p<-expit(parms[1])
    lambda<-exp(parms[2])
    pvec<-1-(1-p)^(0:1)
    pvec<-c(pvec[1],rep(pvec[2],Ninfty))
    gN<-dpois(0:Ninfty,lambda)
    gN<-gN/sum(gN)
    lik<-rep(NA,M)
    for(i in 1:M){
        lik[i]<-sum(dbinom(y[i],J,pvec)*gN)
    }
    -1*sum(log(lik))
}
```

Panel 4.2. Royle–Nichols formulation of the likelihood for the two-state occupancy model. This likelihood function could be used in the **R** wrapper given in Panel 4.1.

close together), then pooling the data in this fashion would typically be a reasonable treatment of the data.

The stations were visited 8 times over 16 days, yielding the following frequencies of detections (464 total stations): (400, 16, 12, 12, 5, 10, 3, 4, 2). Thus, fishers were not detected at 400 sites, only 1 time at 16 sites, twice at 12 sites, etc. The estimates obtained from fitting the two-state occupancy model (MacKenzie et al., 2002) were $\hat{p} = 0.40$ and $\hat{\psi} = 0.14$. The apparent occupancy was 0.138, so there is little adjustment due to detection bias, which might not be too surprising with an array of 6 baited devices, sampled 8 times. Using the estimated p, the power (to detect fishers) of this survey apparatus is $0.98 = 1 - (1 - 0.40)^8$.

Fitting the Poisson abundance mixture model yields an individual detection probability estimate of $\hat{r} = 0.392$, and a mean local abundance of $\hat{\lambda} = 0.162$. The implied occupancy is $\hat{\psi} = 0.15$. Note how similar the density is to the estimated ψ from the non-abundance model. As we noted in the introduction to this chapter, this is because density is low so that $\lambda \approx 1 - \exp(-\lambda)$ in the vicinity of $\lambda = 0$. The distribution of N under the fitted model has most of the mass on $N \leq 3$. In fact, about 85 percent of the sites are unoccupied, 13.8 percent of the sites have a single individual, 1.1 percent have 2 and fewer than 0.1 percent have 3 or more. There appears to be very little abundance-induced heterogeneity in these data.

In many instances, occupancy type models may be ideal for carnivore survey data because abundances are low and individuals often have well-defined territories. Thus, $N \in \{0, 1\}$, or approximately so, as we have just inferred under a particular model. In such cases, occupancy provides an almost-sufficient description of the data, and models of occupancy or occurrence probability translate approximately into models for abundance.

4.2 PREDICTION OF LOCAL ABUNDANCE

In the classical treatment of mixed models, parameters of the random effects distribution are estimated by integrated likelihood in which the random effects are removed from the likelihood (as described in Section 4.1.4). Yet, in some cases, there is interest in obtaining estimates – predictions – of those random effects. In the present context of inference about abundance, we may wish to obtain predictions of individual local population sizes (e.g., see (Dorazio et al., 2005)). *Best Unbiased Prediction* (BUP) is the classical method of predicting/estimating the random effects that were removed from the likelihood by integration (see Chapter 2).

To obtain predictions of unknown quantities, it is necessary to obtain what is referred to (by Bayesians) as the conditional posterior distribution of the quantity for which a prediction is desired. In the present context, that would be $[N_i|y_i, r, \lambda]$. This pmf is obtained by Bayes' rule. Symbolically,

$$[N_i|y_i, r, \lambda] = \frac{f(y_i|N_i, r)g(N_i|\lambda)}{q(y_i|r, \lambda)}.$$

Recall that the denominator on the right-hand side is obtained by the summation operation in Eq. (4.1.8) – i.e., it is the marginal pmf of y_i. As we have conditioned on the observed y_i, it is just a scalar quantity, i.e., some number like 1.6. So, to compute the posterior probability of, e.g., $N_i = 2$, we have:

$$\Pr(N_i = 2|y_i, r, \lambda) = \frac{\frac{J!}{y_i!(J-y_i)!}(1-(1-r)^2)^{y_i}(1-r)^{2(J-y_i)}\frac{e^{-\lambda}\lambda^2}{2!}}{\sum_{k=0}^{\infty}\text{Bin}(y_i|J, p_i(k))\frac{e^{-\lambda}\lambda^k}{k!}}. \quad (4.2.1)$$

The BUP of a particular N_i under any prescribed model will be defined as the conditional expectation of N_i given the data and parameters. That is, $\hat{N}_i = \text{E}(N_i|y_i, r, \lambda)$. Once we compute the conditional posterior probabilities, we evaluate

$$\hat{N}_i = \sum_{k=0}^{\infty} k\hat{\pi}_k,$$

where $\pi_k = \Pr(N_i = k|y_i, r, \lambda)$.

In practice, we estimate this quantity by plugging in the MLEs, resulting in a quantity usually referred to as the *estimated best unbiased predictor* (EBUP). The variance would be the variance of the conditional posterior distribution, but we would expect the 'plug-in' version to understate uncertainty in the prediction because of uncertainty associated with estimating the parameters of the model. Certain procedural adjustments (e.g., parametric boot-strapping, (Dorazio et al., 2005)) can be used to better characterize the uncertainty of predictions.

For the catbird data, we computed the estimated posterior distribution for each of $y = 0, 1, \ldots, 11$. This is shown in Table 4.2. The **R** instructions for carrying out these calculations can be found in the Web Supplement.

4.2.1 What is N?

In the previous examples, we developed models of presence/absence data based on the specification of an underlying model that governs variation in abundance among spatial sample units or patches. Thus, these models have local abundance parameters, N_i, for each spatial sample unit. However, the interpretation of N_i is not necessarily (or even typically) clear. In particular, we might prefer to interpret abundance and associated estimates in terms of density – abundance per unit area. But in most situations (and in all the examples described in this chapter), the sample units are not of known area. Equivalently, the local populations are not closed, because individuals exposed to sampling may reside outside of the putative sample area – the area within which traps or detectors were located randomly or systematically. For example, with a bird point count conducted with a 50 m radius, repeated over several days, there are individuals exposed to sampling on some days, but not on others, due to movement about their home range. The same phenomenon is relevant to trapping grids and other sampling protocols (e.g., capture–recapture), where it has received considerable attention (with various adjustments to obtain 'effective sample area' having been proposed, see Chapter 7). The problem induces an interpretation of N that is, at best, the size of the effective (or exposed or sampled) population of individuals (that is, the number of individuals exposed to sampling). This interpretation arises under the 'random temporary emigration' model of Kendall (1999).

Despite this conceptual problem – the lack of clarity in interpreting N – we prefer the conceptual formulation of these models in terms of abundance because they are relevant to the biological context of the sampling. That is, while we might be able to ignore the problem of "what is N?" by fitting ordinary regression or GLM models to observational data, we prefer the explicit process formulation of the model that we introduced in Section 4.1.

4.3 MODELING COVARIATE EFFECTS

As in Chapter 3, we distinguish between covariates that might influence local abundance, N, and those which influence the observation process via the parameter r.

If we have covariates that are thought to influence abundance, we might naturally choose to model these effects on the mean of N. For example, if

$$N_i \sim \text{Poisson}(\lambda_i)$$

then

$$\log(\lambda_i) = \beta_0 + \beta_1 x_i$$

for some covariate x_i that describes local variation in habitat or some other landscape feature. In Section 4.4.2 we discuss abundance models other than the Poisson. We can similarly model covariate effects on the mean parameter of these other models. Fitting such models can be accomplished by extending the **R** code in Panel 4.1, as we demonstrate shortly.

Table 4.2. Posterior distribution of N given y for the catbird data for $y = 0, 1, \ldots, 11$, and the posterior mean.

y	$N = 0$	1	2	3	4	5	6	7	8	≥ 9	E[$N\|y$]
0	0.94	0.058	0.002	0	0	0	0	0	0	0	0.062
1	0	0.933	0.064	0.002	0	0	0	0	0	0	1.069
2	0	0.861	0.132	0.008	0	0	0	0	0	0	1.148
3	0	0.728	0.247	0.024	0.001	0	0	0	0	0	1.298
4	0	0.531	0.400	0.064	0.005	0	0	0	0	0	1.544
5	0	0.315	0.526	0.141	0.016	0.001	0	0	0	0	1.861
6	0	0.150	0.555	0.249	0.042	0.004	0	0	0	0	2.196
7	0	0.058	0.480	0.359	0.090	0.012	0.001	0	0	0	2.520
8	0	0.019	0.349	0.435	0.163	0.030	0.003	0	0	0	2.847
9	0	0.005	0.217	0.451	0.251	0.064	0.010	0.001	0	0	3.187
10	0	0.001	0.116	0.402	0.333	0.120	0.025	0.003	0	0	3.543
11	0	0	0.053	0.309	0.382	0.192	0.053	0.010	0.001	0	3.916

In practice, we may be able to identify explicit covariates that influence detection probability across samples. In the occupancy models described in Section 3.3, we referred to such covariates as observation covariates, and we modeled their influence on the detection probability parameter p. In the abundance mixture models, such factors should influence the parameter r. To accommodate these more general models, we require an encounter history representation of the model. This is analogous to the formulation described in Section 3.6. The fundamental unit of observation is the encounter history, and we can obtain the integrated likelihood by averaging the probability of a particular encounter history over the possible states, which are indexed by N_i, as before. To achieve this extension, let p_{ij} be the probability of detection at site i and replicate $j = 1, 2, \ldots, J$. We suppose that

$$p_{ij} = 1 - (1 - r_{ij})^{N_i} \tag{4.3.1}$$

with

$$\text{logit}(r_{ij}) = \alpha_0 + \alpha_1 x_{ij}$$

for the case where a single detection covariate, x, is available. In this case, x might be the time-of-day that sample j was conducted at site i, the amount of effort spent sampling, or some other factor likely to influence detection probability.

We can use the expression for p_{ij} in Eq. (4.3.1) to construct the site-specific encounter history probabilities, conditional on N_i. For example, suppose we observe $\mathbf{y}_i = (1, 1, 0)$ from some site sampled $J = 3$ times, then the conditional-on-N encounter probability is

$$\Pr(\mathbf{y}_i | \alpha_0, \alpha_1, N_i) = p_{i1} p_{i2} (1 - p_{i3}). \tag{4.3.2}$$

It is understood here that p_{ij} are functions of the parameters to the right of the '|', but we will be concise and not use the more explicit (and somewhat awkward) representation that we used previously, in which we denoted the dependence as $p_{i1}(r_{i1}, N_i)$. Now we can obtain the unconditional probability of encounter, by averaging Eq. (4.3.2) over possible states of N_i, according to:

$$\Pr(\mathbf{y}_i | \alpha_0, \alpha_1, \lambda) = \sum_{N_i=0}^{\infty} p_{i1} p_{i2} (1 - p_{i3}) \Pr(N_i | \lambda).$$

4.3.1 Analysis of the Swiss Willow Tit Data

We return to the Willow tit data from the Swiss Breeding Bird Survey that were analyzed at various points in Chapter 3.

The **R** specification of the likelihood for the model containing covariates is shown in Panel 4.3. An **R** wrapper that optimizes the likelihood function is provided in the Web Supplement. As done with some of the examples in Chapter 3, the likelihood for the most general model of interest is computed and coefficients are set to 0, except for a particular model that is identified by the vars= reference in the function call. The results of fitting a suite of models having covariates in both r_{ij} and also the quadrat-specific covariates in λ_i (forest cover, elevation, and route length) are given in Table 4.3. As we saw in the analysis of the occupancy model in Chapter 3, the model containing all of the landscape covariates is favored. However, we find under this model that the detection covariates are also included in the best model; whereas, under the simple occupancy model, none of the detection covariates were in the best model. We will return to these data in Chapter 8.

4.4 OTHER MODEL EXTENSIONS

Here we consider extending the model to allow for a non-binomial relationship between p and N, and also alternative models for the local abundance distribution. We also apply these extensions to the catbird data.

4.4.1 Non-binomial Detection

In some cases we might expect that detections of individuals are not independent, and so the binomial-induced function relating p_i to N_i in Eq. (4.1.1) might be inappropriate. In general, faster or slower increases in apparent detection as a function of N could occur. We leave it to the reader to describe plausible mechanisms that yield non-binomial relationships between p and N. One possible model that we consider here is the following:

$$\Pr(y = 1|N_i) = p_i = 1 - (1 - r)^{\theta N_i}, \qquad (4.4.1)$$

where the parameter θ allows for a more or less rapid increase in detection relative to the binomial assumption. While this function is intuitively appealing, because it yields a direct comparison of the effect of local population size relative to that expected under a binomial assumption (i.e., as if detection of individuals are independent Bernoulli trials), it is not generally a valid model in the sense that θ may not be uniquely identifiable. As an example, under the complementary log-log link (recall Eq. (4.1.5)), Eq. (4.4.1) becomes

$$\text{cloglog}(p) = \log(\log(1 - r)) + \log(\theta) + \log(N_i). \qquad (4.4.2)$$

```
RNlikcov<-function(parms,vars=c("a0","b0") ){
  ones<-rep(1,M)
  tmp<-c(0,0,0,0,0,0,0,0,0)
  nam<-c("a0","b0","intensity","date1","date2","elev1","elev2","forest","length")
  names(tmp)<-nam
  tmp[vars]<-parms
  b0<-tmp[1];a0<-tmp[2];a1<-tmp[3];a2<-tmp[4]
  a3<-tmp[5];b1<-tmp[6];b2<-tmp[7];b3<-tmp[8];b4<-tmp[9]

  rvec<-expit(a0*ones + a1*intensity + a2*date + a3*date2)
  lamvec<-exp(b0*ones + b1*elev + b2*elev2 + b3*forest + b4*length)
  lik<-rep(NA,M)

  for(i in 1:M){
    gN<-dpois(0:Ninfty,lamvec[i])
    gN<-gN/sum(gN)
    dvec<-ymat[i,]
    naflag<-is.na(dvec)
    PMAT<-  1-outer( (1-rvec[i,]),0:Ninfty,"^")
    LIK<- t((PMAT^dvec)*((1-PMAT)^(1-dvec)))
    LIK[,naflag]<-1
    likvec<-apply(LIK,1,prod)
    lik[i]<-sum(likvec*gN)
  }
 -1*sum(log(lik))
}
```

Panel 4.3. R construction of the likelihood for the Royle–Nichols model with covariates influencing both detection probability and spatial variation in abundance. The default model has constant λ and constant p. More complex models are specified using the vars= option when the likelihood is passed to nlm.

Table 4.3. Estimates of abundance-mixture model parameters for the territory apparent presence/absence data of willow tits from the Swiss BBS. Maximum likelihood estimates are tabulated for parameters associated with heterogeneity in the probability of detection per territory (α_0, intensity, date1, date2) and with heterogeneity in territory abundance (β_0, elev1, elev2, forest, length).

Model Rank	α_0	Intensity	date	date^2	β_0	elev	elev^2	forest	length	AIC
		Detection parameters				Abundance parameters				
1	−0.471	0.490	−0.083	0.012	0.237	1.911	−0.947	0.588	0.278	414.37
2	−1.16	0.419	−0.064	−0.038	0.465	1.167	–	0.922	0.147	437.42
3	−0.936	0.247	0.203	−0.051	0.863	–	–	0.587	−0.132	502.81
4	−0.856	–	–	–	1.080	–	–	–	–	525.96
5	−0.842	0.278	–	–	0.949	–	–	–	–	526.64
6	−0.843	0.239	0.150	–	0.928	–	–	–	–	527.95
7	−0.842	0.253	0.163	−0.077	0.994	–	–	–	–	529.68
8	−0.851	0.218	0.164	−0.078	1.016	–	–	–	−0.092	530.99

We see that the parameter θ contributes to the intercept in this formulation, and therefore trades off with a function of r. While this precise result does not hold with the logit link function, it does suggest that we should be concerned with the identifiability of the additional parameter. Fortunately, consideration of the complementary log-log (cloglog) link suggests a parameterization that should yield identifiability of the extra-binomial parameter. Namely, in order to have θ be multiplicative with $\log(N_i)$ in Eq. (4.4.2), we require

$$\Pr(y = 1|N_i) = p_i = 1 - (1 - r)^{N_i^\theta}. \tag{4.4.3}$$

We refer to this model as the quasi-binomial detection model (Royle, 2008b). The likelihood specification in **R** is shown in Panel 4.1. We will fit this model to data shortly. Other models for detection might be reasonable but we have not explored or cataloged the possibilities. Royle and Link (2006) describe a general form of dependence, and we discuss the case of 'no dependence' in Section 4.5.

4.4.2 Alternative Abundance Models

We have so far only considered the case where local abundance is assumed to be Poisson. However, many choices for g are possible. One that is usually used to accommodate overdispersion relative to the Poisson is the negative binomial. A convenient parameterization is that based on the mean, μ,

$$g(N|\alpha,\mu) = \frac{\Gamma(N+\alpha)}{\Gamma(\alpha)N!} \left(\frac{\alpha}{\alpha+\mu}\right)^{\alpha} \left(\frac{\mu}{\alpha+\mu}\right)^{N}, \qquad (4.4.4)$$

where $\mathrm{E}(N) = \mu$ and $\mathrm{Var}(N) = \mu + \mu^2/\alpha$, and $\alpha > 0$. It is often described in terms of the 'overdispersion' parameter, $\epsilon = 1/\alpha$. The negative binomial is often motivated as an overdispersed Poisson, where μ is the mean of the limiting Poisson distribution that occurs as $\epsilon \to 0$ (i.e., no overdispersion). The **R** code in Panel 4.1 can be modified to fit this model by using the **R** function for evaluating the negative binomial pmf, dnbinom.

We usually try to avoid using the negative binomial model because it has an unrealistic variance/mean relationship – the variance/mean ratio increases with the mean. This doesn't seem sensible at high abundance levels for most vertebrate species. Secondly, the negative binomial tries to account for mass at 0 by increasing overdispersion, which yields a concomitant increase in mean. As a result, it is often the case, especially in small samples, that the negative binomial mixture can yield unstable MLEs, which can be diagnosed by eigenvalues of the hessian near 0 (see Royle and Nichols (2003) and Kéry et al. (2005)), or, frequently, μ tending to extraordinarily large values (and, conversely, r near 0.0). This is unfortunate because the negative binomial seems to be the default model for counts that exhibit overdispersion relative to the Poisson. However, the preference for the negative binomial model appears to be based more on phenomenological evidence and it does not appear to have any biological or broader scientific basis.

```
lik<-function(parms,Ninfty=100){
    r<-expit(parms[1])
    lambda<-exp(parms[2])
    alpha<-exp(parms[3])
    pvec<-1- (1-r)^( (0:Ninfty)^(1+alpha) )
    gN<-dpois(0:Ninfty,lambda)
    gN<-gN/sum(gN)
    lik<-rep(NA,M)
    for(i in 1:M){
        lik[i]<-sum(dbinom(y[i],J,pvec)*gN)
    }
    -1*sum(log(lik))
}
```

Panel 4.4. Royle–Nichols model with quasi-binomial detection. This likelihood function could be used in the **R** wrapper given in Panel 4.1. Here, $1 + \alpha = \theta$ in Eq. (4.4.3).

Table 4.4. Results of five abundance-mixture models fitted to the catbird data. Models are ordered by AIC. Model 1 (row 1) is the Poisson RN model; Model 2 (row 2) is the Poisson with quasi-binomial detection, having exponent $\theta = \alpha + 1$; Model 3 is the Poisson/log-normal mixture, having $E(N) = \exp(\mu + .5\sigma^2)$; Model 4 is the negative binomial RN model (having overdispersion ϵ); Model 5 is the two-state occupancy model of MacKenzie et al. (2002).

N Model	detection	r or p	$E(N)$	ψ	extra parm	AIC
Po	bin	0.179	0.546	0.421	–	135.94
Po	qb	0.164	0.547	0.421	$\alpha = 0.29$	137.57
Po/log-Norm	bin	0.153	0.655	0.426	$\sigma = 0.72$	137.67
NB	bin	0.149	0.674	0.426	$\epsilon = 0.67$	137.68
Two-state	const	0.230	0.515	0.403	–	138.52

We consider one additional model of abundance, the Poisson/log-normal mixture which assumes that N is Poisson, with a log-normally distributed mean. That is, we place a normal distribution on $\log(\lambda)$. The observation component of the model doesn't change, so $f(y_i|p(N_i))$ is binomial having sample size J and parameter $1 - (1 - r)^{N_i}$, but the abundance model is described by the two components $N_i \sim \text{Po}(\lambda)$ and $\log(\lambda) \sim \text{N}(\mu, \sigma)$. Subsequently, we provide an analysis by integrated likelihood, by evaluating

$$q(y|r, \mu, \sigma) = \sum_{N=0}^{\infty} f(y|p(N)) \left\{ \int_{\phi} g(N|\phi) h(\phi|\mu, \sigma) d\phi \right\}, \qquad (4.4.5)$$

where $\phi = \log(\lambda)$. We give the **R** implementation of this for the catbird data using the function `integrate` in Panel 4.5. As with the negative binomial, this model can be unstable with sparse data and, in such cases, sensitive to choice of `Ninfty`. Caution is recommended.

4.4.3 Example: Analysis of the Catbird Data

We return to the catbird example. Previously we fitted the Poisson abundance mixture under binomial detection, and also the two-state model of MacKenzie et al. (2002). Here we include the Poisson/log-normal mixture, the negative binomial model, under binomial dependence and also the Poisson abundance model under the quasi-binomial dependence Eq. (4.4.3). The parameter estimates under the five models are given in Table 4.4. The Poisson/log-normal mixture is favored by log-likelihood, just slightly over the Poisson model with quasi-binomial detection, and negative-binomial mixture. Because models 2 to 4 have 3 parameters each, a strict adherence to AIC would slightly favor the simpler Poisson model with binomial detection. The estimate of θ for the quasi-binomial detection model indicates a slightly more rapid increase in detection probability with N than expected under the binomial detection model. We conclude that there is not much heterogeneity in abundance among point-count locations at the level of a BBS route. This is not surprising in surveys of breeding birds that do not flock or aggregate.

4.5 FUNCTIONAL INDEPENDENCE BETWEEN P AND ABUNDANCE

The models introduced in Section 4.1 apply to the case wherein site-specific detection probabilities are functionally *dependent* on local abundance (e.g., binomial or what we referred to as quasi-binomial dependence). In this case, information

```
lik<-function(parms,Ninfty=100){
  r<-expit(parms[1])
  mu<-parms[2]
  sigma<-exp(parms[3])
  uy<-unique(y)
  # Compute marginal probabilities Pr(N=k)
  il<-rep(NA,Ninfty+1)
  for(k in 0:Ninfty){
      il[k+1]<-
        integrate(
         function(x){ dpois(k,exp(x))*dnorm(x,mu,sigma) },
          lower=-Inf,upper=Inf)$value
        }

  # compute marginal probabilities of the observations
  proby<-rep(NA,J+1)
  for(k in uy){
      proby[k+1]<- sum(dbinom(k,J,1-(1-r)^(0:Ninfty))*il)
  }

  -1*sum(log(proby[y+1] ))
}
```

Panel 4.5. R construction of the likelihood for a Royle–Nichols type model with a Poisson/log-normal abundance model. This likelihood function could be used in the R wrapper given in Panel 4.1.

about the local abundance distribution $g(N|\theta)$ derives directly from the apparent heterogeneity in p (among sites) induced by variation in abundance. That is, sites with high local population sizes yield higher net detection probabilities, and *vice versa*. Here, the situation in which p is functionally *independent* of N is considered. Under the functional independence case, it is not obvious whether we can obtain information about $g(N|\theta)$. We explore that question here.

In most practical sampling situations, the assumption of functional independence is probably not very appealing or even reasonable. However, certain situations might suggest such a model. For example, it might be a sensible model when sampling highly aggregated species such as insects or amphibians (e.g., 'calling surveys' of vocal anurans (Weir et al., 2005)). Another situation is that involving rare or territorial species such as carnivores. In such cases, local density may be very low, so that local abundance is essentially constant. Thus, heterogeneity in apparent detection probability among sites is negligible, diminishing the importance of the main proposition (i.e., binomial detection) underlying the Royle and Nichols (2003) model. One might also *induce* functional independence by design, i.e., under certain sampling protocols. As an example, one might sample constant-area sample units until a species is first detected (disregarding the time until detection). Then, the influence of variation in local abundance may be mitigated to some extent.

Intuitively, under functional independence, the Royle and Nichols (2003) model reduces to the 'two-state' model described in Chapter 3 (i.e., that of MacKenzie et al., 2002). This is insightful for two reasons. First, it demonstrates that the latter model is not free of assumptions pertaining to local abundance. Indeed, the functional independence assumption appears to us to be somewhat stringent since p is constant for all $N > 0$. Secondly, in the case of functional independence, it is the case that the local abundance distribution, $g(N|\theta)$, cannot be uniquely estimated under the standard design involving replication on homogeneous sample units. To resolve this issue, some auxiliary information about $g(N|\theta)$ is needed. One possibility is to make use of variable area sample units in which sample units are intentionally chosen to have different but known sampling areas. Under this design, the local abundance distribution can be uniquely estimated under functional independence between p and N. It is also the case that variable area sample units allow one to estimate occupancy in the absence of *any* replication of sample units. That is, under the models described by MacKenzie et al. (2002) and many subsequent papers, it was asserted to be necessary to have at least $J = 2$ replicates at some sites. However, it is not necessary to have replication under the variable area sampling design proposed here, when p is functionally independent of N.

4.5.1 Abundance and Occupancy under Independence of p and N

In the general form of the model (see Section 4.1.4), the pmf of y is expressed as the mixture on p:

$$\Pr(y|p, N) = \sum_{N=0}^{\infty} \Pr(y|p(N)) \; \Pr(N).$$

For the case where p is independent of N, the observation model is not a function of N for occupied sites, i.e., $\Pr(y|p, N) = \Pr(y|p)$ whenever $N > 0$; when $N = 0$, $\Pr(y = 0|N = 0) = 1$. Thus we have:

$$
\begin{aligned}
\Pr(y|p, N) &= \sum_{N=0}^{\infty} \Pr(y|p(N)) \; \Pr(N) \\
&= \left\{ \sum_{N=1}^{\infty} \Pr(y|p(N)) \; \Pr(N) \right\} + \Pr(y|N = 0, p) \Pr(N = 0) \\
&= \left\{ \sum_{N=1}^{\infty} \Pr(y|p) \; \Pr(N) \right\} + I(y = 0) \Pr(N = 0).
\end{aligned}
$$

Note that, because $\Pr(y|p)$ is no longer a function of N, we can move that term outside of the summation sign to produce

$$
\begin{aligned}
&= \Pr(y|p) \left\{ \sum_{N=1}^{\infty} \Pr(N) \right\} + I(y = 0) \Pr(N = 0) \\
&= \Pr(y|p) \; \Pr(N > 0) + I(y = 0) \Pr(N = 0).
\end{aligned}
$$

The last expression is precisely equivalent to the zero-inflated binomial likelihood (Eq. (3.5.1) in Chapter 3), with $\psi \equiv \Pr(N > 0)$.

The central question of interest is, can we estimate parameters of $g(N|\theta)$ when p is not a function of N? The answer to that depends on how complex g is. Conceptually, we can substitute the expression $\Pr_g(N > 0|\theta)$ into the likelihood, to obtain the MLE of θ. For example, if g is Poisson, then $\psi(\lambda) = 1 - \exp(-\lambda)$ and the likelihood is

$$q(y|p, \lambda) = f(y|p)\psi(\lambda) + I(y = 0)(1 - \psi(\lambda)).$$

Since ψ is a one-to-one function of λ, λ can be estimated by a parameter substitution (plug $\psi(\lambda)$ into the likelihood). Or, by invariance of MLEs to transformation (Chapter 2), we can obtain the MLE of λ by maximizing the likelihood derived from Eq. (3.5.1), and solving the equality $\hat{\psi} = \Pr_g(N > 0|\hat{\theta})$ for $\hat{\theta}$.

Evidently, parameters of g can be estimated for any single-parameter g, e.g., Poisson, geometric, log-series, etc. However, since p is not a function of N, there is no information about the second moment structure of g in the data, so we cannot choose among these single-parameter models. They all yield the same estimate of ψ, and hence the same log-likelihood. In addition, it is evident that the parameters of multi-parameter distributions are not identifiable because an infinity of parameter combinations yields the same value of ψ (heuristically, you can't solve one equation, that relating $\hat{\psi}$ to parameters of g, for two or more parameters). In summary, when p is functionally independent of N, we cannot uniquely identify the local abundance distribution, but we can equate, redundantly, occupancy to mean abundance.

4.5.2 Choice of Link Functions

In MacKenzie *et al.* (2002) and elsewhere, the logit link function has been used to model the relationship between occupancy probability, ψ, and covariates that describe variation in ψ among sample units. The logit link is customary for parameters on the unit interval, and seldom is justification given for this choice (nor are alternatives considered in most cases). However, when considered in the broader context of the relationship between occupancy and abundance, there can be distinct relationships between the choice of link function and the abundance distribution g. That is, under functional independence between p and N, the relationship between ψ and $g(N|\theta)$ might suggest a particular link function.

For example, suppose that N has a geometric distribution having probability mass function

$$g(N|\theta) = (1 - \theta)\theta^{N},$$

where $0 < \theta < 1$. In this case, $\psi = 1 - \Pr(N = 0) = \theta$. Note that the mean of this geometric random variable is $E(N) = \mu = \theta/(1 - \theta)$ and thus $\log(\mu) = \log(\theta/(1-\theta)) = \text{logit}(\psi)$. That is, a logit-linear model for ψ is equivalent to a log-linear model for $E(N)$ when N has a geometric distribution. If N is geometric, and $\log(\mu_i) = \beta_0 + \beta_1 x_i$ for sample units $i = 1, 2, \ldots, M$ where x_i is some covariate, then, evidently, β_0 and β_1 may be estimated by the parameters of the linear-logistic model for ψ from detection/non-detection data alone.

As a second example, suppose that N is Poisson with mean λ. In this case, ψ and λ are related by the so-called *complementary log-log* (cloglog) link function:

$$\log(- \log(1 - \psi)) = \log(\lambda).$$

Thus, for example, if $\log(\lambda_i) = \beta_0 + \beta_1 x_i$, then also

$$\log(- \log(1 - \psi_i)) = \beta_0 + \beta_1 x_i.$$

Analogous to the case with the geometric distribution for N, we can assert that β_0 and β_1 can be estimated by specification of the appropriate linear function for the cloglog transform of ψ. That is, factors that are linear on the log-mean of N *have the same linear effect* on the complementary log-log scale of occupancy (*not* the logit scale as is customarily used).

These simple results hold under functional independence between p and N, but the issue should be considered more generally. For example (see Royle, 2006), under the binomial relationship between p and N, i.e., $p = 1 - (1 - r)^N$, then

$$\log(\log(1 - p)) = \log(\log(1 - r)) + \log(N) \tag{4.5.1}$$

in which case certain assumptions or approximations regarding the distribution of N might support the use of the cloglog link. For example, if abundance was approximately log-normal (Halley and Inchausti, 2002) then this implies an additive normal effect on the complementary log-log transform of p (Eq. (4.5.1)).

In summary, for models in which p is functionally *dependent* on N, variation in N is manifest as variation in p explicitly, and there is a duality between choice of link function and $g(N)$.

4.5.3 Assessing Functional Independence

When there is functional independence between p and N, the RN model reduces to the simpler two-state model (MacKenzie et al., 2002). However, the latter does not arise as a formal constraint on parameters of the former (note that the number of parameters is the same) and so it is not a reduced model per se. Conceivably, the ordinary two-state occupancy model can fit better than the RN model (an example from Royle (2006) is reproduced below). While we cannot use classical likelihood ratio tests in order to evaluate functional independence, we can use alternative inference paradigms, such as model-selection (e.g., based on AIC).

In fact, the simpler two-state model is a reduced model relative to a much broader class of models – the class of models of all forms of dependence between p and N, of which the RN model is only a particular example. Therefore, a comparison of the RN model vs. the functional independence model is not an omnibus conclusion of functional independence. For example, it could indicate *non-binomial* dependence between p and N (e.g., see Section 4.4.1). However, as suggested in Section 4.5.2, the functional form relating p to N and the distribution of N are closely related.

4.6 THE VARIABLE AREA SAMPLING DESIGN

When detection probability and local abundance are functionally independent, the local abundance distribution cannot be uniquely estimated (in particular, all single-parameter models provide statistically equivalent fits). However, this occurs only under a design in which sample units are homogeneous replicates, having the same effective sample area. We can introduce additional information about the abundance distribution by modifying the sample design to include sample units of varying, but known, areas.

While the precise area of sample units might be unknown in many practical settings (e.g., bird point counts), suppose that one could prescribe the sample area (e.g., a quadrat of specified size) or measure it (e.g., a pond or discrete wetland). Consider the negative binomial distribution, having probability mass function given by Eq. (4.4.4). Recall that, under this parameterization, the mean is μ and variance is $\mu + \mu^2/\alpha$ where μ is the mean of the limiting Poisson. Then, for sample units of varying area, the Poisson model is $N_i \sim \text{Poisson}(\mu A_i)$. The corresponding area-scaled negative binomial distribution has mean μA_i and variance $\mu A_i(1 + \mu A_i/\alpha)$.

This construction, conditional on the amount sampled, has been exploited in a few contexts including, recently, by Airoldi et al. (2006) to model word frequencies in documents of varying length. Bissell (1972) used such a negative binomial model having varying 'element sizes.' To clearly indicate the inclusion of variable area sample units, the pmf of local abundance will be denoted by $g(N|\alpha, \mu, A)$, and the resulting likelihood is the product of components having the form:

$$q(y_i|p, \alpha, \mu, A_i) = f(y_i|J, p)\psi(\alpha, \mu|A_i) + I(y = 0)\{1 - \psi(\alpha, \mu|A_i)\},$$

where $\psi(\alpha, \mu|A) = 1 - g(0|\alpha, \mu, A_i) = 1 - \left(\frac{\alpha}{\alpha + A_i\mu}\right)^\alpha$. Heuristically, we may estimate $\psi(\alpha, \mu|A)$ when we have at least two values of A, and then solve the resulting two (or more) equations for μ and α. In practice, we need not have distinct groups of sites (having common A), but we do need to have sufficient variability in A so as to induce variability in ψ among sites. This is an important design issue that seems worth investigating in advance of the survey.

4.6.1 Illustration

It seems that biologists rarely collect data from sample units of varying area, by design. So, in lieu of a real example, we carried out a brief simulation study to evaluate the variable area design concept. Recall that the goal of the variable area design is to provide information for distinguishing between different models, and also to allow for estimation in abundance distributions having more than one parameter.

In that context, we simulated populations having a negative binomial distribution and then sampled those populations under the functional independence model. We then fitted several models to the resulting data, including the correctly specified negative binomial model, and also the Poisson functional independence model. We also fitted the equivalent functional dependence models (i.e., those described by Royle and Nichols (2003)). The basic goal therefore was to assess whether we can accurately choose the proper model (NB functional independence) and whether it estimates the mean abundance unbiasedly, in suitably large samples, and whether the other (incorrect) models fail in this regard.

We generated a hypothetical population of 400 sample units having areas $A = 1, 2, \ldots, 10$ (40 of each). For each sample unit, we generated local abundance according to a negative binomial distribution with mean $\lambda_0 A$, for λ_0 having values 0.10 or 0.20, and $\alpha = 1.0$ or $\alpha = 0.5$. In the former case, the variance of the negative binomial is $\lambda_0 A + (\lambda_0 A)^2$ and in the latter case it is $\lambda_0 A + 2(\lambda_0 A)^2$. The populations were subjected to sampling $J = 5$ times using presence/absence surveys in which probability of detection for occupied sites was $p = 0.60$ or $p = 0.40$. A total then of 8 distinct cases were considered (2 levels of λ_0, 2 levels of α and 2 levels of p). To illustrate typical data that result from this scenario, the detection frequency distributions for 4 simulated data sets are shown in Table 4.5, where we see that most of the sites yield *no* detections. Typically, about 35 percent of sites in realized data sets have at least 1 detection.

For each of the 8 cases, 200 simulated realizations (of the 400 sites) were generated and each of 4 models were fitted to the data: the correct negative binomial abundance model under functional independence; a Poisson abundance model under functional independence; and both the negative binomial and Poisson under functional *dependence*. AIC was calculated for each fit, and the proportion of cases for which AIC for the data-generating model was lower than that of each of the other 3 models is tabulated in Table 4.6. In addition to this, the median of the sampling distribution of λ is presented (skew in the sampling distribution renders the mean a poor characterization of the typical behavior of the MLE, except in large samples). These brief results indicate that we do quite well in choosing the proper negative binomial model. In particular, for the higher level of overdispersion (cases 5 to 8), AIC suggests the correct model over the Poisson model at least 59 percent of the time. Moreover, the functional dependence models are chosen much less frequently, although the higher overdispersion cases, with $\lambda = 0.1$, seem to suggest the Poisson dependence case (Case 5 in particular favors the Poisson/functional dependence model). Finally, the sampling distribution of the MLE is centered approximately at the correct value of λ in all cases (slight positive bias is suggested for the higher level of overdispersion). Not surprisingly, there is a very strong bias when the abundance distribution is improperly specified. The bias is equally extreme under the incorrect functional dependence model. Note

Table 4.5. Sample realizations (number of detections out of $J = 5$ surveys) for negative binomial local populations with overdispersion $\alpha = 1.0$, and mean $0.10A$ for $A = 1, 2, \ldots, 10$ (40 sites of each area). Probability of detection is $p_0 = 0.40$, independent of N.

Realization	Number of detections					
	0	1	2	3	4	5
1	272	44	46	29	7	2
2	282	32	43	30	10	3
3	274	36	47	31	12	0
4	269	32	60	29	7	3

Table 4.6. Simulation results to assess MLEs under functional independence model and the variable-area sampling design. The data-generating local abundance distribution was negative binomial having parameters (mean) $\lambda_0 A$ and 'size' α. The 'Power' figures are the proportion of the 200 simulations that the correct model was favored (by AIC) over the model identified in the column heading. The last 4 columns are the median of the sampling distribution of $\hat{\lambda}_0$.

case	λ_0	p	α	'Power' against alternative				median($\hat{\lambda}_0$)		
				Poi/ind	NB/dep	Poi/dep	NB/ind	Poi/ind	NB/dep	Poi/dep
1	0.1	0.4	1.0	0.28	0.905	0.805	0.096	0.078	0.079	0.080
2	0.1	0.6	1.0	0.32	0.925	1.000	0.098	0.078	0.076	0.076
3	0.2	0.4	1.0	0.65	0.990	1.000	0.205	0.132	0.135	0.135
4	0.2	0.6	1.0	0.68	0.985	1.000	0.200	0.131	0.125	0.125
5	0.1	0.4	0.5	0.59	0.910	0.140	0.116	0.064	0.065	0.067
6	0.1	0.6	0.5	0.62	0.925	0.570	0.113	0.065	0.064	0.064
7	0.2	0.4	0.5	0.87	0.970	0.990	0.217	0.103	0.104	0.106
8	0.2	0.6	0.5	0.88	0.980	1.000	0.216	0.102	0.098	0.099

that the two functional dependence models yield very similar estimated densities because, in a preponderance of cases, the negative binomial MLEs tended to the Poisson boundary (i.e., $\alpha \to \infty$, no overdispersion).

4.7 ESTIMATING OCCUPANCY IN THE ABSENCE OF REPLICATE SAMPLES

An interesting consequence of the variable-area sampling design is that it provides information on occupancy and detection probability in the absence of replicate samples. To motivate this, consider the Poisson abundance model when p is functionally independent of N. In this case, the marginal probability of detection as a function of sample unit area, A, say $p(A)$, is

$$p(A) = \Pr(y = 1|A) = p(1 - e^{-A\lambda}).$$

For the case where local population size has a negative binomial distribution, then

$$p(A) = p \left\{ 1 - \left(\frac{\alpha}{\alpha + A\lambda} \right)^{\alpha} \right\}.$$

As $A \to \infty$, then $\Pr(y = 1|A) \to p$. That is, as area is accumulated by sampling, the net probability of detection approaches the nominal p, the probability of detection for an occupied site.

This variable area-induced heterogeneity in detection probability provides the heuristic basis for distinguishing between distinct models of occupancy and detection probability from a single replicate. Conceptually, suppose we have a group of sites having area A_1 and another group having area A_2. We can obtain sample statistics estimating the parameters $\Pr(y = 1|A_1) = p(1 - \mathrm{e}^{-A_1\lambda})$ and $\Pr(y = 1|A_2) = p(1 - \mathrm{e}^{-A_2\lambda})$, from which we can estimate parameters p and λ (i.e., we can solve the two equations for the two unknowns). Formally, when N is Poisson, the pmf of y is Bernoulli with

$$p(A) = \Pr(y = 1|A_i) = p(1 - \mathrm{e}^{-A_i\lambda})$$
$$1 - p(A) = \Pr(y = 0|A_i) = 1 - p(1 - \mathrm{e}^{-A_i\lambda})$$

and thus the likelihood for observations $y_1, y_2, \ldots, y_M$ is the product of M such terms. Equivalently, for units that can be pooled into groups of constant sample area, then $n(A) =$ the number of detections in the sample units of size A, and $n(A)$ is binomial with parameter $p(A)$.

In the single replicate situation it is perhaps not surprising that the data requirements for obtaining stable MLEs can be more of a challenge to achieve practically. The heuristic consideration is that a change in $p(A)$ as a function of A has to be realized in the face of binomial sampling variability about the curve $p(A)$. This is achieved as p increases, the range of A increases, and as the number of sites surveyed increases (all of which are potentially controllable). Our preliminary evaluations suggest that to distinguish between negative binomial and Poisson models, and to ensure that the MLEs are reasonably stable, might require several hundred or more observations.

4.8 SUMMARY

In this chapter, we introduced hierarchical models describing the relationship between observations of 'presence/absence' and abundance. These models are consistent with the conceptual framework for much of what we do in this book. The first-stage model – the observation model – is composed of a simple Bernoulli model for detection/non-detection data in which the probability of detection is related explicitly to the size of the local population exposed to sampling. The second-stage model – the process model – describes variation in local abundance among sample units. Under this model, heterogeneity (among sites) in apparent detection probability allows for the estimation of features of the local abundance distribution, $g(N)$ and even estimation of specific values of N, i.e., for particular sites. The hierarchical formulation of a model for this phenomenon via explicit observation and process models yields a coherent, rigorous treatment of inference about abundance and occurrence from observational data.

We specifically considered inference about abundance from presence/absence data using two classes of models. The first class is that in which the observation of presence/absence is influenced by local population size N (which varies among units). The Royle and Nichols (2003) model represents a specific form of what we have called *functional dependence*, and we have elaborated on several other possibilities, including other models for N and alternative models relating p to N. The second class of models that we considered arises under functional *independence* between p and N. In this case, the Royle and Nichols (2003) model of abundance-induced heterogeneity reduces to the two-state model described by MacKenzie et al. (2002). In fact, *all* functional dependence models reduce to this simpler model. However, while the abundance distribution cannot be uniquely determined under the standard design in which homogeneous sites are surveyed repeatedly, it is possible to modify the design so as to yield a uniquely identifiable abundance distribution. In particular, if we choose sample units to have variable area, then there is some ability to distinguish between abundance distributions. Much emphasis in contemporary field sampling is placed on standardization of protocols and methods, and so such designs tend to be uncommon. However, there are natural situations for which they might arise (sampling of wetlands or discrete habitat types), and certainly the structure can be imposed *a priori* if there was some benefit to doing so. One further implication of the variable area sampling design is that density can be estimated in the *absence* of replicate samples. As such, there may be some interest in describing sampling protocols that *induce* functional independence.

The models considered in this chapter require various types of model assumptions, such as a local abundance distribution, a probability model for the observations, and relationships between p and N. As always, questions of sensitivity to assumptions could be raised. Our view is that inference about occupancy and abundance is necessarily a model-based enterprise since, given only data on presence/absence, N is a latent variable, the properties of which must be deduced from mean and variance structure of Bernoulli observations. There is no free lunch – the price being, in this case, the specification of models. We do not advocate a particular model for the conduct of inference from presence/absence data. We do, however, advocate model-based inference as a general framework for providing context and interpreting data. While it is tempting (and also convenient at times) to criticize model-based inference as being (potentially) sensitive to model choice, we believe that sensitivity to model assumptions should be put into a broader context (that being that occupancy is, fundamentally, related to the distribution of organisms in space, and is a matter of scale and abundance or density). Alternatively, one could adopt estimators of occupancy that are, on the face of things, lacking in explicit model assumption about abundance (e.g., MacKenzie et al., 2002). It might be tempting to claim that such procedures are 'robust' to choice of $g(N)$. However, the absence of any explicit statement of a model is not equivalent to robustness to potential, unspecified, alternatives. Nor does it imply that their effects are negligible or irrelevant (Dorazio, 2007; and see Little (2004) for a good example in the context of design-based vs. model-based estimation in sampling). While an estimator of ψ can be derived in the absence of model assumptions regarding abundance, an equivalent view is that it arises under functional independence. Whether or not it implies some form of robustness is therefore debatable.

5

INFERENCE IN CLOSED POPULATIONS

A classical problem in statistical ecology is estimating the size of a closed population. In this chapter, we provide an introduction to some basic concepts and models underlying this classical problem, and we describe some of its contemporary manifestations.

A closed population is one that is not subject to additions or subtractions through demographic processes or temporary movement. That is, a closed population is a static collection of individuals, experiencing no mortality, recruitment, emigration, or immigration. As such, there is no such construct in reality. Nevertheless, the literature on inference in closed populations is vast. There are many monographs and books largely devoted to the topic (e.g., Otis et al. (1978) and Borchers et al. (2002) deal exclusively with closed populations). Dozens of papers are published regularly in top journals that deal largely with closed population problems. A literature search of the journal *Biometrics* reveals 24 papers relevant to 'closed populations' from 2000 through 2006, and 52 papers among a handful[1] of popular ecological journals over the same time frame.

There are several reasons for this interest in models of closed populations. First, *abundance* is a fundamental object of inference in many ecological studies, and closed population models provide the simplest 'design' that allows for direct inference about abundance from a sample of marked individuals. Second, the closed population as a *theoretical* mathematical construct has considerable utility as a framework for formulating models and studying analytic procedures, and for developing elaborations of those models and procedures. For example, models with individual covariates are widely studied in the context of closed populations because of the tractability that the closed population system yields. The multinomial distribution that underlies the basic statistical formulation of closed population models is fundamental to nearly all capture–recapture problems. Many methodological innovations in capture–recapture amount to little more than slight modifications of multinomial cell probabilities of existing models. Thus it is useful to

---

[1] Journal of Wildlife Management, Conservation Biology, Ecology, Journal of Animal Ecology, Journal of Applied Ecology.

gain some experience with the basic multinomial model structure that arises under closed population sampling problems. Finally, the notion of a closed population is useful because of the conceptual clarity that closed population systems provide. While almost no biological system is truly closed, the assumption of a closed population, along with a model describing the encounter process, brings some context to animal sampling problems. It provides a basis for the interpretation of observational data that is concise and unambiguous in the model-based framework for inference within which we operate. For example, consider simple counts (e.g., in consecutive years) of $n = 1000$ vs. $n = 2000$, which are supposed to represent counts of individuals in some population. Interpretation of these counts requires some other piece of information that describes the observation process (i.e., related to effort). It seems natural to summarize that additional piece of information by one or more parameters related to probability of detection, and the observed n by an estimate of population size, $\hat{N}$, relevant to some hypothetical closed population.

The main objectives of this chapter are to provide a coarse review of some basic concepts, models, terminology, and literature. We don't aim to provide an encyclopedic coverage of these things (for that, see Williams et al. (2002)). Rather, we aim to cover the main technical concepts and methods that are in widespread use. In keeping with one of the broader themes of this book, we provide coverage of both classical and Bayesian treatments of conventional models for closed populations and the use of **R** and **WinBUGS** to solve these problems. We focus here primarily on the specific inference problem of estimating N and various ways to do that under certain designs and models. One of the major themes that we want to develop in this chapter is the notion that the observation process induces 'error' in the form of bias due to imperfect detection. The second major theme is that the multinomial distribution is integral to inference in closed populations, typically as a model of the observation process. The final theme of this chapter is that Bayesian analysis of closed population models can be accomplished using a method known as 'data augmentation.' In the context of closed populations, data augmentation yields a technical equivalence between closed population models and models of occupancy (Section 5.6). More specifically, multinomial observation models that arise in closed population sampling can be reparameterized as zero-inflated binomial models with the use of data augmentation. This is convenient in some cases (e.g., Chapters 6 and 7), but it also provides a direct technical and conceptual linkage to the models of occurrence or occupancy that constitute one of the other broad methodological themes of this book.

We begin with a brief discussion on the multinomial distribution and then focus attention on estimating the size of a closed population under the simplest model of the detection process. We briefly discuss other multinomial observation models, including removal sampling, sampling with multiple observers, and distance sampling. Finally, we introduce a new method of formulating models for closed populations based on *data augmentation*, which we use repeatedly throughout this book as a framework for analysis of other models.

5.1 FUN WITH MULTINOMIAL DISTRIBUTIONS

We introduced the multinomial distribution in Chapter 2. The multinomial is fundamental to inference in ecology, because it is a natural distribution for modeling frequencies of discrete outcomes. In the case of capture–recapture type models in general and closed population models specifically, there are 3 special characteristics of the multinomial distributions that are shared by virtually all of the situations that we will encounter. First is that the multinomial outcome $\{y_k; k = 1, 2, \ldots, K\}$ resulting from a capture–recapture study is the vector of frequencies of distinct *encounter histories*. For example, if we sample a population of individuals twice, then potential encounter histories are $\{11, 01, 10, 00\}$ representing an individual encountered in both samples (11), only the second sample (01), etc. Note that usually the last encounter history, 00, is the encounter history representing the event 'not encountered'. In a population of N individuals, the frequencies of these 4 possible outcomes have a multinomial distribution with sample size N and probabilities $\boldsymbol{\pi} = (\pi_{11}, \pi_{01}, \pi_{10}, \pi_{00})$. Secondly, the multinomial cell probabilities are *structured*. That is, they are related to one another by a smaller collection of more fundamental parameters that describe the observation process (the detectability of individuals during sampling). Thus, typically, the multinomial cell probabilities don't have any intrinsic meaning (in the population sampling context), but the parameters that comprise those cell probabilities do. For example, in keeping with the previous example of 4 possible encounter histories, we might suppose that the probability of detecting an individual during sampling, p, is constant for both sampling occasions. Then, the multinomial cell probabilities are related to p according to: $\pi_{11} = p^2$, $\pi_{01} = (1 - p)p$, $\pi_{10} = p(1 - p)$, and $\pi_{00} = (1 - p)(1 - p)$. Thirdly, it is almost always the case that the multinomial sample size is unknown. In the case of animal sampling, the sample size is N, the size of the closed population, and N is almost always the object of inference.

5.1.1 Example

The frequencies associated with any finite partition of a continuous random variable y can be rendered as a multinomial (e.g., see Serfling, 1980, 2.7). For example, if y is a Normal$(0, 1)$ random variable, then, for a sample of size n, we can tabulate the frequencies $n_1 = \sum_{i=1}^{n} I(y \leq -1)$, $n_2 = \sum_{i=1}^{n} I(-1 < y \leq 1)$, $n_3 = \sum_{i=1}^{n} I(y > 1)$

Table 5.1. Binomial frequencies in a sample of size 79 with $J = 6$.

k	=	0	1	2	3	4	5	6
n_k	=	11	25	22	13	5	1	2

where $I(\cdot)$ is the indicator function. These frequencies constitute a multinomial sample having cell probabilities $\pi_1 = \Pr(y \leq -1)$, etc.

For a discrete random variable, when conditioned on the sample size n, the observed frequencies have a multinomial distribution related to the cell probabilities equal to the pmf of the random variable. The binomial structured multinomial is relevant to things we do in this chapter and elsewhere. Suppose $y \sim \text{Bin}(J, p)$, i.e.,

$$f(y|J, p) = \frac{J!}{y!(J-y)!} p^y (1-p)^{J-y}.$$

Then $\mathbf{n} = (n_0, n_1, \ldots, n_J)$, where $n_k = \sum_{i=1}^{n} I(y_i = k)$, has a multinomial distribution with cell probabilities $\{\pi_0, \pi_1, \ldots, \pi_J\}$ where

$$\pi_0 = \Pr(y = 0) = (1-p)^J$$
$$\pi_1 = \Pr(y = 1) = Jp(1-p)^{J-1}$$
$$\cdots \quad \cdots$$
$$\cdots \quad \cdots$$
$$\pi_J = \Pr(y = J) = p^J.$$

The point of this is that while the counts are binomial, we can organize them into groups and then deal with the multinomial likelihood of the grouped counts. As an example, a study was conducted that yielded the 79 binomial counts in Table 5.1, based on $J = 6$. The multinomial likelihood can be maximized using the **R** code given in Panel 5.1. It is easy to obtain parameter estimates of structured multinomial distributions.

Sometimes we encounter structured multinomials where the cell probabilities derive from a geometric pmf. Suppose that n ducks are marked (banded) and released and that the annual survival probability is ϕ. Let n_j be the number of ducks that die in year j and let $n_{J+1} = N - \sum_{j=1}^{J} n_j$ be the number surviving after J years. Then, the frequencies $\{n_1, n_2, \ldots, n_{J+1}\}$ are multinomial with cell probabilities

$$\pi_j = (1-\phi)\phi^{j-1}$$

and π_{J+1} for the last cell is $1 - \sum_{j=1}^{J} \pi_j = \phi^J$, which is the probability of surviving after J years. This model structure is common in animal demography, the classical example being band-recovery models, which are just a manifestation of a general survival process. This particular model element is the relevant biological process model; whereas, in practice, we also have to accommodate a component model for the observation process. We'll encounter several examples of that in later chapters of this book.

5.1.2 Shrinking and growing multinomials by conditioning and unconditioning

Suppose we have a multinomial with K cells. We will encounter many situations in which we are interested in constructing a multinomial whose categories are a strict subset of these K categories. In this case, we can reduce the dimension of the original multinomial by *conditioning* on a subset of events. Alternatively, we sometimes have an interest in constructing a multinomial that includes the K categories as a strict subset. In this case, we can expand the size of the multinomial by *unconditioning* on the sample size of the original multinomial.

Suppose $y \sim \text{Bin}(J, p)$ with $J = 5$. There are 6 possible outcomes of y, and so the sample frequencies have a multinomial distribution with cell probabilities structured according to a binomial pmf. Now consider that we are concerned about the multinomial distribution that describes the frequencies of y given that $y > 0$ (this will be the typical usage of conditioning). We want the cell probabilities of this new multinomial, which are $\text{Pr}(y = k | y > 0)$. So, we need to partition the probabilities $\text{Pr}(y = k)$ into components according to the law of total probability:

$$\text{Pr}(y = k) = \text{Pr}(y = k | y > 0)\,\text{Pr}(y > 0) + \text{Pr}(y = k | y = 0)\,\text{Pr}(y = 0)$$

```
silly.fn<-function(){

k <-c( 0, 1, 2, 3,4,5,6)
nk<-c(11,25,22,13,5,1,2)

lik<-function(parms){
   p<-expit(parms[1])
  -1*sum(nk*log(dbinom(k,6,p)))
 }

nlm(lik,c(0),hessian=TRUE)

}
```

Panel 5.1. A simple function to maximize a multinomial likelihood for the data in Table 5.1.

Note that the second part evaluates to 0 for $k = 1, 2, \ldots, J$, since the events $y = k$ and $y = 0$ are mutually exclusive outcomes. Thus, it is the case that

$$\Pr(y = k | y > 0) = \Pr(y = k) / \Pr(y > 0).$$

We see that the conditional probabilities are scaled versions of the original probabilities – each is scaled by the complement of the event that was conditioned on. This turns out to be somewhat useful in many situations, usually in the context of accounting for bias induced by sampling. In that context, the relevant event, or group that is conditioned on, is the group of individuals that appear in the sample – i.e., captured individuals. The resulting distribution in this case is usually a zero-truncated binomial (e.g., see Section 5.2.2). The zero-truncated Poisson is another such distribution that arises in some applications. Its pmf is

$$\Pr(y = k | y > 0) = \frac{\exp(-\lambda)\lambda^y}{y!(1 - \exp(-\lambda))}.$$

The reverse operation to conditioning (let's say unconditioning), expands the dimension of the multinomial. Suppose y is multinomial with K cells and cell probability $\{\pi_k\}$, and sample size N. Suppose further that N is binomial with sample size M and parameter ψ. Then it transpires that y is multinomial with $K + 1$ cells and parameters $\psi \pi_k$ for $k = 1, 2, \ldots, K$ and the last cell, $\pi_{K+1} = 1 - \psi$. This reverse operation comes in handy in Section 5.6 and elsewhere in this book. Obviously, to get rid of the new cell – we condition it away, by dividing the first K cells by the complement of the last, which is ψ. Hence, we are left with the multinomial that we started with.

These two operations that we perform on multinomial distributions are simply manifestations of conditional probability and the law of total probability.

5.2 ESTIMATING THE SIZE OF A CLOSED POPULATION

Suppose we sample a population of size N and observe n distinct individuals. The classical problem associated with closed populations is to estimate the number of individuals in the population based on the sample of size n, i.e., to estimate N. We could equate $n = N$ if we think we did a census. But how would we know? To formalize the answer to this question, we need to describe a *model* that relates n to N that we can analyze statistically.

The essence of almost all contemporary animal sampling methods is the assumption of binomial sampling. We assert that $n \sim \text{Bin}(N, \pi)$, which is to say that individuals that appear in the sample represent a random sample from the population of size N. Under this view, n itself is the outcome of the experiment[2]. One helpful way to conceptualize this is to imagine a latent variable w_i associated with individuals $i = 1, 2, \ldots, N$, such that $w_i = 1$ if individual i appears in the sample, and $w_i = 0$ if not. We assume that the w_i are Bernoulli trials, i.e., they are independent with $\Pr(w_i = 1) = \pi$. The probability that an individual appears in the sample, π, is a function of certain parameters, the precise form of which is to be described shortly. In animal sampling problems, the magnitude of π is largely a function of 'method' – e.g., the level of effort and the manner of sampling. Ideally, we would like π to be close to 1. This is because the MSE of n_k, taken across replicate populations indexed by k, (this could be space, time, or both), has both a sampling component and an environmental component (Section 1.3.2). As π gets close to 1, the sampling component, which is purely nuisance variation, diminishes to zero. This desire to have a high sampling fraction competes with economics – in practical situations, it is often more expensive to achieve something close to a census (i.e., π near 1). This trade-off motivates much of contemporary animal sampling applications. In estimating the size of a hypothetically closed population, the conceptual objective is to estimate π so that we can convert our sample count to a population size estimate. Under the binomial assumption,

$$\mathrm{E}(n) = \pi \times N$$

and thus estimators of N can be motivated heuristically as an 'adjustment' of the observed count, according to:

$$\hat{N} = n/\hat{\pi}.$$

Estimation of the probability of capture, π, requires additional information. There seems to be two conceptually distinct approaches that yield this information. The first is to acquire that information by design – specifically to sample the population repeatedly, in order to obtain J independent observations of the population. This provides direct information about the encounter process. For example, when the sampling occurs under closure, we can be certain that a $(1, 0, 0, 1)$ encounter history represents 2 legitimate non-captures, in which case the probability of detection might be estimated as $2/4$. The second manner by which to acquire information about detection probability is to have a 'magic covariate' – i.e., a covariate that is related to p in a special way. The most common situation here is distance sampling, but that concept applies to other problems in which biased samples arise (e.g., size and length bias). A recent example involving reporting of tornados is Anderson et al. (2007). There are hybrid methods that involve both replication and covariates. These 'individual covariate models' are discussed in Chapter 6.

---

[2] Not a formal experiment in the statistical sense, but in the more abstract sense of being the outcome of a probabilistic mechanism.

5.2.1 The Classic Design based on Replication

Here we consider the classical problem of estimating the size of a population of N individuals that are sampled, with replacement, on J separate occasions. We assume that the probability of detection, p, is identical for all individuals and all occasions and that the observations are independent. This combination of design and assumptions is colloquially referred to as 'Model M_0' in the capture–recapture literature.

We obtain an $n \times J$ matrix of detections $\boldsymbol{Y}$, whose rows correspond to the sequence of detections (an encounter history) observed for each of the n unique individuals in the sample. An example is shown in Table 5.6. These are capture histories of *Microtus* reported by Williams et al. (2002, p. 525) that have been analyzed extensively by others. The data used here include encounter histories of 56 adult males that were sampled on 5 consecutive days (additional details are provided below).

To formalize the solution to our inference problem, we need to specify the joint distribution of the data as a function of the unknown quantities, N and p. Under the assumptions of Model M_0, each binary observation y_{ij} is the outcome of an independent Bernoulli(p) trial; therefore, the encounter history observed for the ith individual may be summarized by the total number of times the ith individual was detected in J attempts: $y_i = \sum_{j=1}^{J} y_{ij}$. To solve the inference problem, we require the joint pmf of the observed vector of detections $\boldsymbol{y} = (y_1, \ldots, y_n)$ and n, given the unknown parameters N and p:

$$[\boldsymbol{y}, n | N, p]. \tag{5.2.1}$$

What is this distribution? To develop some intuition, let's assume counterfactually that we had 'observed' $y = 0$ for the $N - n$ individuals in the population that were not detected and that we only need to estimate the probability of detection. In other words, suppose we knew the size of the population and only want to estimate p. Under the assumptions of Model M_0, $y \overset{iid}{\sim} \text{Bin}(J, p)$; therefore, we can express the joint pmf of $\boldsymbol{y}$, which is assumed to include the $N - n$ 'observations' of $y = 0$, as follows:

$$[\boldsymbol{y}|p] = \prod_{i=1}^{N} \text{Bin}(y_i | J, p). \tag{5.2.2}$$

The information in $\boldsymbol{y}$ can be summarized as 'grouped data' by calculating the total number of individuals detected in k of J sampling occasions as follows:

$n_k = \sum_{i=1}^{N} I(y_i = k)$, where $I(\cdot)$ is the indicator function. We use these summary statistics to rewrite Eq. (5.2.2) as follows:

$$[\boldsymbol{y}|p] = \prod_{k=0}^{J} \{\text{Bin}(k|J,p)\}^{n_k}.$$

Thus, the likelihood function for p is easily expressed in terms of the grouped data. i.e., the product-binomial:

$$L(p|\boldsymbol{y}) = L(p|n_0, n_1, \ldots, n_J)$$
$$= \prod_{k=0}^{J} \pi_k^{n_k}$$

where $\pi_k = \binom{J}{k} p^k (1-p)^{J-k}$ denotes the binomial pmf.

Now, let's consider our original problem where the frequency of zeros, $n_0 = N - n$, is unobserved and we need to estimate both N and p. Notice that the vector $(n_1, n_2, \ldots, n_J)$, i.e., the 'grouped' representation of $\boldsymbol{y}$, contains the observed frequencies of J *mutually exclusive and discrete* outcomes $(\{1, 2, \ldots, J\})$. Because $y = 0$ is the *only* unobservable outcome, we may conclude that the joint distribution of $(n_1, n_2, \ldots, n_J)$ is multinomial with pmf

$$f(n_1, \ldots, n_J | N, \pi_1, \ldots, \pi_J) = \binom{N}{n_1, \ldots n_J} \pi_1^{n_1} \cdots \pi_J^{n_J} \left(1 - \sum_{k=1}^{J} \pi_k\right)^{N-n}.$$

Therefore, the likelihood function for N and p is easily expressed in terms of the grouped data:

$$L(N, p|\boldsymbol{y}, n) = L(N, p|n_1, \ldots, n_J)$$
$$= \frac{N!}{(N-n)!} \pi_1^{n_1} \cdots \pi_J^{n_J} \left(1 - \sum_{k=1}^{J} \pi_k\right)^{N-n}. \qquad (5.2.3)$$

Furthermore, taking the structure of π_k into consideration yields an even simpler expression for the likelihood function:

$$L(N, p|\boldsymbol{y}, n) = \frac{N!}{(N-n)!} p^{\sum_{i=1}^{n} y_i} (1-p)^{J \cdot N - \sum_{i=1}^{n} y_i}. \qquad (5.2.4)$$

Using this likelihood function, it is straightforward to fit Model M_0 in **R**, as we will demonstrate later.

5.2.2 Inference based on the Conditional Likelihood

The joint pmf of the observations $\mathbf{y}$ and n can be factored according to

$$[\mathbf{y}, n|N, p] = [\mathbf{y}|n, p][n|N, p]$$

where $[n|N, p] = \text{Bin}(n|N, 1-(1-p)^J)$. The first component, $[\mathbf{y}|n, p]$ is the conditional (on n) distribution of the observations – and the resulting likelihood is usually referred to as the conditional likelihood[3]. It is common in practice to use the first component of the decomposition to estimate p (note that $[\mathbf{y}|n, p]$ is a function of p but not N). Then, we can use the 'conditional MLE' of p along with n to obtain the conditional estimator of N. That is, under the binomial assumption, $\text{E}[n] = \pi(p)N$ and so $\hat{N} = n/\pi(\hat{p})$. The so-called conditional estimators are pervasive in applications of abundance estimation. In part, this is because of their heuristic justification. Early work by Sanathanan (1972) established the asymptotic equivalence of conditional and unconditional estimators of N.

It remains to identify this conditional distribution $[\mathbf{y}|n, p]$. It is tempting to declare this to be a binomial. But note that y has a binomial distribution if N were known. In that case, the y_i are *iid* realizations of a binomial random variable having sample size J and parameter p. The important distinction between the observations $y_1, \ldots, y_n$ and a strict binomial sample is that, in the former case, we do not have the zero counts. That is, all of the observations are ≥ 1. Thus, the sample obtained from a capture–recapture study could be viewed as a biased sample relative to a hypothetical 'known-N' binomial sample.

We have previously encountered a similar situation in Chapter 3, where we were exposed to the *zero-inflated binomial*, which arose by mixing excess zeros in with binomial observations. The context was a site occupancy type of problem, in which the observed zeros were composed of truly unoccupied sites, and sites that are occupied but where the species was not detected. Here we have the conceptual opposite problem – we have binomial observations but with *no* zeros. The corresponding modified binomial distribution for this case is a zero-truncated binomial. Thus, while the complete data likelihood would be based on a binomial pmf, $\text{Pr}(y) = \text{Bin}(y|J, p)$, we seek the likelihood based on the conditional probabilities $\text{Pr}(y = k|y > 0)$. We obtain this by taking the mass of the zero cell and distributing it over the remaining cells (see Section 5.1.2).

---

[3]Whereas then $[\mathbf{y}, n|N, p]$ is colloquially referred to as the unconditional likelihood. It is, more properly, the joint distribution of $\mathbf{y}$ and n.

This can also be accomplished by application of Bayes' rule. Let w be a Bernoulli random variable indicating capture ($w = 1$) or not ($w = 0$). Then, we seek the conditional distribution $[y|w = 1]$. Bayes' rule yields

$$\Pr(y|w = 1) = \frac{\Pr(w = 1|y)\Pr(y)}{\Pr(w = 1)}; \quad \text{for } y = 1, 2, \ldots, J.$$

Note that $\Pr(w = 1|y) = 1$ (by definition of $w = 1$ as being 'captured') for the support of y. Thus, $\Pr(y|w = 1) = \Pr(y)/\Pr(w = 1)$ where $\Pr(y)$ are binomial probabilities. Finally, $\Pr(w = 1) = \Pr(y > 0) = 1 - (1 - p)^J$ (under M_0, but in general it is the zero-cell of the multinomial, π_0). Thus,

$$\Pr(y = k|w = 1) = \text{Bin}(k|J, p)/(1 - (1-p)^J).$$

The likelihood of the n observations is therefore multinomial, having J cells, and cell probabilities specified by $\Pr(y = k|y > 0)$.

5.3 EXAMPLES

Here we consider several examples that seem to be typical applications of closed population models. These examples also highlight some deficiencies of the closed population models and indicate the need for more general models.

5.3.1 Analysis of the Microtus Data

The *Microtus* data were collected by J.D. Nichols at Patuxent Wildlife Research Center, Laurel, MD during 1981. The sampling was carried out using a 10×10 grid of traps baited with corn. Considerable detail on these data are available in Williams et al. (2002, Chapter 19), and they have been analyzed in many journal articles. We consider here data of adult males, from the first sampling period beginning June 27.

For the *Microtus* data, for which a few of the encounter histories were shown in Table 5.6, the encounter frequency distribution is $n_k = (12, 8, 9, 12, 15)$ for $k = 1, 2, 3, 4, 5$, respectively. Construction of the log-likelihood in **R** is shown in Panel 5.2. Execution of this function and some of the output are also given in Panel 5.2. The estimates are $\hat{p} = 0.60$, and for the number of individuals not captured, n_0, the MLE occurs on the boundary $\hat{n}_0 = 0.00$. This is a sensible result, given the high detection probability of this sampling apparatus (baited traps arranged in a grid). The probability of an individual being captured at least 1 time is $1 - (1 - 0.60)^5 = 0.99$. We conclude that Dr. Nichols has a keen ability to capture *Microtus*. In Williams et al. (2002) some additional models were fitted, and there is considerable discussion of the use of these data for the estimation of density. We provide some further analysis of these data in Section 5.6 and in subsequent Chapters.

5.3.2 Horned lizard surveys

These data are from a capture–recapture survey of flat-tailed horned lizards (*Phrynosoma mcallii*; Figure 5.1) in southwestern Arizona and originate from studies to evaluate monitoring strategies for the species (Young and Royle, 2006; Royle and Young, 2008). The specific data set consists of capture/recapture data from a single 9 ha plot (300 m × 300 m). There were 14 capture occasions over 17 days (14 June to 1 July 2005). A total of 68 individuals were captured 134 times. The distribution of capture frequencies was $(34, 16, 10, 4, 2, 2)$ for 1 to 6 captures respectively, and no individual was captured more than 6 times. The species is notoriously difficult to sample, due to their habit of burying themselves in the sand to avoid detection (Figure 5.2). The low detection probability is indicated by the sparse encounter histories.

Model M_0 was fitted to these data yielding $\hat{p} = 0.117$ (0.0122) and $\hat{n}_0 = 14.024$ (5.414). To estimate the density of this species, we might use an estimator of the form:

$$\hat{D} = \frac{68 + 14.024 \text{ individuals}}{9 \text{ ha}} = 9.114 \text{ individuals/ha}$$

having SE $= 0.602$. We discuss the interpretation of such density estimates in the following example and also in subsequent analyses of these data.

```
lik0<-function(parms){
    p<-  expit(parms[1])
    n0<-  exp(parms[2])
    N<-sum(nvec) + n0
    cpvec<-dbinom(0:5,5,p)
    -1*(lgamma(N+1) - lgamma(n0+1) + sum(c(n0,nvec)*log(cpvec) ))
}

nvec<- c(12,8,9,12,15)
out<-nlm(lik0,c(-1,1),hessian=TRUE)
```

Panel 5.2. R specification of the likelihood for Model M_0, and its execution for the *Microtus* data.

Figure 5.1. Left: Flat-tailed horned lizard (FTHL) (*Phrynosoma mcallii*) in all its glory. Right: Quality habitat of the flat-tailed horned lizard. *Photos courtesy of K.V. Young.*

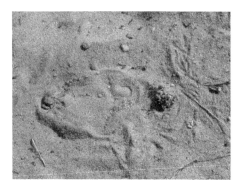

Figure 5.2. Figures illustrating FTHL elusiveness. Left: typical lizard 'dug-in'. Right: this area was thoroughly searched and lizard in the middle was missed by surveyors who nearly stepped on it (note footprints) before eventually noticing it. *Photos courtesy of K.V. Young.*

5.3.3 Tiger camera trapping data

Closed population models are commonly used to to estimate densities of carnivores from arrays of camera traps. Tigers and other large cats have unique stripe or spotting patterns that allow individual animals to be identified uniquely (e.g., Karanth, 1995; Karanth and Nichols, 1998, 2000; Trolle and Kéry, 2003; Karanth et al., 2006). We provide an illustration of this application using data from the Nagarahole reserve in the state of Karnataka, southwestern India. For an analysis of these data, see Royle et al. (2008). Further analyses are described in Chapters 6 and 7. The particular data used here were collected during 12 sampling events from 24 January to 16 March 2006. The survey included 120 trap stations, consisting of two camera-traps each.

There were 45 tigers observed, having detection frequencies (out of $J = 12$) $(30, 10, 3, 1, 1)$ for $k = 1, 2, 3, 4, 5$, respectively (no individual was observed more than 5 times). We obtain $\hat{n}_0 = 25.30$, and $\hat{p} = 0.081$. Thus, $\hat{N} = 70.30$. Unlike with the *Microtus* example given previously, tiger capture rates are very low and hence, in this case, a large fraction of the population appears to have gone undetected. As a result, the precision of the estimates also appears to be low ($\text{SE}(\hat{p}) = 0.015$ and $\text{SE}(\hat{N}) = 10.22$).

5.3.4 On Estimating Density

This situation, in which a population is sampled by a spatial array of traps (or, in the case of the lizard example, a prescribed area is surveyed), is typical of situations in which closed population models are applied for estimating density. We obtain an estimate of population size, $\hat{N}$, and divide by the sample area. The main difficulty in interpreting results is that the population is not strictly closed – or rather, N is not well-defined in the sense that the effective sample area of the grid is unknown, in most practical situations. If the spacing of the traps is sufficiently close (so that most individuals are exposed to multiple traps), then the effective sample area is likely to be larger than the nominal area covered, physically, by the 10×10 grid of traps (in the *Microtus* example). Another important consideration in such spatial capture–recapture problems is that there is heterogeneity in detection probability induced by unequal exposure to traps. For example, individuals having home ranges near the boundary of the trapping grid experience less exposure to trapping than individuals near the center, on average.

These problems are pervasive in almost all spatial capture–recapture problems (and in other contexts), and many ad hoc solutions have been proposed. See Karanth and Nichols (1998); Trolle and Kéry (2003); Karanth et al. (2004) for context and relevant discussion. We address these considerations further in Chapters 6 and 7.

5.4 AN ENCOUNTER HISTORY FORMULATION

In the previous section, the model was formulated in terms of the frequencies, $\{n_k\}$, the total number of individuals detected $k = 1, \ldots, J$ times. A more general formulation of the model is in terms of encounter history frequencies. This is necessary when detection is assumed to vary over time so that the individual

Table 5.2. Encounter history frequencies when $J = 3$. Unique encounter histories are assigned an integer index from 0-7 for convenience.

frequency	$j = 1$	$j = 2$	$j = 3$
n_0	0	0	0
n_1	1	0	0
n_2	0	1	0
n_3	0	0	1
n_4	1	1	0
n_5	1	0	1
n_6	0	1	1
n_7	1	1	1

Bernoulli trials do not have a constant success probability. In the absence of individual effects on p, the individual encounter histories (e.g., Table 5.6) can be pooled into groups of unique encounter histories, indexed by h a unique combination of zeros and ones. For convenience (and consistency), we will reference these encounter histories by an integer subscript k. Thus, let n_k be the number of individuals having encounter history k where $\{n_k; k > 0\}$ are the *observable* encounter history frequencies, and n_0 is the unobservable encounter history frequency. Thus, for sampling that occurs on J occasions there are 2^J possible encounter histories (including that corresponding to 'not encountered'). As an example, for $J = 3$, the possible encounter histories are shown in Table 5.2.

As before, the nominal objective is to estimate the frequency of zeros, n_0. The pmf of the vector $\mathbf{n} = (n_1, n_2, \ldots, n_K)$ is multinomial, having the form:

$$[\mathbf{n}|N, \boldsymbol{\pi}] \propto \frac{N!}{(N - n)!} \left\{ \prod_{k=1}^{7} \pi_k^{n_k} \right\} (1 - \pi.)^{N-n}$$

where $n = \sum_{k=1}^{7} n_k$ is the number of unique individuals observed, $\pi. = \sum_{k=1}^{7} \pi_k$ is the net probability of encountering an individual in at least one of the J samples, and $\pi_0 = (1 - \pi.) = $ probability of not encountering an individual (at all).

The multinomial cell probabilities, $\{\pi_k\}$, are probabilities of observing each distinct encounter history k, and are functions of parameters that describe the detection process, the precise form depending on the hypothetical model under consideration. A useful model is that in which the detection probabilities vary over time, a model that is referred to as 'Model M_t' in the jargon of capture–recapture. Under this model, each potential encounter history corresponds to a multinomial outcome, and the multinomial cell probabilities correspond to the probabilities of J independent Bernoulli trials. For example,

$$\pi_1 = \Pr(\mathbf{y} = (1, 0, 0)) = p_1(1 - p_2)(1 - p_3)$$

where p_t is the probability of encountering an individual in period t.

In some situations we may have time-specific covariates that might influence p, and it would be conventional to model those according to

$$\text{logit}(p_j) = \alpha_0 + \alpha_1 x_j$$

where x_j might be effort, weather, etc. Such models are used sometimes with longer-term studies, often of banded birds. They are structurally similar to models having covariates, which we address in Section 5.6.4 and Chapter 6.

Operationally, it is more convenient and mathematically equivalent (but computationally less efficient) to treat each observation as its own cell and compute n cell-probabilities. An **R** function that computes the log-likelihood function for each observation is shown in Panel 5.3. The data and **R** function for fitting this model are provided in the Web Supplement.

5.4.1 Model variations

Model extensions can be obtained by simple modifications of the multinomial cell probabilities. This gives rise to a sequence of closed population models that are usually fit in succession, followed by a suite of goodness-of-fit tests and AIC calculations, to arrive at the best model. Otis et al. (1978) recognized 8 such

```
lik<-function(parms){
    pvec<-  expit(parms[1:5])
    n0<- exp(parms[6])
    N<-nind + n0
    loglik<-rep(NA,nind)
    for(i in 1:nind){
      yvec<-Y[i,]
      loglik[i]<- sum(yvec*log(pvec) + (1-yvec)*log(1-pvec))
    }
zerocell<- n0*sum(log(1-pvec))
fact<-lgamma(N+1)-lgamma(n0+1)
-1*(fact + zerocell + sum(loglik))
}
```

Panel 5.3. R specification of the likelihood for Model M_t using the *Microtus* data. Here, the **R** object Y is the encounter history matrix.

models, including M_0, M_t, and various other modifications. The suite of models continues to grow as variations on the basic theme and new models are proposed (e.g., see Yang and Chao (2005), for general models of behavioral response). These closed population models (and many that have not been invented yet) are essentially multinomial observation models derived by combining and recombining the $2^J - 1$ observable encounter histories. We consider models of behavioral response and the Yang and Chao (2005) variation for the *Microtus* data in Section 5.6.

5.5 OTHER MULTINOMIAL OBSERVATION MODELS

Much effort has been directed toward developing sampling protocols that allow population size to be estimated in the presence of imperfect detection but that do not require the formal capture and marking of individuals. In addition to classical capture–recapture, the situation just described, protocols based on double or multiple observer sampling (and variations) have become popular in many animal sampling problems. Another popular class of protocols are so-called 'removal' sampling methods. Removal sampling was originally applied to exploited populations where individuals are physically removed from the population in successive samples. Information about encounter probability (removal probability) is obtained from the apparent decrease in individuals removed over time.

Such protocols have recently become popular in bird sampling (Nichols et al., 2000; Farnsworth et al., 2002; Rosenstock et al., 2002) based on 'point counts', and other situations where sampling is carried out at a large number of spatially referenced sample units. These protocols yield a multivariate count statistic **y** that has a multinomial observation model

$$\mathbf{y}|N, \boldsymbol{\pi} \sim \text{Multin}(N, \boldsymbol{\pi}).$$

The differences among protocols are manifest in parameterization of the multinomial cell probabilities $\boldsymbol{\pi}$.

5.5.1 Pooling Robustness of the Multinomial

A result that is of some utility in the context of spatial sampling of wildlife populations is the following, which has to do with aggregating count statistics. If $y_1 \sim \text{Bin}(N_1, p)$ and $y_2 \sim \text{Bin}(N_2, p)$ then $y_1 + y_2 \sim \text{Bin}(N_1 + N_2, p)$. The multinomial equivalent is also true, i.e., if you add up multinomials with the same cell probabilities but different sample sizes, their total is a multinomial with the same cell probabilities and sample size being the sum of the sample sizes of the constituent multinomials.

Table 5.3. Data and observation model structure for a double-observer sampling protocol where both observers make independent observations.

Observations	Multinomial cell probabilities
y_{11}, seen by both observers	$\pi_{11} = p_1 p_2$
y_{10}, seen by first but not second	$\pi_{10} = p_1(1 - p_2)$
y_{01}, seen by second but not first	$\pi_{01} = (1 - p_1)p_2$
y_{00}, not seen by either observer	$\pi_{00} = (1 - p_1)(1 - p_2)$

This is useful in spatial sampling problems that yield sparse samples at many units. In such cases, it is convenient to aggregate data across sample units. Denote the aggregated count statistic as $\mathbf{y} = (y_1, y_2, y_3, y_4)$. The observation model for the aggregated counts is multinomial of the form

$$f(\mathbf{y}|N, p) = \frac{N!}{\left\{ \prod_{k=1}^{K} y_k! \right\} (N - y.)!} \left\{ \prod_{k=1}^{K} \pi_k^{y_k} \right\} (1 - \pi.)^{N - y.}$$

where N is the 'total' population exposed to sampling across all sample units and the π_k is structured by the protocol employed. For example, under a removal protocol then $\pi_k = p(1 - p)^{k-1}$.

Thus, multinomial observation models are *pooling robust* in the sense that no matter what the distribution of site-specific abundances (the N_i parameters), the observation model of the pooled count statistics has the same form, with the same multinomial cell probabilities. This allows us to pool the count statistics spatially and to estimate the parameters of the detection process and the total abundance N. However, in pooling the data, information about spatial variation is lost. Also, the pooled count statistic is not generally the sufficient statistic for $N = \sum_i N_i$ (Royle, 2004a).

5.5.2 Multiple Observer Sampling

The observation model that results from multiple observer sampling is equivalent to capture–recapture model M_t except that replicate samples are obtained (usually simultaneously) using 2 or more observers. In the case of two observers, suppose they make independent observations, then the encounter history frequencies and corresponding multinomial cell probabilities are as shown in Table 5.3. Early references on this protocol include Magnusson et al. (1978) and Cook and Jacobson (1979).

Practical implementation requires some attention. Typically, observers have to enumerate individuals in a manner that establishes their identity unambiguously.

This is because the observers must reconcile the lists after sampling in order to obtain the y_{11} frequency. For example, it is common for observers to sketch a map of the area sampled, and then reconcile the maps after sampling. Independence of observations is crucial, and this typically might be a practical field limitation. For example, if an observer is more likely to detect an individual as a result of another observer's detection (e.g., the former detects it as a result of the latter's scribbling on paper and intense focus on a bush 50 m away from the point of observation). A variation has been proposed (Nichols et al., 2000) in which the observations are dependent. One observer serves as a checker, records the first observer's data, and adds birds to the list that the first observer missed. During the survey, the observers switch roles to yield identifiability of both observer's detection probabilities. This protocol seems more practical in many field situations.

Note that the objective of these multiple observer methods is to induce dependence in the observed counts via the multinomial observation structure. An alternative approach to data collection with multiple observers is to obtain simultaneous *independent counts* resulting in data without the reconciled count y_{11}. For these data, the hierarchical models described in Royle (2004c) are appropriate. We address such models (in part) in Chapter 8. We expect that this type of multiple observer protocol should be more widely useful than existing multiple observer methods.

5.5.3 Example: Aerial Waterfowl Survey

Here we consider counts of mallard ducks (*Anas platyrhynchos*) collected by the U.S. Fish and Wildlife Service during the 2005 annual waterfowl population survey in the northeastern United States and eastern Canada (Koneff et al., 2008). Sample units in this fixed-wing survey are 18 mile linear segments (Smith, 1995), and 121 such segments were sampled using two observers (front seat, back seat) with detection probabilities p_1, and p_2, respectively. A total of 162 groups (or clusters) of birds were observed. The detection history frequencies of each group size, aggregated over all 121 segments, are given in Table 5.4. The segments are 18 miles long and 0.25 miles wide, so the total sampled area is 544.5 mi^2. The main focus of this study was to assess the efficacy of the aerial surveys, as characterized by the detection probability parameter. Longer term, parties seek an operational approach to adjust observed counts for imperfect sampling, or differential efficacy of platforms (e.g., helicopters vs. fixed-wing aircraft). In this type of survey, we might also wonder whether the size of the group affects its detectability. We note that the data used here deviate slightly from Koneff et al. (2008), and they also consider slightly different models.

For now, we ignore the group size (but see Chapter 6), and focus attention on the pooled counts (pooling all group sizes). The **R** instructions for obtaining the

Table 5.4. Mallard double-observer detection history frequencies from 121 aerial segments. In the detection history (x, y) position x represents the pilot and y represents the backseat observer.

detection history	Group size							total
	1	2	3	4	5	6	7	
$i = 1$ (0,1)	10	6	1	2	0	0	0	19
$i = 2$ (1,0)	52	19	4	1	1	2	0	79
$i = 3$ (1,1)	42	13	3	4	0	0	2	64

MLEs of p_1 (pilot), p_2 (back-seat observer) and $n_0 = N - n$, are given in Panel 5.4. Executing those commands yields $\hat{p}_1 = 0.776$ $\hat{p}_2 = 0.450$ and $\hat{n}_0 = 22.19$. It appears that the front-seat observer has a higher detection probability, which might be expected because that observer has a better field of view and more time to record individuals (Koneff et al., 2008).

To obtain an estimate of total population size, we need to obtain an estimate of the average group size, μ_{gs}. Using the sample observations (under the hypothesis that group size does not influence detectability), we estimate $\hat{\mu}_{gs} = 1.623$ having a standard error of 0.089. Then, an estimate of the density of individuals per square mile is $\hat{D} = ((162 + 22.19) * \mu_{gs})/544.5 = 0.549$ ducks/mile^2. It remains to characterize the uncertainty associated with this estimate. Note that the SE($\hat{n}_0$) = 8.07. The estimated density is the product of two estimated quantities, $\hat{N}\hat{\mu}_{gs}/$area. An unbiased estimate of the exact variance of the product (Goodman, 1960) is

$$\text{Var}(\hat{N}\hat{\mu}_{gs}) = (\hat{N}^2)\text{Var}(\hat{\mu}_{gs}) + (\hat{\mu}_{gs})^2\text{Var}(\hat{N}) - \text{Var}(\hat{N})\text{Var}(\hat{\mu}_{gs}) = 439.76.$$

Thus, the SE of $\hat{D}$ is approximately 0.039. This is a valid approximation when there is no group size bias. We address this issue in Chapter 6.

In general we would like to preserve the group size information in the analysis, since it seems possible that detection probability depends on group size. In addition, there is sometimes intrinsic theoretical interest in studies of flocking behavior or group structure, and factors that influence group size or structure (Silverman et al., 2001; Silverman, 2004). One natural way to account for group size is to formulate a stratified multinomial model, one in which each group size represents a distinct segment of the population, having its own population size parameter, i.e., let $N_k; k = 1, 2, \ldots, K$ be the number of groups of size k in the population. Then, we can construct the observation model as a product-multinomial, one multinomial for each group size (e.g., see Pollock et al., 1984), and then we focus on estimating the parameters $N_k; k = 1, 2, \ldots, K$. The problem with this approach is the sparsity of the data – typically several of the observed group sizes are poorly represented (as in Table 5.4). This basic problem motivates a simple hierarchical solution in which additional model structure is imposed on group size (that is, a prior distribution for N_k). If x_i is the group size of the ith observation, then assume that x has pmf $g(x|\theta)$ for $x = 1, 2, \ldots$. The resulting hierarchical model can be analyzed by likelihood methods (Royle, 2008b). Alternatively, the model can be formulated as a model having an 'individual covariate', which we consider when we return to these data in Chapter 6.

5.5.4 Removal Sampling

Removal sampling, in which individuals are sequentially removed from a population, is commonly employed as a means of estimating population size in exploited populations. An early reference from fisheries is Zippin (1956). If the population is closed, then successive removals should decline over time provided 'catchability' (a fisheries term for capture efficiency) and sampling effort remain constant with each removal.

The data that arise under the removal sampling protocol are the counts of individuals removed from the population during each sampling event. For example, if $J = 3$, then the data consist of the counts $y_j =$ the number of individuals removed during removal j. Under a removal protocol, the observations are realizations of a geometric random variable. As such, the removal counts have a structured multinomial distribution with geometric cell probabilities

$$\pi_k = (1 - p)^{k-1} p^k$$

```
y<-c(64,19,79)
nind<-sum(y)

lik<-function(parms){
    p<-  expit(parms[1:2])
    n0<- exp(parms[3])
    N<-nind + n0
    pvec<-c(p[1]*p[2],(1-p[1])*p[2],p[1]*(1-p[2]),(1-p[1])*(1-p[2]))
    -1*(lgamma(N+1)-lgamma(n0+1) +  sum(c(y,n0)*log(pvec)))
  }

nlm(lik,c(0,0,1),hessian=TRUE)
```

Panel 5.4. R instructions for fitting the double-observer model to the aerial waterfowl survey data.

for the cell corresponding to the number of individuals removed in removal sample k. The last cell probability, corresponding to 'not removed', is $\pi_{J+1} = (1 - p)^J$. It is natural to formulate the model in terms of p_j where p_j is a function of effort (or perhaps some other covariates). Without additional information the capture probabilities cannot be time-specific and unrestricted (Otis et al., 1978).

Removal sampling was developed originally for the case in which individuals are physically removed from the population. As such, it is often used in the context of exploited populations or species that are difficult to mark, such as insects (e.g., Russell et al., 2005, apply it to native bees). The protocol is now routinely applied to situations in which removal is temporary. For example, removal of fishes (Dorazio et al., 2005) or salamanders (Jung et al., 2005) from a stream or pond and put, temporarily, into a bucket. After several removal samples, the individuals are returned to the stream. Recently the basic concept has also been applied to bird point counting (Farnsworth et al., 2002; Royle, 2004a). In this case, individuals are not removed at all (even temporarily). Instead, the method in this case is a kind of mental removal wherein the observer reports the first detection of each individual into successive intervals (e.g., of 3 minutes). Application of this method is based on the supposition that one can note a bird and then ignore it subsequently so that it is not counted again.

5.5.5 Distance Sampling

Distance sampling (Buckland et al., 2001) is a common method of estimating the density of wildlife populations. It often involves a moving platform (such as an aircraft or boat) with observers recording distances to individuals. Typically, the observations are continuous distance measurements, but it is often applied to the case where observations are recorded into distance classes. This is the case in sampling forest birds using point counting protocols in which it is almost impossible to record distance, because birds are usually detected aurally. However, there is some belief that accuracy can be achieved in recording into distance classes, or that the effect of error is mitigated in such cases. When data are collected into distance classes, the observation model is multinomial having $K + 1$ cells (Buckland et al., 2001), where K is the number of distance classes. Royle et al. (2004) present a hierarchical model for modeling spatial variation in bird abundance from data collected using a distance sampling protocol. That model is a special case of those described in Chapter 8.

We discuss distance sampling in more detail in Chapter 8.

5.6 DATA AUGMENTATION

In this section, we present an alternative parameterization of closed population models. In the simple context of Model M_0, this approach provides a formal, technical linkage between closed population models and models for occupancy (Section 3.7). In Royle et al. (2007a), we provided a formal description of this linkage, illustrating how data augmentation may be used to reparameterize a multinomial model with unknown sample size to a model having excess zeros. The model of the augmented data is a zero-inflated version of the complete-data model where N is known. In fact, the resulting models prove to be equivalent to the occupancy models considered in Chapter 3. In effect, the parameter N of the closed population models is replaced, under data augmentation, by the parameter ψ. This duality between closed population size estimation and site occupancy models is enormously useful for analyzing models with individual random effects and covariates (Chapter 6) and certain generalizations of these models (Chapter 7). This duality can also be applied in more complex models involving animal communities (Chapter 12) and open population models (Chapter 10).

5.6.1 Heuristic development

We previously noted that if N was known, the data from a closed population study would comprise simple binomial counts having parameter p, under the assumptions of Model M_0. The difficulty is that the size of the data set, N is not known, and we must account formally for the fact that the sample is biased by the absence of observations wherein $y = 0$. The classical solution(s) were described in Section 5.2. Here we employ the method of data augmentation, which allows us to reformulate the problem to one in which the size of the data set *is* known.

Under data augmentation, we literally augment the observed data set (of size n) with a large number, say $M - n$, of 'all 0' encounter histories. We will assume that we can be certain that M is much larger than N. We refer to these augmented zeros as 'pseudo-individuals.' Under the simple assertion that M is sufficiently large (in a sense to be described shortly), it can be demonstrated that the pmf of the augmented data is a zero-inflated version of the pmf of the random variable when N is known. The new inference problem in this situation is to partition the pseudo-individuals into fixed zeros and sampling zeros (individuals that are members of the population that was sampled). We recognize this objective, and the data structure, as being precisely equivalent to that of estimating the number of occupied sites (Section 3.7).

This duality between models of occupancy and closed population size was first suggested (to the best of our knowledge) by Nichols and Karanth (2002) – who applied the problem in reverse. They were interested in estimating the number of occupied sites. To achieve that, they tossed out the all-zero encounter histories and estimated N using the classical model M_0, defining $\psi = N/M$. In our work (Royle et al., 2007a), we have formalized this duality and applied it in the opposite direction to simplify the analysis of models where the zero frequency is not observed. In the following sections, we discuss some of the technical issues justifying this approach and provide several examples.

5.6.2 Bayesian Motivation by Analysis of M_0

Consider here the classical closed population situation in which a population of N individuals is sampled J times, leading to observed encounter histories on $n \leq N$ individuals. As before, the pmf of the observed encounter frequencies is multinomial (and, more generally, so is the pmf for observed encounter *history* frequencies):

$$f(n_1, n_2, \ldots, n_J | \pi_1, \pi_2, \ldots, \pi_J, N) \propto \frac{N!}{n_0!} \left(\pi_0^{n_0} \pi_1^{n_1} \pi_2^{n_2} \pi_3^{n_3} \right)$$

where, in the simplest case, π_k are binomial probabilities.

In contemplating a Bayesian approach to the analysis of this problem, we require prior distributions for the model parameters p and N. Customary, or at least sensible priors would be proper uniform priors: $p \sim U(0,1)$ and $N \sim Du(0, M)$ for M reasonably large. (Du indicates the discrete uniform distribution). In practice, we are not too concerned about misspecifying M to be too small because we can always change our minds and rerun the analysis. That is, this uniform prior is not a manifestation of specific prior information but rather the manifestation of the *absence* of prior information. Thus, if the choice of M does influence the result, which can be diagnosed by high posterior mass near M, then we should start over with a larger choice of M.

Our strategy here is to reformulate the likelihood under this prior specification and to observe that this model specification is equivalent to a model of occupancy wherein the zeros are observed. To accomplish this goal, first note that an alternative representation of the prior for N is to assume that $N \sim Bin(M, \psi)$ with $\psi \sim U(0,1)$. That is, if we remove ψ by integration, by evaluating the integral:

$$[N|M] = \int_0^1 Bin(N|M, \psi) \, U(\psi|0, 1) \, d\psi.$$

The resulting prior distribution, $[N|M]$, is $Du(0, M)$. We need not actually carry out this integration. Instead, we might as well focus attention on the joint posterior distribution of N, and ψ which has the form

$$[N, p, \psi | \boldsymbol{n}] \propto f(n_1, n_2, \dots, n_J | \pi_1, \pi_2, \dots, \pi_J, N) \, \text{Bin} \, (N | M, \psi)$$

which a Bayesian might analyze using methods of Markov chain Monte Carlo, with ψ having the $U(0, 1)$ prior.

Alternatively, we might also choose to remove N from the likelihood by integration, and this yields precisely the zero-inflated binomial likelihood that forms the basis of likelihood analysis of the models described in Chapter 3. To see this, we apply the basic operation of 'growing a multinomial distribution' (Section 5.1.2). The result is that we have a new multinomial with cell probabilities $\psi \pi_k$ for $k = 1, 2, \dots, J$ and the new last cell is $1 - \psi$. The resulting likelihood does not contain N, but does contain the parameter ψ, which is the probability that an individual on the list of size M is a member of the population of size N that was exposed to sampling.

Some might be skeptical about conditioning on M, an arbitrary constant which induces the existence of a bunch of zero 'pseudo-individuals'. However, M as the upper limit of a uniform prior is sensible, and this prior *induces* the population of pseudo-individuals.

5.6.3 Implementation

We could implement data augmentation directly using the binomial prior on N. However, we hierarchicalize the model one-step further, which will prove useful in later chapters of the book.

Note that the binomial prior on N can be constructed by introducing a set of latent indicator variables z_i $(i = 1, 2, \dots, M)$ with prior $z_i \overset{iid}{\sim} \text{Bern}(\psi)$. We suppose that observation i from the augmented list is included in the population with probability ψ. That is, the latent z variables determine which individuals of the superpopulation of size M are members of the exposed population of size N. Under this construction, $N \equiv \sum_{i=1}^{M} z_i$ which has the required $\text{Bin}(M, \psi)$ prior. Now, analysis of the model requires attention to the z_i variables.

With the introduction of the collection of latent variables $\{z_i\}$, the data structure is summarized in Table 5.5. Note that $z_1, \dots, z_n$ are observed $(=1)$ and $y_i \equiv 0$ for $i > n$. The augmented values $z_{n+1}, \dots, z_M$ are unobserved. The objective is, in a sense, to impute these missing values. Under this formulation, population size is now a derived parameter $N = \sum_{i=1}^{M} z_i$.

For the augmented data, the model is very simple, being composed of two components. The observation model is

$$y_i \sim \text{Bin}(J, z_i \, p)$$

Table 5.5. Some encounter histories for the data augmented version of the *Microtus* data. For model M_0, only the individual totals y_i are required. Note that the augmented pseudo-data for $i = n+1, n+2, \ldots, M$ are zeros. z_i is the binary population inclusion indicator variable.

individual	t1	t2	t3	t4	t5	y_i	z_i
1	0	1	1	0	0	2	1
2	1	0	1	1	0	3	1
3	1	1	1	0	1	4	1
$\vdots$	$\vdots$	$\vdots$	$\vdots$	$\vdots$	$\vdots$	$\vdots$	$\vdots$
	1	1	1	1	1	5	1
	0	0	0	0	1	1	1
	1	0	1	0	1	3	1
$n-2$	1	1	1	1	1	5	1
$n-1$	1	1	1	1	1	5	1
n	1	1	1	0	1	4	1
$n+1$	0	0	0	0	0	0	?
	0	0	0	0	0	0	?
	0	0	0	0	0	0	?
	0	0	0	0	0	0	?
	0	0	0	0	0	0	?
$\vdots$	$\vdots$	$\vdots$	$\vdots$	$\vdots$	$\vdots$	$\vdots$	?
M	0	0	0	0	0	0	?

whereas the augmented state process is obtained by the introduction of a set of latent indicator variables

$$z_i = \begin{cases} 1 \text{ if individual } i \text{ is a member of the population} \\ 0 \text{ if not} \end{cases}$$

$$z_i \sim \text{Bern}(\psi).$$

By now, this should look familiar, as the model is precisely equivalent to the hierarchical formulation of the occupancy model described in Chapter 3. That is, this is a model of site occupancy, where a site in this case is an individual from the augmented data set consisting of the original n observations, and an additional $M - n$ all zeros. The **WinBUGS** model specification is given in Panel 5.5. (which is the same as Panel 3.5 from Chapter 3 except that, here, we compute N as the sum of the latent indicator variables). Note also that we described the MCMC algorithm for this model in Chapter 3. For the simple case, it requires only sampling from distributions of known form (beta distributions and Bernoulli trials). We find Bayesian analysis of models requiring prior distributions on ψ (and similar parameters later) to be considerably more convenient than models having integer population size parameters.

Table 5.6. Some encounter histories for the adult male *Microtus* data from Williams et al. (2002). y_i denotes the total number of days that the ith individual was detected.

\multicolumn{5}{c}{Sampling day}					y_i
1	2	3	4	5	
0	1	1	0	0	2
1	0	1	1	0	3
1	1	1	0	1	4
1	1	1	1	1	5
0	0	0	0	1	1
1	0	1	0	1	3
1	1	1	1	1	5
1	1	1	1	1	5
1	1	1	0	1	4
0	1	0	1	0	2

5.6.4 Analysis of the Microtus Data

For illustration, we provide a reanalysis of the *Microtus* data under model M_0, using the model specification given in Panel 5.5. The **R** instructions for carrying out this analysis are provided in the Web Supplement.

We augmented the data set with $M - n = 50$ all-zero encounter histories. The posterior distributions for p and N are shown in Figure 5.3, where we see that the posterior mass of N piles up around 0, consistent with our previous finding that the MLE of N occurred on the boundary $N = n$. In fact, $\Pr(N \leq 58) \approx 0.986$ (recall that $n = 56$ individuals were captured). The posterior mean of p is 0.63, and the $(2.5, 97.5)$ percentiles of the posterior are $(0.570, 0.687)$.

```
model {
psi~dunif(0,1)
p~dunif(0,1)
for(i in 1:M){
    z[i]~dbin(psi,1)
    mu[i]<-z[i]*p
    y[i]~dbin(mu[i],J)
}
N<-sum(z[1:M])
}
```

Panel 5.5. WinBUGS model specification for the reparameterization of Model M_0 that is induced by data augmentation.

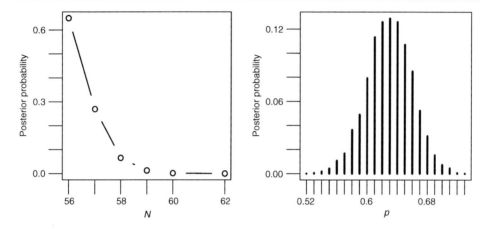

Figure 5.3. Posterior of N and p for the *Microtus* data under model M_0, analyzed in *WinBUGS* using data augmentation.

5.6.4.1 Model Extension: Behavioral Effects

One good reason to use data augmentation for the analysis of closed population models is that it allows us to avoid dealing with the unknown sample size parameter N. Instead, models are converted to zero-inflated logistic-regression type models, and most model extensions can be implemented without additional difficulty (beyond that required for Model M_0). We illustrate that point here using the *Microtus* data, following Royle (2008a).

We noted previously that the traps were baited with corn. This raises the possibility of what is usually referred to as a behavioral response where we might expect detection probability to increase after initial capture because individuals will return for an easy meal (it could also work the other way around – capture being the *Microtus* equivalent of being abducted by aliens).

We consider here the classic behavioral response 'Model M_b' (Otis et al., 1978), as well as a model suggested recently by Yang and Chao (2005) allowing for a short-term behavioral response (what they referred to as an ephemeral response). The classical model (the 'persistent' behavioral response model) supposes that previous capture might increase or decrease an individual's probability of capture. If we let $x_{ij} = 1$ if an individual was captured prior to sample j, then the model can be formulated as a logistic-regression type model for the Bernoulli observations y_{ij}, provided that N is known. The ephemeral response of Yang and Chao (2005) extends this model to include an autoregression term in the model, so that individual detection probability is influenced by capture in the previous sample. This broader

Table 5.7. Analysis of the Microtus data under the behavioral response model of Yang and Chao (2005).

parameter	mean	sd	2.5%	median	97.5%
N	57.740	1.880	56.000	57.000	63.000
α_0	0.095	0.255	-0.434	0.103	0.568
α_1	0.231	0.339	-0.429	0.230	0.900
α_2	0.639	0.384	-0.114	0.639	1.383
ψ	0.321	0.036	0.254	0.320	0.394

class of models has the form

$$\text{logit}(p_{ij}) = \alpha_0(1-x_{ij}) + \alpha_1 x_{ij} + \alpha_2 y_{i,j-1}.$$

The Markovian autoregressive structure is indicated by the regression on $y_{i,j-1}$ and x_{ij} is the indicator of previous capture. We have chosen a parameterization of the intercept here in which α_0 is the logit-scale probability of detection for individuals that have not previously been captured, and α_1 is that for individuals that have been captured.

As before, it is reasonably straightforward to describe the model in **WinBUGS** (Panel 5.6). An **R** function for organizing the data and executing **WinBUGS** is provided in the Web Supplement. We recommend caution when implementing certain classes of models in **WinBUGS** (see our discussion of this in Section 6.4.2). This model is an example where we sometimes realize very poor mixing (or no mixing) of the Markov chains when certain prior distributions have infinite support. In this case, the Markov chain for the model in which mean parameters have a `dnorm(0, .001)` mixes very poorly in some cases. One solution which seems to resolve the problem is to use priors having bounded support. In the present case, we used U($-5, 5$) priors for the regression parameters (as implemented in Panel 5.6).

The results of fitting the behavioral effects model to the *Microtus* data are given in Table 5.7. It would appear that there is not a strong persistent behavioral response (the difference between α_0 and α_1 is small). There is some evidence for a slight ephemeral response – α_2 is indicated to be positive and most of the posterior mass is above 0 for that parameter.

5.7 SUMMARY

In this chapter, we have given a brief introduction to certain concepts associated with inference about population size in closed populations. In real life, there aren't really any strictly closed populations. Yet, closed population models are widely used in practice for a number of reasons. First, they form the basis of a large number of extensions that are of great practical relevance. Secondly, there are instances where few options are available due to the extraordinary difficulty of obtaining sufficient data. This is common in the sampling of certain carnivores and rare species (Karanth and Nichols, 1998; Trolle and Kéry, 2003). Thirdly, closure is as much a feature of the observation process as the demographic process and so some populations might be practically closed over short periods in which case closed population models provide reasonable estimates of abundance or density. In this sense, closure is an important design consideration.

On a conceptual level, studying closed population methods allows us to gain experience with some fundamental concepts that have much broader applicability. The main concepts that are most often emphasized on the topic are (1) detection bias, or detection probability, is a fundamental element of the observation process and (2) the multinomial observation model is fundamental to inference. The main novel concept that we presented in this chapter is the use of *data augmentation* to facilitate Bayesian analysis of closed population models. The basic idea behind data augmentation is to introduce latent structure that yields a useful *reparameterization* of closed population models. In the present case, we add a bunch of all-zero encounter histories. We recognize that the likelihood of the augmented data is a

```
model {

psi~dunif(0,1)
a0~dunif(-5,5)
a1~dunif(-5,5)
a2~dunif(-5,5)

for(i in 1:(nind+nz)){
  z[i]~dbin(psi,1)
  for(j in 1:J){
    logit(p[i,j])<-a0*(1-prevcap[i,j]) + a1*prevcap[i,j]  + a2*lagY[i,j]
    mu[i,j]<-p[i,j]*z[i]
    y[i,j] ~ dbern(mu[i,j])
  }
}
N<-sum(z[1:(nind+nz)])
}
```

Panel 5.6. **WinBUGS** model specification for the data-augmentation reparameterization of the Yang and Chao (2005) model with persistent and ephemeral behavior effects.

zero-inflated version of the 'complete data' likelihood – which has a representation (at the observation level) as a logistic regression model. The resulting likelihood does not have N as a parameter but rather, in its place, the parameter ψ, the probability that an individual on the augmented list corresponds to an individual in the population exposed to sampling. The resulting models are little more than zero-inflated binomial (logistic) regression models, and consequently very easy to analyze. Our point in pursuing the use of data augmentation is that such zero-inflated models are easier to analyze in some cases (and especially in **WinBUGS**). The zero-inflated binomial model that results has a very simple structure (see Panel 5.7), and this proves to be immensely useful later on, even for very complex models. We adopt data augmentation for analysis of models with individual effects in the following chapter (Chapter 6); also for the analysis of Jolly–Seber type models in Chapter 10 and community models in Chapter 12.

Detection bias and multinomial observation models provide some of the practical motivation for discussing methods for closed populations. But why is this chapter relevant to the broader theme of the book, hierarchical models? The main relevance is motivational – that we often will be confronted with situations in practice in which there are spatial, temporal, species, or some other group structure in our data. In these situations, we have closed population models that must be 'linked together' by additional structure relating N across groups. This linkage is achieved with hierarchical models. For example, in the waterfowl example, the unit of observation was a group of individuals. The inference problem suggests the need for additional model structure beyond the simple double-observer likelihood. In many sampling problems, data are collected according to standard protocols (i.e., described in this chapter) at a large number of spatial sample units (e.g., transects in the case of the waterfowl survey, point count locations in our bird surveys, stream segments in salamander surveys). Important extensions of closed population models that

---

Observation model:

- $[y_i|z_i] = \mathrm{Bin}(J, z_i \cdot p)$

State model:

- $z_i \sim \mathrm{Bern}(\psi)$

Priors:

- $p, \psi \sim U(0,1) \times U(0,1)$

Panel 5.7. Hierarchical formulation of Model M_0 under data augmentation.

allow for the inclusion of spatial structure in the model are described in Chapter 8. Finally, we also introduced a couple of spatial capture–recapture problems that are typical of the kind of sampling problem in which closed population methods are often applied. The *Microtus* data, the tiger camera trapping data, and the horned lizard survey data are all typical examples of spatial capture–recapture data. For these problems, the closed-population framework gets us half-way to where we want to be. That is, we are able to estimate the size of the population that is exposed to sampling but, because animals move freely onto and off of the plot, we do not know the effective sample area of the plot. A hierarchical extension of these models allows us to deal with this problem. This is the topic of Chapter 7.

6

MODELS WITH INDIVIDUAL EFFECTS

While the simple closed population models described in Chapter 5 are useful in limited situations, the class of closed population models can be generalized considerably to include individual effects. Such models are enormously useful in many different contexts.

We recognize three basic classes of individual effects models. First are the *individual heterogeneity* models, in which there exists unstructured heterogeneity in the form of a latent random effect. Specifically, we suppose that p is an individual-specific latent effect, a random variable, having pdf $g(p|\theta)$. Analysis of this class of models was pioneered by Burnham and Overton (1978, 1979) and is usually now referred to as 'Model M_h' although it is actually a very large class of models. Note that we previously described a model of this phenomenon (heterogeneity in p) in Chapter 4. In that case, heterogeneity in p was induced by spatial variation in abundance. The second type of models are the so-called *individual covariate* models in which the heterogeneity among individuals is structured by measurable covariates, e.g., body mass of individual, or some other trait that influences detection probability (or, in the case of open systems, a trait that influences survival). Such models are often colloquially referred to as 'the Huggins–Alho model' due to Huggins (1989) and Alho (1990), although it would be more accurate to refer to the Huggins–Alho 'model' as an estimation procedure since its development seems to be very much in the spirit of classical design-based sampling. They propose what is usually referred to as a 'conditional estimator' of N, based on Horvitz–Thompson concepts of unequal probability sampling (e.g., Thompson, 2002, Chapter 6). A final class of individual covariate models are what we refer to as *spatial capture–recapture* models. A special case is the classical distance sampling problem, which has a precise linkage to the individual covariate models. We focus on the first two types of individual effects models in this chapter. Spatial capture–recapture models are described in Chapter 7.

Individual effects models are a hot topic right now. Historically, estimation of N in the presence of individual heterogeneity (i.e., models of the first type) was dominated by Burnham's 'jackknife estimator', which is distinctly not model-based. Over the last 10 years or so there has been a paradigm shift to more distinctly model-based views (Norris and Pollock, 1996; Coull and Agresti, 1999; Pledger, 2000; Dorazio and Royle, 2003). Concurrent to this, our colleague W. Link has been actively working to refute the whole concept of individual heterogeneity models (Link, 2003) and also the foundations of the jackknife estimator (Link, 2006). The active research in this area has generated considerable enthusiasm among statisticians at conferences and meetings but alas, not so much among ecologists. Similarly, individual covariate models have been dominated by what is viewed as being a non-parametric procedure (i.e., the Huggins–Alho procedure), and there has been an increasing focus on model-based formulations. This is a natural evolution as certain types of problems require that additional information be brought to bear on the problem, either in the form of auxiliary data or additional model structure (e.g., individual covariates in open systems (Bonner and Schwarz, 2004)). A certain degree of synthesis and unification of the 3 types of individual effects models arises in the presence of a fully model-based framework. This is a central theme of this chapter and also of Chapter 7.

What do we have to offer on the topic of individual effects models? We offer a conceptual unification of these individual effects models as hierarchical models. They are not so different conceptually, technically, or mathematically, but their treatment historically, in the literature and in their application, has been very diffuse. Within the hierarchical modeling framework, these different classes of models have concise representations as a sequence of observation and process model components. Secondly, we offer a unification of the manner in which inference is achieved under these various models using data augmentation. We first introduced the concept of data augmentation in Chapter 5, but its utility and importance was not self-evident at that point. Instead, it was a statistical curiosity more than anything, formally linking models of occurrence to models of closed population sampling. In the realm of individual effects models, the benefit of data augmentation is realized because data augmentation renders these models as zero-inflated logistic-regression models, which are analyzed easily by Bayesian methods. Using data augmentation, inference under models with individual effects is no more difficult than under model M_0 and other closed population models that were considered in Chapter 5.

6.1 INDIVIDUAL HETEROGENEITY MODELS

Here we consider the class of models that is colloquially referred to as 'Model M_h' in the literature (Otis et al., 1978). Interest in these models is motivated by the possibility that detection probability varies by individual. That is, that *heterogeneity* exists in detection probability, hence the 'h' in M_h, but we cannot describe an explicit causal mechanism for that heterogeneity.

Heterogeneity in detectability arises for many important reasons – behavior, breeding status social status, or physical characteristics. We consider a problem here where estimating species richness is formulated as a closed population model with heterogeneity. In that case, heterogeneity arises due to variations in song, physical appearance, and other characteristics of *species*. Another factor that leads to heterogeneity in studies that involve spatially organized trapping arrays is the situation of an individual's home range relative to the traps themselves. Here, we introduce several examples and address specific extensions of individual effects models for this phenomenon in Chapter 7.

There are two important factors that motivate interest in heterogeneity models. First is that the effect of heterogeneity in detection probability leads to biased estimators of N, under the nominal model in which heterogeneity is excluded. The second factor motivating interest in heterogeneity models is that certain biological and observational mechanisms induce (or suggest) the presence of heterogeneous detection probabilities. While the first concern is of practical importance, whenever the object of inference is abundance or density, the second concern is interesting within a hierarchical modeling framework, as it often will suggest the construction of some component of the hierarchical model (as in Chapters 4 and 7).

6.1.1 Bias Induced by Heterogeneity

That the presence of heterogeneity biases the ordinary estimator of population size is intuitively obvious, as individuals with high detection probabilities appear in the sample in greater proportion their occurrence in the population at large. Thus, the 'average p' of captured individuals is greater than the population averaged p. This can be formalized as follows. Suppose that $p \sim g(p)$, having mean μ_p and variance σ_p^2. Then the pdf of p conditional on capture is, by Bayes' rule:

$$[p|y = 1] = \frac{[y = 1|p]g(p)}{[y = 1]}.$$

Substituting $\Pr(y = 1|p) = p$ and $\Pr(y = 1) = \int pg(p)\mathrm{d}p = \mu_p$ yields

$$[p|y = 1] = \frac{p}{\mu_p}g(p)$$

which has expected value

$$\mathrm{E}[p|y = 1] = \int p\frac{p}{\mu_p}g(p)\mathrm{d}p = \int \frac{p^2}{\mu_p}g(p)\mathrm{d}p = \frac{\sigma_p^2}{\mu_p} + \mu_p.$$

Since $\sigma_p^2/\mu_p > 0$, this establishes the result.

Estimators of N appropriate for M_h were developed first by Burnham (1972), who considered a beta-binomial mixture model and a 'non-parametric jackknife' estimator (Burnham and Overton, 1978, 1979). The jackknife estimator is in widespread use, and probably is the *de facto* standard because it has been implemented in popular software packages, such as MARK (White and Burnham, 1999) and COMDYN (Hines et al., 1999). As such, there is a casual equivalence made between 'Model M_h' and the application of the jackknife estimator to data. In fact, the so-called 'Model M_h' is a very broad class of models for describing variation in detection probability among individuals. At this time, a large number of different parametric models have been applied to the problem of modeling heterogeneity. For example, Coull and Agresti (1999) developed a model in which the logit-transformed p's have a normal distribution. Pledger (2000) considered a model in which variation in p is described by a finite-mixture of point masses. Dorazio and Royle (2003) considered a beta-binomial mixture and compared various classes using case studies and simulations. We will describe several of these models subsequently.

6.1.2 Model construction

Technically, model M_h is only slight extension of the basic null model M_0, to include an individual random effect. The key idea is the existence of variation in p among *individuals*, and so models are constructed by modeling p_i as a random effect or latent variable. For a population of size N subjected to J samples, then, supposing that N were known, the heterogeneity model is described by

$$y_i|p_i \sim \mathrm{Bin}(J, p_i); \quad i = 1, 2, \ldots, N$$

and

$$p_i \sim g(p|\theta),$$

where g is some probability density function for p. The various flavors of model M_h correspond to the particular choice of g. Some of the more prominent versions are described subsequently.

Using standard likelihood methods of inference, we can attack this problem by integrated likelihood. In this case, as with basically every closed population model, the observation model that arises under any choice of g is a structured multinomial. In particular, in a study consisting of J survey periods, then the *observed* values of y_i are the integers $1, 2, \ldots, J$. Conditioning on N introduces n_0 as the unobserved detection frequency and the resulting pmf for the vector $\mathbf{n} = (n_1, n_2, \ldots, n_J)$ of observed detection frequencies (where $n_k = \sum_{i=1}^{n} I(y_i = k)$) is the multinomial:

$$[\mathbf{n}|\theta, N] = \frac{(n + n_0)!}{n_0! \prod n_k!} \pi_0^{n_0} \pi_1^{n_1} \ldots \pi_J^{n_J}.$$

This is basically where we started with M_0 as well. The difference is that the cell probabilities are not simple binomial cell probabilities, but rather they are the marginal or average probabilities, where the averaging takes place over the prescribed random effects distribution g, as follows:

$$\pi_k = \Pr(y = k) = \int_0^1 \text{Bin}(k|J, p)\, g(p|\theta)\, dp.$$

The technical framework here is precisely equivalent to that of the abundance-induced heterogeneity models described in Chapter 4, where we considered integrated likelihood in several contexts. In the present context, p is a random effect, and in the integrated likelihood, the random effect is removed from the conditional likelihood (the likelihood that is conditional on the random effect) by integration.

These cell probabilities, π_k, are analytic for a few cases, including the beta and so-called finite-mixture, both of which will be described subsequently. For other cases, e.g., the logit-normal, we have to do this integration numerically, which is not difficult, as we will demonstrate below.

6.2 FLAVORS OF M_h

Burnham's jackknife estimator is still the most widely used estimator for Model M_h. This procedure is viewed as being non-parametric because no particular $g(p)$ is explicitly prescribed. However, the construction of the estimator implies a particular relationship among moments and, unfortunately, there may not be *any* g that possesses this relationship, or at least the existence of such a class is not obvious (Link, 2006). Thus, its interpretation as a non-parametric procedure may be questionable.

Alternatively, it is possible to construct a number of model-based estimators by prescribing a $g(p)$ and providing standard methods of parametric inference. This model-based framework described in the preceding section provides a precise mathematical rendering of the individual heterogeneity model. A number of specific classes of g have been considered in some detail in the literature, and we review some of these now.

6.2.0.1 Finite-Mixtures

After Burnham's jackknife estimator, it was quite a few years before anyone devised an alternative mousetrap. Norris and Pollock (1996) devised what they referred to as a 'non-parametric MLE' of N in the presence of heterogeneity. What they proposed is commonly referred to as a finite-mixture or latent-class model for p. Under this model, each individual p_i may belong to one of C classes, but the class membership is unknown. That is, the potential values of p_i, the *support points*, are $p_i \in \{p_c; c = 1, 2, \dots, C\}$. They have *mass* $g_c = \Pr(p = p_c)$ where $\sum_{c=1}^{C} g_c = 1$. Pledger (2000) gives a general treatment of these models.

For example, suppose the existence of two latent classes. In this case, the marginal probability of encountering a bird k times is, by the law of total probability,

$$\pi_k = \Pr(y = k|p_1, p_2, g_1) = \mathrm{Bin}(k|J, p_1)g_1 + \mathrm{Bin}(k|J, p_2)(1 - g_1).$$

This is the discrete analog of the marginalization operation that we mentioned in the previous section. There is only a minor conceptual tweak here as we went from a continuous random variable to a discrete random variable. As in similar applications that we have encountered previously, the pmf of the observations is a structured multinomial with cell probabilities π_k. This model is easy to analyze because of the simple form of the cell probabilities. In particular, all cell probabilities can be computed in one **R** instruction:

```
cellprobs<- dbinom(0:J,J,p1)*g1 + dbinom(0:J,J,p2)*(1-g1)
```

The likelihood can be completely described and maximized in only 1 or 2 more instructions, given the basic capability to maximize a multinomial likelihood (see Chapter 5).

The finite-mixture models represent $g(p)$ as a discrete pmf with arbitrary (but discrete) support. This is the sense in which the model is 'non-parametric'. However, all of the support points and their masses are estimated, and the number of support points is unknown. Thus, the model is highly parameterized and so it is unclear whether there are advantages that arise from being 'non-parametric' in this case.

Pledger (2000) generalized the basic framework and formalized likelihood inference across broad classes of models under an encounter history formulation of the model. For example, a model with time effects and individual heterogeneity can be described by distinct encounter histories, with detection probability specified by

$$\mathrm{logit}(p_{ij}) = \alpha_i + \beta_j.$$

In this case, β_j are fixed effects, whereas α_i are assumed to vary according to a latent class model.

Note the relationship between this and the abundance-induced heterogeneity models described in Chapter 4. The latter have a large number of classes (theoretically an infinite number of classes), but the support points and their masses are constrained by the assumption of an abundance distribution.

6.2.0.2 Logit-Normal Mixtures

A continuous mixture that is in widespread use for mixed models in many contexts throughout applied statistics is the logit-normal mixture, in which $\eta_i = \text{logit}(p_i) \sim \text{N}(\mu, \sigma^2)$. Coull and Agresti (1999) adapted this model for N estimation problems. As before, classical analysis can be achieved by integrated likelihood in which the random effect is removed from the conditional likelihood (the likelihood that is conditional on the random effect) by integration. In the present case, this requires that we calculate the following integral:

$$\Pr(y|J, \mu, \sigma^2) = \int_{-\infty}^{\infty} \text{Bin}(y|J, p(\eta)) \, g(\eta|\mu, \sigma) \, d\eta,$$

where $g(\eta|\mu, \sigma)$ is a normal density. This would have to be carried out numerically, but this is not difficult using **R** (and most software), as we demonstrate in the subsequent example.

6.2.0.3 Beta-Binomial Mixture

Burnham (1972) initially considered models in which p had a beta distribution, denoted by $p_i \sim \text{Be}(a, b)$. This is an interesting model because the beta prior is conjugate for the binomial parameter p, and hence it yields some analytic tractability (Dorazio and Royle, 2003) under a model where the only source of variation in detection probability is that due to individual heterogeneity. In this case, the detection frequencies are multinomial, where the cell probabilities are structured according to a beta-binomial pmf. That is, if $y|p \sim \text{Bin}(J, p)$ with $p \sim \text{Be}(a, b)$ then, marginally, y is Beta-Binomial with probabilities

$$\pi_k = \Pr(y = k) = \frac{\Gamma(J + 1)}{\Gamma(y + 1)\Gamma(J - y + 1)} \frac{\Gamma(a + y)\Gamma(J + b - y)}{\Gamma(a + b + J)} \frac{\Gamma(a + b)}{\Gamma(a)\Gamma(b)}$$

and thus, conditional on N, the observed frequencies $\mathbf{n}_k$ have a multinomial distribution with cell probabilities $\{\pi_0, \pi_1, \ldots, \pi_J\}$.

The multinomial log-likelihood derived from the beta model for p can be described and maximized in only a few lines of **R** code. For this, we make use of the `lgamma` function in **R** for computing $\log(\Gamma(arg))$.

6.2.0.4 So many Models M_h, so little data

An interesting aspect of the N-estimation problem in the presence of heterogeneity is that conventional goodness-of-fit statistics, such as deviance, cannot be relied on for selecting a model. For example, Coull and Agresti (1999) showed that, if capture probabilities are relatively low or vary greatly among individuals, the logistic-normal and latent-class models both may fit the observed data reasonably well, but provide substantially different inferences about N. This basic result was demonstrated also by Dorazio and Royle (2003) among several classes of models. The reason for this behavior is that model-based estimates of N actually correspond to extrapolations for the capture histories of unobserved individuals (Fienberg, 1972), so it is not surprising that the extrapolated values are sensitive to model structure. These results suggest that prior knowledge of potential sources of heterogeneity in individual rates of capture and data-based criteria must both be considered when selecting models for estimating N. Our view, which we stated in Dorazio and Royle (2003, 2005b), is that continuous mixtures should normally be favored on biological grounds because the causes of heterogeneity are many and varied (that is, unless there is some a priori belief in latent classes (which is not implausible)). Secondly, continuous mixtures have the advantage of parsimony. That is, they cost relatively less (in terms of parameters) to model heterogeneity than finite-mixture models.

The problem of selecting among different classes of heterogeneity models was further elaborated by Link (2003). He demonstrated that, in some cases, different choices of g can fit the data identically (or very nearly so), but imply wildly different values of N. Essentially, when you broaden the class of models to include all possible distributions for p, then N is not identifiable across classes. The effect of misspecification of heterogeneity models is related to the degree of heterogeneity and the mean detection probability. As Link noted, differences among mixtures are more pronounced as the mass of $g(p)$ is concentrated near zero. One might view low mean detection or high levels of heterogeneity as suggesting that the species of interest cannot be reliably, or effectively, sampled. This should be viewed as a biological sampling issue to be considered in survey design prior to data collection, not a statistical issue to rectify after the fact by considering complex models of the detection process or by engaging in conventional model selection strategies aimed at choosing among different mixture distributions (Dorazio and Royle, 2005b; Royle, 2006). Consistent with the broader philosophical framework within which we operate (i.e., parametric inference), we should not be overly apprehensive of picking a model and carrying out inference under that model, provided that the model is useful for its intended purpose.

That N is non-identifiable in this general sense, across all possible classes of models, does not diminish its central importance in ecology and conservation biology where estimation of N or density is important in certain problems. We know heterogeneity must arise in certain situations, due to biological mechanisms that we can describe. Thus, there will be cases where some effort should be made to account for it explicitly by a model. In this regard, there are active attempts to get around this problem by broadening the class of models. For example, Morgan and Ridout (2008) considered an extended class of models by mixing distributions of continuous and discrete support. Dorazio et al. (2008) developed a class of models based on the Dirichlet process (DP) prior, which is designed to account for a modeler's uncertainty in the latent distribution of detectability. Estimates of N based on the DP prior therefore should be robust to errors in model specification. Mao (2004) developed a purely non-parametric procedure for estimating a lower bound for N that generally exceeds the trivial bound of $N = n$. However, the applicability of estimates of lower bounds in ecological problems seems unclear.

6.2.1 Example: Flat-tailed Horned Lizard Data

We return to the flat-tailed horned lizard data introduced in Section 5.3.2. Recall that these data originate from samples carried out on a 9 ha plot of dimension 300 m × 300 m, with 14 capture occasions over 17 days. A total of 68 individuals were captured 134 times. The distribution of capture frequencies was (34, 16, 10, 4, 2, 2) for 1 to 6 captures respectively, and no individual was captured more than 6 times. We fitted the basic null model ('M_0') in which p is constant, yielding $\hat{N} = 82.02$ (SE = 5.414) $\hat{p} = 0.117$ (SE = 0.0122) and the AIC for this model was -242.661.

We noted two general problems that arise in spatial capture–recapture surveys. First, that the definition of N is rendered somewhat ambiguous by movement of individuals. That is, while it may be that N represents the size of some actual population of individuals that were exposed to sampling (a 'super-population', Kendall et al., 1997; Kendall, 1999), we do not know the area over which those individuals were drawn. Secondly, the proximity of an individual's home ranges relative to the area being surveyed induces heterogeneity in detection probability. This provides the heuristic justification for consideration of heterogeneity models for such data. It still doesn't provide a firm basis for interpretation, but it does suggest the mechanism that will yield heterogeneous detection probabilities.

For this purpose, we fitted the logit-normal version of the heterogeneity model (Coull and Agresti, 1999). The **R** code is shown in Panel 6.1, along with instructions to execute it for these data. The function `Mhlik` computes the integrated binomial cell probabilities, making use of the **R** functions for computing the binomial and normal densities, as well as the function `integrate`. The estimates obtained under this model were $\hat{N} = 106.8$ (SE $= 20.8$), $\hat{\mu} = -2.56$, and $\hat{\sigma} = 0.795$. The corresponding AIC was -247.3902. This model is somewhat favored by AIC, relative to the null model. The interested reader should be able to modify this **R** code to fit the beta-binomial or finite-mixture model.

We see that the heterogeneity model yields quite a substantial increase in the estimated abundance, from about 82 under M_0 to about 107 under the logit-normal mixture version of M_h. One view of this difference is that we believe the latter estimate to be more relevant because heterogeneity *must* be induced by the spatial nature of the system – the variable exposure of individuals to detection by the trapping apparatus. We do pay a price for entertaining this more complex model – that being the increased uncertainty in N that results from fitting the heterogeneity model.

6.2.2 Estimating Tiger Abundance

We return now to the tiger camera trapping data introduced in Section 5.3.3. Recall that 45 individuals were captured and, under the null model, we have $\hat{N} = 70.4$ and $p = 0.0805$. This model has an AIC score of -160.095.

We fitted the logit-normal flavor of Model M_h as implemented in Panel 6.1, which yields $\hat{N} = 111.69$, and $\hat{\mu} = -3.27$ and $\sigma = 0.89$. The AIC for this model is -159.650. The estimate of N is somewhat imprecise, as indicated by the very flat profile likelihood (Figure 6.1). The profile likelihood can be constructed by recursive calls to a modified version of the function in Panel 6.1, in which N is fixed. As N is varied, and the likelihood maximized as a function of the remaining two parameters, the resulting likelihood minimum *as a function of* N is the profile likelihood. The profile likelihood has end points on either side of the MLE of n_0 that yield an increase of 1.92 negative log-likelihood units. For the tiger data, the 95 percent profile likelihood lower bound is in the vicinity of 60, but the profile likelihood does not achieve a difference of 1.92 to the right of the MLE within the range considered in Figure 6.1.

We note that the heterogeneity model is only slightly better than the constant-p model, judging by the AIC scores. The very low detection probability for tigers, and the resulting sparse data, makes it difficult to detect heterogeneity due to movement. The effect of movement is to shrink individual p's toward the origin by a variable degree depending on an individual's relative exposure to traps. We might improve on our estimate of N by considering the movement mechanism explicitly. We'll revisit these data in the next chapter where we will consider models that are based on explicit exposure/movement mechanisms.

6.3 INFERENCE ABOUT SPECIES RICHNESS

A natural application of heterogeneity models, and one of increasing interest, is that of estimating species richness, or the number of distinct species in a community of animals (Burnham and Overton, 1979; Bunge and Fitzpatrick, 1993; Boulinier et al., 1998; Cam et al., 2002c,b; Kéry and Schmid, 2006; Kéry and Plattner, 2007). Species identity is a natural (taxonomic) mark that may be observed in repeated samplings of a community. Natural communities of animals generally include common species that are easily detected, rare species that are difficult to find, and an assortment of species at intermediate levels of abundance whose

```
---------- data ---------------------------
nx<-c(34, 16, 10, 4, 2, 2,0,0,0,0,0,0,0,0)
nind<-sum(nx)
J<-14
---------- definition of likelihood ----------
Mhlik<-function(parms){
  mu<-parms[1]
  sigma<-exp(parms[2])
  n0<-exp(parms[3])

 il<-rep(NA,J+1)
 for(k in 0:J){
   il[k+1]<-integrate(
    function(x){
      dbinom(k,J,expit(x))*dnorm(x,mu,sigma)
      },
      lower=-Inf,upper=Inf)$value
    }
 -1*(lgamma(n0+nind+1) - lgamma(n0+1) + sum(c(n0,nx)*log(il)))
}

---------- minimization of -likelihood -------

nlm(Mhlik,c(-1,0,log(10) ),hessian=TRUE)
```

Panel 6.1. R construction of the integrated likelihood for the logit-normal version of a closed population model with heterogeneous detection probabilities. Additional R instructions specify the data and minimize the objective function for the lizard data.

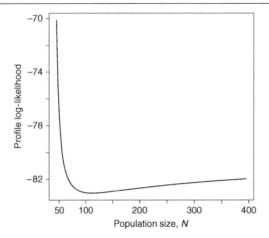

Figure 6.1. Profile likelihood of N for the tiger data. The profile likelihood is minimized at $\hat{N} = 111.69$.

detectabilities vary greatly. Behavioral differences among species also contribute to their variable rates of detection.

There are two fundamentally different designs used in the analysis of species richness from such surveys. A design that is strictly analogous to sampling of closed populations would have J replicate samples drawn of the same community, e.g., at a location or prescribed area. An alternative design that is very commonly used for practical reasons (i.e., the absence of replicates) is that in which spatial subsamples are regarded as replications. That is, we have J distinct spatial samples within some larger region of interest. This yields a collection of spatial encounter histories that are treated in precisely the same manner as if the replicates were true replicates. The spatial subsampling design is described in Williams et al. (2002, pp. 557–558). In either case, we have the $n \times J$ matrix of detection/non-detection data that will be regarded, by the model, as being identical to encounter histories in capture–recapture. Dorazio and Royle (2005a) addressed an important extension of this design, and corresponding model, to the case where true replicate samples are made at a number of spatial locations. This situation allows for inference about both occurrence probability and detection probability. We discuss this further in Chapter 12.

Here we consider the North American Breeding Bird Survey data from 2001, taken from Dorazio and Royle (2003). These data are roadside point counts of birds, made at $J = 50$ 'stops' that are regarded as replicate samples of the landscape being sampled by the route, as described above. This particular route is located in Maryland. For 2001, a total of $n = 71$ species were *observed* in the sample, and

Table 6.1. North American Breeding Bird Survey data from a route in Maryland. Frequencies are number of species detected k times for $k = 1, 2, \ldots, 50$.

# detections	frequency (# species)																	
1-18:	15	7	5	2	5	6	3	2	1	2	4	4	0	1	0	2	3	1
19-36:	1	1	1	0	0	0	1	1	0	1	0	1	0	0	0	0	0	1

no species was detected at more than 36 stops. The number of species detected $1, 2, \ldots, 36$ times is given in Table 6.1.

This model can be fit in **R** using the `integrate` function, as demonstrated by Panel 6.1. The estimates for these data were $\hat{\mu} = -2.44$, $\hat{\sigma} = 1.49$ and $\hat{n}_0 = 11.53$ so that $\hat{N} = 82.3$[1]. The 95 percent profile likelihood interval for these data is $(73.69, 104.29)$.

6.3.1 Spatial Subsamples as Replicates

The classical design for estimating N in closed populations (Chapter 5) is that in which a population is sampled J times. Implicit in the conceptual formulation is that the population occupies a single location or area. Often, as we have seen in the previous examples (voles, lizards, tigers), we may only be able to describe this area approximately (e.g., a 9 ha plot, or a polygon around camera trap locations). However, in the present application of closed population models to estimating species richness, we do not have this situation of replicate samples of a population that is (conceptually) static. Rather, we have used spatial samples as replicates so that, potentially, each sample exposes a different 'population' of species. In the BBS data, each route consists of 50 stops, and we view these stops as replicates of the population of species that is exposed to sampling by the collection of 50 routes. The use of spatial subsamples as replicates is common in this context, having been used by almost all of the applications of which we are aware (Burnham and Overton, 1979; Boulinier et al., 1998; Cam et al., 2002b,c; Dorazio and Royle, 2003).

Given this distinct spatial sampling context, it is natural to question whether this has any relevance to inference or interpretation of N. The spatial sampling should be relevant because there is a component of variation that is now strictly due to spatial sampling (occurrence or not at the level of a subsample) and also a component of variation due to detectability *at occupied subsamples*.

---

[1] Logit-normal estimates for these data are slightly different than those reported in Dorazio and Royle (2003, Table 3) which we believe are in error.

One view that yields a clear interpretation of model parameters is that in which we can assert that species occur, or not, on all subsamples along the route. That is, if the route is occupied, then all 50 stops are also occupied and *vice versa*. In this case, variation in p is due to variation in detectability of species alone. A consequence of this view is that N is the number of species *present* on the 50 particular spatial subsamples (not the region from which those subsamples were randomly chosen). This view is consistent with the closed population sampling view from which the method derives. We (Dorazio and Royle, 2005a) incorrectly stated that the use of this spatial subsampling design implies $\psi = 1$. It does not. The view that all spatial replicates are occupied (and $\psi = 1$) is only one possible interpretation of the design. An alternative and prevailing view is that, by neglecting the occurrence process in the model, there is an implicit confounding of a subsampling occurrence probability for species i, say ψ_i, with detection probability. That is, if occurrence of a species is independent across spatial subsamples then, when spatial subsamples are used as replicates, this induces a confounding of conditional detection probability, say $p_i^{(c)}$, with ψ_i so that the $p_i = p_i^{(c)}\psi_i$. Here $p_i^{(c)}$ is the normal 'conditional on capture' detection probability (i.e., that contained in the models of Chapter 5). In this case, the view that N is the size of the community that was exposed to sampling by the spatial subsamples (Boulinier et al., 1998; Cam et al., 2002c) appears to be justified.

It is possible to formulate models, under more general sampling designs, in which one can obtain explicit information about both subsample occurrence probabilities and also conditional-on-occurrence detection probabilities (Dorazio and Royle, 2005a). The design in this case has formal replicate samples at each spatial sample unit. Applications of this design to modeling community structure can be found in Dorazio et al. (2006); Kéry et al. (2008); Kéry and Royle (2008a,b). These models are the topic of Chapter 12.

6.4 BAYESIAN ANALYSIS OF HETEROGENEITY MODELS USING DATA AUGMENTATION

Analysis of capture–recapture models with individual effects poses a number of technical challenges. Importantly, the population size, N, is unknown and, in a Bayesian context, each time a new N is sampled from the posterior distribution, the dimension of the parameter space (the individual effect) also changes. Here, we provide a framework for the analysis of individual effects models based on data augmentation.

We have previously introduced the concept of data augmentation for the reparameterization of closed population models. The method has proved to be generally applicable to multinomial type observation models with N unknown. In such cases, we can augment the data set with a number, say $M - n$, of all-zero encounter histories, and then proceed with an analysis of the resulting model, which is reparameterized to a zero-inflated logistic-regression model. Here we adopt data augmentation to aid in the analysis of heterogeneity models, a class of models which provided some of the motivation of Royle et al. (2007a). In the case of individual effects models, each individual in the augmented list of size M has associated with it a latent variable, the individual effect. This is viewed as a complete data problem for purposes of the MCMC, and updating these individual effects is no more difficult than in any generalized linear mixed-model (GLMM) type of problem.

The hierarchical model for the augmented data set has an observation model consisting of two components. First, for the observables, we have:

$$y_i | p_i, z_i \sim \text{Bin}(J, z_i\, p_i), \quad \text{for } i = 1, 2, \ldots, M$$

and, for the parameters p_i,

$$\eta_i \equiv \text{logit}(p_i) \sim \text{N}(\mu, \sigma^2)$$

and, the 'process model,' induced by data augmentation,

$$z_i \sim \text{Bern}(\psi).$$

It might appear awkward to refer to this component as a process model since it is the model component for 'made-up' data (at least in part), and therefore is not much of an interesting process. But the model represents the prior distribution on N which will, in more complex situations, be a meaningful process model. And also, this model component is analogous to the occupancy process that data augmentation induces.

6.4.1 Example: Estimating Species Richness

We return to the North American Breeding Bird Survey data used for estimating avian species richness along a BBS route. The **WinBUGS** model specification is given in Panel 6.2, along with the **R** instructions for organizing the data *and* executing **WinBUGS** from within **R**.

It is also not difficult to implement an MCMC algorithm in **R**, which we have done. We only sketch out the algorithm here, but an implementation is available in the Web Supplement. We suppose a customary set of prior specifications in which we specify a diffuse normal prior on μ, and an inverse-gamma prior on $\tau = 1/\sigma^2$. Further, we suppose the natural $U(0, 1)$ prior for ψ. Then, each iteration of the MCMC algorithm is composed of the following steps.

Table 6.2. Results of fitting logit-normal heterogeneity model for estimating species richness from the BBS data using both a **WinBUGS** implementation (see Panel 6.2) and also a native **R** implementation. **WinBUGS** results are based on $\approx$105K iterations whereas **R** results are based on $\approx$500k iterations.

	WinBUGS					Native **R**				
	mean	sd	2.5%	median	97.5%	mean	sd	2.5%	median	97.5%
N	89.560	12.000	75.000	87.000	120.000	89.293	11.945	75.000	87.000	120.000
μ_p	−2.680	0.431	−3.724	−2.613	−2.051	−2.670	0.431	−3.721	−2.601	−2.043
σ_p	1.670	0.296	1.218	1.624	2.361	1.659	0.298	1.207	1.616	2.369

(1) For each $i = 1, 2, \ldots, M$, where M is the size of the augmented data set, obtain z_i from a Bernoulli distribution having probability $\Pr(z_i = 1 | y_i = 0)$ which can be obtained from application of Bayes' rule.

(2) Draw ψ from a beta distribution having parameters $1 + \sum_i z_i$ and $1 + M - \sum_i z_i$.

(3) μ and $1/\sigma^2$ are sampled from standard distributions (normal and gamma, respectively).

(4) Updates of each η_i can be obtained by use of a random walk Metropolis–Hastings-type algorithm (Section 2.4.4).

The results of fitting the model in **WinBUGS** are summarized in Table 6.2, based on 105K iterations. Note that the posterior of N is highly skewed and so the posterior means, even the medians, are not so similar to the MLE reported earlier. For comparison, the mode of the posterior of N was 83, having only slightly less posterior probability than $N = 84$, the two values that straddle the MLE. The posterior distribution of N is shown in Figure 6.2.

6.4.2 Of Bugs in BUGS

A vast majority of models described in this book can be analyzed using methods of classical statistics (usually variations of integrated likelihood). In addition, Bayesian analysis is straightforward for most of these models, and this is facilitated by the use of data augmentation. We have implemented most examples described in the book using both classical likelihood-based methods, and Bayesian analysis by MCMC, using native **R** programs that we have written. Despite this, we have adopted a strong focus on the use of **WinBUGS** for the Bayesian analysis of hierarchical models. There are several reasons for this. First, freedom from having to develop software on a case-by-case basis allows one to focus on model development for the problem in question, not on 'code development.' Secondly, writing **R** code (or code in some other language) requires experience and specific technical knowledge of computing *and* MCMC *and* probability calculus, which we think many ecologists

```
nx<-c(15, 7, 5, 2, 5, 6, 3, 2, 1, 2,
   4, 4, 0, 1, 0, 2, 3, 1, 1, 1,
   1, 0, 0, 0, 1, 1, 0, 1, 0, 1,
   0, 0, 0, 0, 0, 1, 0, 0, 0, 0,
   0, 0, 0, 0, 0, 0, 0, 0, 0, 0)
J<-length(nx)
nind<-sum(nx)
y<-rep(1:50,nx)
nz<-250
y<-c(y,rep(0,nz))

sink("winbugsmodel.txt")
 cat("
   model {
     psi~dunif(0,1)
     mu.p~dnorm(0,0.001)
     tau~dgamma(.001,.001)
     sigma<-sqrt(1/tau)
     for(i in 1:(nind+nz)){
        z[i]~dbin(psi,1)
        eta[i]~dnorm(mu.p,tau)I(-20,20)
        logit(p[i])<-eta[i]
        mu[i]<-p[i]*z[i]
        y[i] ~ dbin(mu[i],J)
   }
 N<-sum(z[1:(nind+nz)])
}
",fill=TRUE)
sink()

data <- list ("y","nind","nz","J")
zst<-c(rep(1,nind),rbinom(nz,1,.25))
inits <- function(){
  list (mu.p=rnorm(1),tau=runif(1,0,1),z=zst)
}
parameters <- c("N","mu.p","sigma")
out <- bugs (data, inits, parameters, "winbugsmodel.txt")
```

Panel 6.2. **WinBUGS** model specification and its execution from **R** using **R2WinBUGS**. Note that, in the model specification, the normal random effects prior is truncated to have compact support (see Section 6.4.2). These **R** instructions can be copied directly into an **R** script and executed. nz is the number of all-zero encounter histories with which augment the data set.

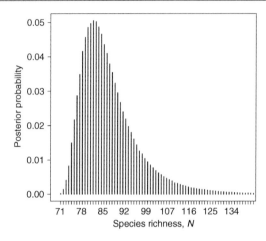

Figure 6.2. Posterior of N for the 2001 Breeding Bird Survey data from a Maryland route. The posterior distribution was computed by Markov chain Monte Carlo in **R** using the algorithm described in Section 6.4.1.

will not have. Some might argue that one should know what's going on 'under the hood'. Of course that is a sensible argument, but the reality of it is that almost no one does and the same rule is not applied when it comes to learning conventional likelihood procedures (e.g., how do SAS or MARK or Excel do what they do?).

However, the use of **WinBUGS** is not a panacea. It is very easy to find models, parameterizations of models, and specifications of models for which **WinBUGS** fails. And, importantly, it is not always obvious what constitutes failure or why failure was realized.

Bayesian analysis of models by Markov chain Monte Carlo is notoriously sensitive to the parameterization of the model. Devising an efficient MCMC algorithm depends very much on finding suitable or 'good' parameterizations of the model under consideration. Apparently innocuous changes can have substantial effects on the mixing of the Markov chains. Parameterization of models in **WinBUGS** is an important issue that should be considered whenever mixing appears to be a problem. In addition to the issue of parameterization, we also must consider the 'specification' of the model – that is, how we describe it using **WinBUGS** pseudo-code. A given parameterization of a model has a non-unique representation in **WinBUGS**, and not all representations work equally well. In some models where **WinBUGS** initially fails (or chooses an algorithm that yields a poorly mixing Markov chain), slight, subtle, or apparently innocuous reformulations of the description cause

WinBUGS to choose alternative algorithms that yield dramatically improved mixing. We will provide several examples of this in the context of applications.

One class of models where we have experienced repeated difficulties is in logistic-regression type models with priors having unbounded support. We find that such priors often result either in poor mixing, or no mixing at all (but not always). A few modifications to the model specification seem to (usually) resolve the problem. First, we can change the MCMC setting for 'log concave' to `UpdaterSlice` in the `Methods.odc` file which can be found in the directory `WinBUGS14\Updater\Rsrc` of your **WinBUGS** installation. Second, a less direct construction of a model seems to do the trick in some instances. Thirdly, truncate the normal prior on the interval $[-B, B]$ or, where appropriate, use a uniform prior with sufficiently large B so as to not induce truncation of the posterior distribution. We conjecture that these last two suggestions trick **WinBUGS** into changing algorithms, perhaps to a more general but less efficient algorithm that can properly sample from the distributions in question. Another problem area in this class of models involves the use of logit and log functions on the left-hand side of an assignment. For example, in some models `logit(p) <- a + b*x` can produce either errors ('traps'), poor mixing, and other problems. Whereas the reformulation of the same operation using two instructions, `lp <- a + b*x` and `p <- 1/(1+exp(-lp))`, seems to always work.

6.5 INDIVIDUAL COVARIATE MODELS

In the previous sections of this chapter, we introduced models containing unstructured variation or individual heterogeneity in the form of an individual random effect. While we believe that these models are useful in some situations, the result of Link (2003) adds some very relevant context to their utility (that is, we cannot hope to sort through a list of models and choose the best one using AIC or any other data-based method of model selection). As such, one should think about the context of the situation to identify a suitable class of models and stick to it for the purposes of conducting inference. In the absence of information that might suggest a model, our view is that a priori selection of the class of models is a necessary and reasonable course of action. This may bother some, but the alternative is to consider no model of heterogeneity – which is probably unrealistic in many situations. An alternative approach to dealing with heterogeneity is to identify specific factors that cause heterogeneity – i.e., individual-specific covariates that influence detection probability. These 'individual covariate' models have proved to be enormously useful in many problems and such models are the subject of the remainder of this chapter.

Table 6.3. Classes of individual covariate models with respect to the status/treatment of the individual covariate.

model class	n individuals	n_0 individuals
behavioral response	observed	observed
heterogeneity models	unobserved	unobserved
individual covariates	observed	unobserved

Models with individual covariates are very similar to the unstructured heterogeneity models described previously. The main distinction (of course, of some importance) is that in the individual covariate models, we *observe* the covariate for the n individuals, and so the covariate is 'missing data' only for the $N - n$ unobserved individuals. Conversely, in Model M_h, the 'covariate' is missing for all N individuals. In individual covariate models, we do have information about the covariate distribution by virtue of some covariate values being observed. Thus the identifiability problem raised by Link (2003) may be less practically relevant. In Chapter 5 we considered another class of models in which covariates varied by individual – those being the 'behavioral response' models. For these models, the individual covariate is known for *all N* individuals (it is 1 for previous capture and 0 otherwise). A nice conceptual benefit of adopting a hierarchical formulation of closed population models is that the relationship between the models becomes transparent. They only differ as to what set of covariate values is latent or unobserved (i.e., none, some, or all). The relationship in this regard among the various models is summarized in Table 6.3.

The general design framework considered here is consistent with that for sampling closed populations. We suppose a population of size N is sampled on J occasions and yields a sample of n unique individuals, captured y_i times. A covariate, x, thought to influence p, is measured on each individual:

$$\text{logit}(p_i(x_i)) = \alpha_0 + \alpha_1 x_i.$$

We adopt a logit-linear relationship between p and x in this case but other link functions could just be used. We saw in Chapter 5 how easy it was to solve such problems when x_i represented a behavioral response. In the present case, the main technical challenge is that we don't know covariate values for the individuals that did not appear in the sample. We have thus observed a biased sample of individuals – biased toward those individuals with values of x_i that are favorable to being captured. In statistics, this is a problem of *non-ignorable missingness*. The missingness is non-ignorable '....if the failure to observe a value depends on the value that would have been observed' (Ibrahim et al., 2005). In the present context, the probability that the covariate is missing (i.e., that the individual is not captured) depends on the value of the covariate through the detection probability. The treatment we adopt here (see Royle, 2008a) is consistent with the contemporary treatment of such problems in statistics, but not with their classical treatment in capture–recapture.

6.5.1 Background

Considerable attention has been paid to this problem of individual covariates in capture–recapture models (see Williams et al. (2002) pp. 300–302 and Pollock (2002) for reviews).

Three basic strategies have been suggested for modeling individual covariates. One approach suggested by Pollock et al. (1984) requires stratification of individuals into a finite number of discrete classes, yielding K strata with stratum population sizes N_k. Under this approach, the collection of N_k parameters is the object of inference. This approach is ideally suited for discrete covariates, especially those that take on a small number of values. For example, in Section 5.5.3, we introduced data from a waterfowl survey where the covariate was the size of the group of individuals. We could define N_k to be the number of groups of size k and apply this stratified estimator. One shortcoming is that the parameter dimension increases with the cardinality of the covariate, or number of strata in the case of a continuous covariate. Avoiding this dimensionality increase by specification of fewer strata yields a poorer approximation of the covariate distribution, and the issue is compounded as additional covariates are considered. These issues motivate the hierarchical modeling solution that we advocate. By specifying an additional model for the N_k parameters, estimation proceeds without any difficulty.

The stratification approach of Pollock et al. (1984) has not been widely adopted. Instead, two philosophically divergent solutions for attacking inference in individual covariate problems have been applied. By far, the most popular is the so-called Huggins–Alho procedure (Huggins, 1989; Alho, 1990). Sometimes this is referred to as the Huggins–Alho *model* which is confusing and conceptually misleading as its origins are strictly design-based, and while such procedures do have a model-based justification, the consequences of that are never addressed in applications. Thus, the Huggins–Alho procedure represents a specific solution to a fairly broad problem, one that has a myriad of potential solutions. This procedure is based on a generalized Horvitz–Thompson estimator for unequal probability sampling (e.g., Thompson, 2002, Chapter 6). In this context, the sample inclusion probabilities are functions of the individual covariates, and these can be estimated by maximum likelihood. Under this approach, N is a derived parameter, its estimation being based on estimates of individual detection probabilities, and classical methods of asymptotic inference are employed. This has a formulation that is similar to the notion of conditional estimators, and it is often referred to as such, but the usage of the term in this context is somewhat ambiguous as the Huggins–Alho estimator does not arise under a factorization of the joint likelihood, as virtually all other instances of the conditional estimator do.

A third approach, colloquially referred to as the 'joint likelihood approach,' (or 'full likelihood' see Borchers et al. (2002)) is based on specification of a probability distribution for the covariates, say $g(x)$, and then developing the joint likelihood of the parameters, including the parameters of $g(x)$, based on the joint distribution of (y_i, x_i) for individuals $i = 1, 2, \ldots, N$. This approach is consistent with the broader hierarchical modeling view that we advocate, and we emphasize it in this chapter. We thus refer to this formulation of the problem as the hierarchical (or model-based) formulation, whether one adopts a classical likelihood analysis, or a Bayesian one.

The hierarchical modeling approach, when the missing covariate is removed from the likelihood by integration, has a basic formulation as a multinomial observation model with unknown population size, N. Thus, the model is amenable to efficient Bayesian analysis by data augmentation, which is the general strategy for inference that we employ for these models. We provide two examples. In the first, we revisit the meadow vole data, where the covariate *body mass* is thought to explain heterogeneity in detection probability. The second example is the waterfowl survey survey data described in Section 5.5.3, in which the effect of a discrete covariate (group size), is evaluated.

6.5.2 Model Formulation

Suppose a closed population of size N is sampled on J occasions, yielding a sample of n unique individuals, where individuals $i = 1, 2, \ldots, n$ were captured y_i times. For clarity, in the initial development here, we assume that detection probability does not vary over the J occasions, but the following development extends directly to a formulation in terms of capture histories (which we illustrate in both examples).

We consider logit-linear functions relating detection probability to some measured covariate, x_i,

$$\text{logit}(p_i) = \alpha_0 + \alpha_1 x_i. \qquad (6.5.1)$$

We suppose that the covariate x has probability distribution (or pmf) $g(x|\theta)$. In the first example below we consider a continuous covariate, in which case we choose g to be normal. In the second example, x is discrete and we consider a Poisson pmf in that case.

The observations consist of the (y_i, x_i) pairs for $i = 1, 2, \ldots, n$ as well as the sample size itself, n. To proceed with inference about N, and possibly the parameters of Eq. (6.5.1), we seek the joint distribution of these observables. That is, $[\mathbf{x}_n, \mathbf{y}_n, n | \boldsymbol{\alpha}, \theta]$ where $\mathbf{x} = (x_1, \dots, x_n)$, $\mathbf{y} = (y_1, \dots, y_n)$ and $\boldsymbol{\alpha} = (\alpha_0, \alpha_1)$. A solution can be achieved by specifying the conditional distribution of y and x for each individual conditional on the event that the individual appeared in the sample (Royle, 2008a), which can be derived by an application of Bayes' rule.

To proceed here, note that if N were known, then $[y|x]$ is binomial and $[x] = g(x|\theta)$ is prescribed. In this case, the joint distribution of all (y_i, x_i) pairs is

$$[\mathbf{y}_n, \mathbf{y}_0, \mathbf{x}_n, \mathbf{x}_0] = \prod_{i=1}^{N} \text{Bin}(y_i | p(x_i)) g(x_i | \theta),$$

where $\mathbf{y}_0$ is the vector of 'data' corresponding to uncaptured individuals, i.e., $\mathbf{y}_0 = (y_{n+1}, \dots, y_N)$ and similarly for $\mathbf{x}_0$. Since (y_i, x_i) are observed only for $i = 1, 2, \dots, n$, consider partitioning the joint distribution as follows:

$$[\mathbf{y}_n, \mathbf{y}_0, \mathbf{x}_n, \mathbf{x}_0] = \prod_{i=1}^{n} \text{Bin}(y_i | p(x_i)) g(x_i | \theta) \prod_{i=n+1}^{N} \text{Bin}(y_i | p(x_i)) g(x_i | \theta).$$

In the second part, we don't know x_i, but we know that $y_i = 0$. So let us resolve this problem as we always do in such cases by removing the unobserved x's by integration, to yield the marginal probability of observing $y = 0$. That is,

$$\Pr(y = 0) = f(0 | \boldsymbol{\alpha}, \theta) = \int_x \text{Bin}(0 | p(x)) g(x | \theta) \mathrm{d}x.$$

Finally, note that n constitutes a binomial sample having probability $1 - f(0 | \boldsymbol{\alpha}, \theta)$. As such, the 'joint likelihood' is:

$$L(N, \boldsymbol{\alpha}, \theta | \mathbf{y}_n, \mathbf{x}_n, n) = \left\{ \prod_{i=1}^{n} \text{Bin}(y_i | J, p(x_i; \boldsymbol{\alpha})) \right\} \left\{ \prod_{i=1}^{n} g(x_i | \theta) \right\}$$

$$\times \frac{N!}{n!(N-n)!} f(0 | \boldsymbol{\alpha}, \theta)^{N-n}. \qquad (6.5.2)$$

We can 'condition on n' which yields

$$L_c(\boldsymbol{\alpha}, \theta | \mathbf{y}_n, \mathbf{x}_n) = \frac{\left\{ \prod_{i=1}^{n} \text{Bin}(y_i | J, p(x_i; \boldsymbol{\alpha})) \right\} \left\{ \prod_{i=1}^{n} g(x_i | \theta) \right\}}{(1 - f(0 | \boldsymbol{\alpha}, \theta))^n}. \qquad (6.5.3)$$

We do not provide a classical likelihood analysis of the examples below, but analysis of either the conditional or unconditional models can be achieved without difficulty in **R** by use of the `integrate` function (or with the `adapt` library for doing adaptive quadrature). Note that the unconditional likelihood here requires specification of $g(x)$. As such, this standard conditional formulation is inconsistent with the conventional Huggins–Alho procedure.

6.5.3 Bayesian Estimation by Data Augmentation

Here we adopt a Bayesian formulation of the individual covariate model based on data augmentation introduced in Chapter 5. The subsequent analyses follow from Royle (2008a).

As with previous applications of data augmentation, the basic strategy is to augment the observed data set with a fixed, known number of all-zero capture histories and to model the augmented data set as a zero-inflated version of the complete-data model using an unknown, but estimable, zero-inflation parameter. In the present context, we introduce the zero observations ($y_{n+1} = 0, y_{n+2} = 0, \ldots, y_M = 0$), and a set of latent covariate values $\{x_i\}_{i=n+1}^{M}$. As always, use of data augmentation requires us to assume a set of latent indicator variables $\{z_i\}_{i=1}^{M}$ which are observed ($z_i = 1$) for $i = 1, 2, \ldots, n$ and unobserved for $i = n+1, \ldots, M$. We suppose that $z_i \sim \text{Bern}(\psi)$ where ψ is the inclusion probability. That is, the probability that an element of the augmented population (of size M) is a member of the sampled population (of size N). The object of inference is the population size, N, which is a derived parameter $N = \sum_{i=1}^{M} z_i$.

Thus, the model for the augmented data consists of the following 3 components: Conditional on x_i, the observation model is

$$[y_i | p(x_i), z_i] = \text{Bin}(y_i | J, z_i \, p(x_i))$$

with

$$\text{logit}(p_i) = \alpha_0 + \alpha_1 x_i.$$

The operative 'process model' has components

$$[x_i] = \text{N}(\mu_x, \sigma_x^2)$$

and, due to data augmentation,

$$z_i \sim \text{Bern}(\psi).$$

Estimation and inference are straightforward using conventional Markov chain Monte Carlo (MCMC) methods in either **R**, or **WinBUGS** (subject to caveats mentioned previously). As a technical matter, the difficult component of implementing the MCMC algorithm is sampling each 'missing' x_i from its full-conditional distribution, which does not generally have a convenient form. We can, however, draw samples of x_i using a Metropolis–Hastings type algorithm in which we sample from the full-conditional proportional to

$$[x | y = 0] \propto (1 - p(x))^J g(x | \theta)$$

which is not difficult to do efficiently in most cases. The MCMC algorithm is otherwise fairly standard. We avoid the technical details of implementing this in **R**, and instead consider implementation in **WinBUGS** in the following examples.

Table 6.4. Estimates under a model with an individual covariate (body mass) for the *Microtus* data. Results are based on approximately 100K Monte Carlo draws from the posterior.

parameter	mean	sd	2.5%	median	97.5%
N	60.17	3.65	56.0	59.0	69.0
α_0	0.59	0.14	0.323	0.591	0.874
α_1	1.03	0.18	0.672	1.013	1.400
μ	−0.12	0.18	−0.501	−0.107	0.199
ψ	0.33	0.04	0.262	0.333	0.418
σ	1.09	0.13	0.878	1.075	1.391

6.5.4 Example: *Microtus* Trapping Data

We return to the *Microtus* data described in Chapter 5. Body mass (grams) was measured on each individual at the time of first capture, and we consider here whether there is an effect of body mass on detection probability. In Section 5.6.4, we considered the modified behavioral response model of Yang and Chao (2005). Before expanding on this model we consider the simplest individual covariate model, which has no temporal effects in detection probability. Thus, the detection model is

$$\text{logit}(p_i) = \alpha_0 + \alpha_1 x_i.$$

In this analysis (and all subsequent ones), the body mass covariate for the observations was standardized to have mean zero and unit variance. The model is precisely that described in the previous Section (Section 6.5.3). Body mass in the population of mice is assumed to be normally distributed with mean μ and variance σ^2, i.e., $x \sim \text{N}(\mu, \sigma^2)$.

The **WinBUGS** model specification is given in Panel 6.3. **R** instructions for data organization, and executing **WinBUGS** are provided in the Web Supplement. We note that this is an example of a model that sometimes doesn't mix properly (see Section 6.4.2). The prior for the covariate can be truncated using the following specification: `mass[i]` $\sim$ `dnorm(mu, tau)I(-10,10)`. This seems innocuous since we standardized the covariate to have mean 0 and unit variance. The estimates under this covariate-only model are shown in Table 6.4. A strong positive effect of body mass on detection probability is indicated, although the posterior of N is concentrated near the number of observed individuals (56) due to the high mean probability of detection. Note that the posterior mean of μ is −0.12, indicating that the body mass of the population of individuals is slightly less than individuals that appeared in the sample (which have standardized mean body mass 0). This is consistent with the positive effect of body mass.

Next, we consider expanding the modified behavioral response model of Yang and Chao (2005) (Section 5.6.4) to include the body mass covariate, so that we have

$$\text{logit}(p_{ij}) = \alpha_0(1 - x_{1,ij}) + \alpha_1 x_{1,ij} + \alpha_2 y_{i,j-1} + \alpha_3 x_{2,i}.$$

The covariate $x_{1,ij}$ is an indicator of previous capture, as before, and $x_{2,i}$ is the body mass covariate. Note that we have used a parameterization of the model in which α_0 is the mean for individuals that have not previously been captured, and α_1 is the intercept for previously captured individuals.

This model was fitted in **WinBUGS** using the specification described in Panel 6.4. We augmented the data set with $M - n = 100$ observations of $y = 0$, and corresponding 'missing' covariate values. We require prior distributions for the structural parameters of the model. For these, we adopt the customary default priors, diffuse normal priors for the mean parameters and gamma priors for the precision parameters. The **R** instructions for passing the data to **WinBUGS** using the **R2WinBUGS** function `bugs` are provided in the Web Supplement. We refer the reader back to the discussion of **WinBUGS** sensitivity to parameterization and model specification. We have resolved poor mixing in this case by truncation of the normal prior for the individual covariate. However, alternative implementations are possible.

```
model {
    psi~dunif(0,1)
    mu~dnorm(0,.001)
    tau~dgamma(.001,.001)
    sigma<-sqrt(1/tau)
    a0~dnorm(0,0.1)
    a1~dnorm(0,0.1)
    for(i in 1:(nind+nz)){
        mass[i]~dnorm(mu, tau)
        z[i]~dbern(psi)
        logit(p[i])<- a0 + a1*mass[i]
        muY[i]<-z[i]*p[i]
        y[i]~dbin(muY[i],J)
    }
    N<-sum(z[1:(nind+nz)])
    }
```

Panel 6.3. WinBUGS model specification for analysis of a simple individual covariate model for the *Microtus* data.

We note that the ephemeral response is weak (Table 6.5). On the other hand, the persistent behavioral response is indicated to be strong in the presence of the body mass covariate, and there is a large effect of body mass. The posterior distributions of both α_1 (the behavioral effect) and also of α_3 (the body mass coefficient) are concentrated above 0. Thus, the results indicate a large increase in detection probability once an individual is captured, and also that heavier individuals have higher probabilities of a capture. The conclusion, evidently, is that fat mice simply cannot control themselves around free corn. Secondarily, we note that when these two effects are included in the model, the posterior mass of N shifts away from the lower bound, n (the number of individuals captured). This is the intuitive effect, since the high apparent detection probability is due to the higher recapture rate of previously caught individuals, and the propensity for heavier individuals of being trapped.

The estimated population mean and standard deviation of the body mass covariate indicate that the measured covariate on the observed individuals

```
model {

psi~dunif(0,1)
tau~dgamma(.001,.001)
sigma<-sqrt(1/tau)
mu~dunif(-10,10)
a0~dunif(-10,10)
a1~dunif(-10,10)
a2~dunif(-10,10)
a3~dunif(-10,10)

for(i in 1:(nind+nz)){
    mass[i]~dnorm(mu,tau)I(-10,10)
    z[i]~dbin(psi,1)
    for(j in 1:J){
        logit(p[i,j])<- a0*(1-prevcap[i,j]) + a1*prevcap[i,j]
        + a2*lagY[i,j] + a3*mass[i]
        muY[i,j]<-z[i]*p[i,j]
        y[i,j] ~ dbern(muY[i,j])
    }
}
N<-sum(z[1:(nind+nz)])
}
```

Panel 6.4. WinBUGS specification of individual covariate model for the *Microtus* data. We find that poor mixing can arise in some cases when the prior distributions for logit-normal effects are unbounded. As shown here, the problem is resolved by truncating the prior distribution of the body mass covariate.

represents only a slightly biased sample. Whereas the sample mean and standard deviation of body mass were 41.79 and 11.96, respectively, the estimated (posterior mean) population mean and variance (upon back-transforming the posterior means of the estimates for μ and σ given in Table 6.5) are obtained by solving $(x - 41.79)/11.96 = -0.186$ which yields the population value $E[x] = 39.57$, and by solving $\text{Var}((x - 41.79)/11.96) = 1.117^2$, which yields the population standard deviation of $SD(x) = 13.36$. Thus, the population distribution of body mass is slightly lower and more variable, than the sample, as expected. The probability of capture for an individual that is s standard deviations from the mean body mass was evaluated for $s = (-2, -1, 0, 1, 2)$, for a hypothetical individual that was not previously captured. The **R** instructions for doing this, and the results, are:

```
>  px<- expit(0.245 +1.013*c(-2,-1,0,1,2))
>  netp<- 1-(1-px)^5
>  netp
  [1] 0.5408940 0.8512751 0.9836848 0.9994690 0.9999928
```

Individuals whose body mass is 2 standard deviations below the mean have only about a 54 percent chance of being captured during a study composed of $J = 5$ periods. Thus, considerable heterogeneity in p is induced by dependence of p on the covariate body mass. Now, consider the same calculation for an individual that *was* previously captured. This yields,

```
>  px<- expit(0.583 + 1.013*c(-2,-1,0,1,2))
>  netp<- 1-(1-px)^5
>  netp
  [1] 0.6536425 0.9183586 0.9940995 0.9998640 0.9999985
```

The interpretation of this is as follows: The probability of capture for an individual 2 standard deviations below the mean body mass increases by about 20 percent after initial capture. Due to the high mean capture probability (α_0), there is little effect of initial capture for individuals whose body mass is much above the mean (they have capture probabilities near 1).

6.5.5 Example: Group Size in Waterfowl Surveys

In this example, we consider the aerial survey data on mallard ducks (*Anas platyrhynchos*) which were introduced previously in Section 5.5.3. There were a total of 162 groups of waterfowl observed. As noted previously, sampling was conducted using two observers (front seat, back seat) in a double-observer sampling protocol. This yields 3 observable encounter histories of the form $\{(0,1), (1,0), (1,1)\}$,

Table 6.5. Posterior summary statistics from a model with both persistent and ephemeral behavioral responses and the body mass covariate (α_3) fitted to the *Microtus* data.

node	mean	sd	MC ERR	2.5%	median	97.5%
N	63.670	6.459	0.068	56.000	62.000	81.000
α_0	0.245	0.287	0.002	−0.343	0.254	0.787
α_1	0.583	0.378	0.002	−0.148	0.579	1.342
α_2	0.176	0.427	0.002	−0.678	0.178	1.005
α_3	1.013	0.184	0.001	0.666	1.007	1.388
μ	−0.186	0.208	0.002	−0.659	−0.166	0.164
σ	1.117	0.143	0.001	0.889	1.099	1.449

indicating whether the group was seen by the back seat observer only, the front seat observer only, or both observers. The model for such data is equivalent to a capture–recapture study with $J = 2$ periods, and detection probabilities p_1 (for the front-seat observer), and p_2 (back-seat observer). We are interested in estimating the density of ducks over the survey region, while allowing for the possibility that the observers have different detection probabilities, and also that detection probability varies by group size.

In what follows, the population group size distribution is assumed to be $1 +$ Poisson(λ), a right-shifted Poisson distribution, having mean $1 + \lambda$. The model for the probability of detection for group i was

$$\text{logit}(p_{ik}) = \alpha_k + \beta x_i \qquad (6.5.4)$$

for observer $k = 1, 2$ and observation $i = 1, 2, \ldots, 162$.

Data augmentation proceeds by adding a large number of $(0, 0)$ encounter histories to the data. For this analysis, 125 zero encounter histories were added, which proved to be adequate in the sense that the posterior mass of N was concentrated away from the upper limit (see Figure 6.3). To implement this in **WinBUGS**, we must zero-inflate a multinomial observation having 4 cells. Thus, we reorganize the data set into individual observations of the 4-dimensional multinomial random variable, with corresponding covariate 'group size'. In **WinBUGS** we will describe this by y[i,1:4] $\sim$ dmulti(mu[i,1:4],1) where y[i,1:4] is a vector of length 4, having a single 1 in the cell corresponding to the encounter history for that observation (there are probably other ways to approach the formulation of this model in **WinBUGS**). We organized the cell probabilities of the multinomial to correspond to the encounter histories $\{(0, 1), (1, 0), (1, 1), (0, 0)\}$, having probabilities $\{(1-p_{i1})p_{i2}, p_{i1}(1-p_{i2}), p_{i1}p_{i2}, (1-p_{i1})(1-p_{i2})\}$. Thus, the last cell corresponds to uncaptured individuals and the augmented data are vectors of the form $(0, 0, 0, 1)$. The zero-inflation of this multinomial model for this individual effects model is shown in Panel 6.5. The data file and **R** commands for fitting the model in **WinBUGS** are provided in the Web Supplement.

The quantities of interest are the total number of groups exposed to sampling and also the total number of individuals. These are both derived parameters, being $N_g = \sum_{i=1}^{M} z_i$, and $N_{\text{ind}} = \sum_{i=1}^{M} z_i x_i$, respectively. Density is also a derived parameter, being

$$D = \frac{N_{\text{ind}}}{544.5}.$$

These quantities are defined as deterministic nodes in Panel 6.5.

Posterior summaries of model parameters based on 200000 Monte Carlo draws from the posterior distribution are given in Table 6.6. The posterior mass of the group size coefficient, β, was slightly negative, but concentrated in the vicinity of zero. Thus, there does not appear to be an effect of group size on detection probability, at least of the form specified by Eq. (6.5.4). The estimated average detection probabilities are ordered as anticipated, $p_1 = 0.794$ and $p_2 = 0.501$, for front and back seat observers, respectively. Finally, the posterior distribution of the total number of groups, N_g, is shown in Figure 6.3. Posterior summaries of both N_g and N_{ind} are given in Table 6.6.

We note that one can solve this particular problem using classical likelihood methods (Royle, 2008b). Because the individual covariate is discrete, the likelihood for the observed group-size frequencies is multinomial with probabilities that depend on p_{ik} in Eq. (6.5.4), as well as parameters of the group size model $g(x)$. The parameters of this model can be estimated easily by maximizing this multinomial likelihood. However, the quantity of interest is a function of several of those parameters – the mean size of unobserved groups and the number of unobserved groups. Thus, one would have to conjure up an approximate variance of this quantity and also take into account the variation in group size, number of groups, and estimation variance. In contrast, one convenient aspect of Bayesian analysis by MCMC is that we can obtain posterior summaries of derived parameters merely by summarizing the appropriate function of those parameters. In the present case, the last several lines of Panel 6.5 are:

```
for(i in 1:(nind+nz)){
  tmp[i]<- 1 + gs[i]
  tmp2[i]<-tmp[i]*z[i]
}
Nind<-sum(tmp2[1:(nind+nz)])
D<-Nind/544.5
```

Thus, for each unobserved group, a posterior draw of the number of individuals is made (this is 1 + gs[i] where gs[i] $\sim$ dpois(mu.gs)), and these are summed up for each group that is a member of the exposed population (i.e., for which z[i]=1).

Table 6.6. Posterior summaries of model parameters for the aerial waterfowl survey data. Based on 200K posterior samples.

parameter	mean	sd	$q_{0.025}$	$q_{0.50}$	$q_{0.975}$
β	-0.194	0.198	-0.612	-0.183	0.168
λ	1.829	0.153	1.577	1.813	2.186
α_1	1.611	0.473	0.697	1.604	2.550
α_2	0.237	0.490	-0.695	0.224	1.230
p_1	0.799	0.052	0.686	0.805	0.887
p_2	0.510	0.076	0.363	0.510	0.660
N_g	196.400	17.250	174.000	193.000	243.0
N_{ind}	544.900	74.990	459.500	524.900	753.7
D	0.999	0.140	0.844	0.963	1.394

Finally, a posterior draw of density, D, is obtained as a linear function of the number of individuals – by dividing the latter by the surveyed area.

6.6 SUMMARY

In this chapter, we described two widely used classes of closed population models containing individual effects. The first of those comprises models containing unstructured heterogeneity in the form of a random effect. This class of models is commonly referred to as 'Model M_h', and its popularization originates from 'Burnham's jackknife estimator' (Burnham and Overton, 1978). This estimator is widely used by ecologists to estimate species richness and population size when detection probabilities vary among individuals. Interest in heterogeneity models continues to expand, with subsequent contributions by Chao (1987), Norris and Pollock (1996), Pledger (2000), and many other recent papers. Despite considerable interest in such models, there is also quite a bit of ongoing controversy over the preferred formulation, interpretation and even their basic utility (Link, 2003; Mao, 2004; Pledger, 2005; Dorazio and Royle, 2005b; Holzmann et al., 2006; Link, 2006). Our view is consistent with our broader adoption of parametric inference – that being that it is reasonable to choose a single class of models and to conduct inference about parameters under that class. The more relevant issue should be not whether your model is correct (none is), but rather, it should be whether the model and analysis provide meaningful context to the problem at hand – whether it is useful for some intended purpose. While we are certain that the debate over the sensitivity to model choice and its implications will continue, so too will the use of these models for describing variation in detection probability, which arguably is pervasive in biological systems.

```
model {

psi~dunif(0,1)
logmu.gs~dnorm(0,.001)
mu.gs<-exp(logmu.gs)
mu1.p~dnorm(0,.001)
mu2.p~dnorm(0,.001)
beta~dnorm(0,.001)

 for(i in 1:(nind+nz)){
    z[i]~dbin(psi,1)
    gs[i]~dpois(mu.gs)
    logit(p1[i])<- mu1.p + beta*(1+gs[i])
    logit(p2[i])<- mu2.p + beta*(1+gs[i])

    cp1[i]<- (1-p1[i])*p2[i]
    cp2[i]<- p1[i]*(1-p2[i])
    cp3[i]<- p1[i]*p2[i]
    cp4[i]<- (1-p1[i])*(1-p2[i])

    mu[i,1]<-z[i]*cp1[i]
    mu[i,2]<-z[i]*cp2[i]
    mu[i,3]<-z[i]*cp3[i]
    mu[i,4]<-z[i]*cp4[i] + (1-z[i])

     ncap[i,1:4]~dmulti(mu[i,1:4],1)
 }
 # mean detection probability for observers 1 and 2
   logit(p1bar)<-mu1.p + beta
   logit(p2bar)<-mu2.p + beta
 # population size of groups is a derived parameter
   Ng<-sum(z[1:(nind+nz)])
   for(i in 1:(nind+nz)){
     tmp[i]<- 1 + gs[i]
     tmp2[i]<-tmp[i]*z[i]
   }
 Nind<-sum(tmp2[1:(nind+nz)])
 D<-Nind/544.5
}
```

Panel 6.5. WinBUGS model specification for the double-observer sampling data with an individual covariate (group size). The situation yields a multinomial likelihood, and analysis is carried out by data augmentation. N_g and density (D) are derived parameters – functions of the $z[i]$ indicator variables.

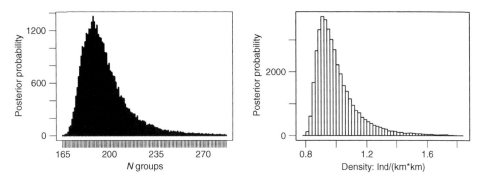

Figure 6.3. Posterior distribution of the total number of mallard groups on the surveyed sample units.

The second methodological focus of this chapter was on the individual covariate models. The essential concept motivating these models is that individuals have unequal probabilities of appearing in the sample, and we can identify measurable covariates that help explain variation in detection probability among individuals. The Huggins–Alho method (Huggins, 1989; Alho, 1990) is the standard method of inference for N in such problems. This is an intuitive solution to the estimation problem, as it is based on classical methods of unequal probability sampling similar in concept to Horvitz–Thompson estimation. The method does not seem sufficiently generic or flexible when considering other classes of models with individual covariates (e.g., open population models, and see Chapter 7). We have adopted what is sometimes referred to as the 'joint likelihood' approach – this approach specifies a model for the covariate in addition to an observation model that is conditional on the covariate value for each individual. Then, the covariate values for individuals that are not encountered are treated as missing values. This approach is consistent with the broader hierarchical modeling theme expressed in this book and also with more contemporary statistical approaches to the treatment of missing data.

Both classes of individual effects models described in this chapter have natural formulations as hierarchical models, despite the fact that their historical treatment was inconsistent even with parametric inference. In the case of M_h the practice was dominated by Burnham's jackknife estimator, and in the case of individual covariates, practice was dominated by the Huggins–Alho procedure. Both are heuristically motivated, but ad hoc solutions, to the respective problems. Conversely, both classes of models become unified within a hierarchical framework in which the model is fully specified in terms of a model for the observations, and a model for the unobserved (M_h) or partially observed (individual covariate models) individual effect. Bayesian analysis of these models is straightforward using *data augmentation*, which yields convenient reparameterizations as zero-inflated versions of the complete-data ('known-N') models.

7

SPATIAL CAPTURE–RECAPTURE
MODELS

Much of the theory and methodology underlying inference about population size is concerned with the situation where the population of interest is well-defined in the sense that one can randomly sample individuals that are associated with some location, area or place and, usually, uniquely identify them. However, a problem that we have encountered in several examples in previous chapters is that even when a spatial sample area can be precisely delineated, movement of individuals onto and off of the sample unit induces a form of non-closure. This non-closure has a direct effect on our ability to interpret the estimates of N that result from closed population models.

Individuals within populations are spatially organized; they (frequently) have home ranges or territories, or some sense of 'place' within which they live. The juxtaposition of this place relative to the sampling or trapping apparatus has important implications for sampling, modeling, estimation, and interpretation. In particular, this juxtaposition induces two general problems. First is that the effective sample area is unknown or, equivalently, the population size of individuals for which the sampled area applies is not known. We clarify this point in Section 7.3 in the context of precise models for animal movement and trapping. The second problem is that this juxtaposition of individuals with traps induces heterogeneity in capture probability, as a result of variable exposure to capture among individuals. Certain individuals, e.g., those with home ranges on the edge of a trapping grid, might experience little exposure to the trapping grid, perhaps only coming into contact with 1 or 2 traps. Conversely, individuals whose territories are located squarely in the center of a trapping grid might come into contact with many traps. As such, these individuals should experience higher probabilities of capture than individuals of the former type.

In this chapter we consider a class of models that we call spatial capture–recapture models. These models apply to the situation in which a classical capture–recapture type of study is conducted but each individual also has a location of capture associated with it. Thus, we observe a *spatial encounter history* for each individual.

That is, in addition to the usual encounter history composed of binary observations indicating encounter (or not), we also have, for each encounter, a spatial coordinate indicating the individual's location. As an example, suppose a population is sampled $J = 5$ times; then a spatial encounter history for individual i might be of the form $\mathbf{h}_i = (0, 0, \mathbf{x}_{i3}, 0, \mathbf{x}_{i5})$, where $\mathbf{x}$ is, normally, a two-dimensional vector of coordinates[1]. As before, a 0 indicates non-capture. Intuitively, there is more information here than relates simply detection probability. We can decompose the spatial encounter history into two distinct pieces of information. First is information about detection probability, in which we could express the previous encounter history as $\mathbf{y}_i = (0, 0, 1, 0, 1)$. Second, is information about where individual i lives – its home range, or territory. This comes from the coordinates $\mathbf{x}_{i3}$ and $\mathbf{x}_{i5}$.

While statistical inference from spatial sampling of individuals has a long history in the ecological literature, the field remains diffuse both conceptually and methodologically with no unifying theoretical framework that ties these methods together through a common model formulation and estimation/inference framework. Historically, ecologists have used auxiliary information on location to estimate the mean or maximum distance moved to adjust the effective sample area, or used various other heuristic or ad hoc 'adjustments' (Wilson and Anderson, 1985a,c; Karanth and Nichols, 1998; Trolle and Kéry, 2003). By ad hoc we mean that such procedures lack an explicit and precise model description or they adopt an informal approach to estimation. Formalization of the use of this information requires the precise definition of a model – the linkage of $\mathbf{x}$ to some notion of territory or home range, and the linkage of observations $\mathbf{y}$ to $\mathbf{x}$.

We present models that make use of spatial information to account explicitly for the heterogeneity induced by sampling, the movement of individuals, and to make inferences about the absolute density of individuals (i.e., the number of individuals per unit area). We do this by the construction of hierarchical models that admit explicit notions of location (a territory or home range) and movement of individuals in relation to the traps or area searched. In the development of these models, we introduce an individual random effect that is related to each individual's 'place' – which we call the individual's activity center. We treat these activity centers as unobserved or partially observed latent variables, which are dealt with in the usual way, by integration of the conditional distribution of the observations given the random effects, over the prior distribution of the latent effect in question. As such, these models are variations of models with individual covariates (Chapter 6). We formalize that linkage in this chapter. We provide a unifying framework for Bayesian inference under these models using data augmentation.

---

[1]The exception to this case will be distance sampling, in which the covariate is a deterministic function of location, but we will retain the usage of x to represent the (potentially) observable covariate in that case.

The most prominent variation of these spatial capture–recapture models is ordinary distance sampling (Buckland et al., 2001). While distance sampling is not formally capture–'recapture' – as these are not typically recaptures – the basic model structure of distance sampling makes it a particular case of a much broader class of spatial capture–recapture models. In the basic case, we have a spatial capture history corresponding to $J = 1$ observation, i.e., a single coordinate. To develop some concepts, we consider a variation of distance sampling in which the location of each individual is measured with error. This can be dealt with by a design modification in which replicate samples are obtained (e.g., using 2 observers). Under this extension, the distance sampling model is more directly relevant to the broader class of spatial capture–recapture models.

The second sampling situation that we consider in this chapter consists of a slightly different protocol and design, in which a prescribed area is subjected to a search of uniform intensity. The search is repeated J times to yield spatial encounter histories on n individuals. The formalization of a model for observations and an underlying movement and distribution process allows for the estimation of absolute density from spatial capture–recapture data (Royle and Young, 2008). An analysis of this situation using basic probability calculus yields a precise (and intuitive and interesting) linkage to the temporary emigration concept (Kendall et al. 1997). Finally, we consider the classical design/protocol in which an array or grid of traps is used in a capture–recapture study. This is typical of small-mammal studies (Parmenter et al., 2003), and it is now common to sample carnivore populations using arrays of camera traps (Karanth, 1995; Karanth and Nichols, 1998; Trolle and Kéry, 2003; O'Brien et al., 2003; Wallace et al., 2003; Karanth et al., 2004; Kawanishi and Sunquist, 2004; Karanth et al., 2006) in which individuals are identified uniquely by spotting or striping patterns, or by genetic tags in which individuals are identified using their DNA from scat (Creel et al., 2003) or so-called 'hair snares' (Woods et al., 1999). While these are not areal sample units, we can conceptualize the problem as a process model in which the probability of capture of an individual depends on the distance between an individual's intrinsic location and each trap.

7.0.1 A Hierarchical Model for Temporary Emigration

We formalize the movement process of individuals relative to a spatial sampling apparatus by the specification of a simple hierarchical model. Suppose a population of size N is subjected to repeated sampling, and let z_{ij} be an indicator of whether individual i is physically exposed to sampling during survey j. We assume that z_{ij} are *iid* Bernoulli variables, having probability ϕ. This is the 'random' temporary emigration model of Kendall et al. (1997). Now suppose we observe $y_{ij} = 1$ with probability p if $z_{ij} = 1$. Otherwise, if $z_{ij} = 0$, then we observe $y_{ij} = 0$ with probability 1. Thus, the observation model is a two-part model, with the first component (operative if $z_{ij} = 1$) corresponding to the realization of a binomial random variable and the second component being the addition of structural zeros that arise when $z_{ij} = 0$ (the individual not being exposed to capture). It is natural to formulate the observation model as a conditional model in which $y_{ij}|z_{ij} \sim \text{Bern}(p\,z_{ij})$ and $z_{ij} \sim \text{Bern}(\phi)$. The parameter $1-\phi$ is sometimes called the temporary emigration probability. The parameter ϕ reflects exposure to sampling, so we might also use the term *exposure probability*.

In the basic design in which a single observation is made for each of $j = 1, 2, \ldots, J$ samples, we do not obtain any information on ϕ independent of p and so, marginally, $y_{ij} \sim \text{Bern}(p\,\phi)$. That is, p and ϕ are confounded. However, just as the 'robust design' yields information on state-transitions in classical multi-state models (Kendall et al., 1997; Kendall and Nichols, 2002), the spatial information in spatial capture–recapture problems provides information about the exposure or availability state of individuals. Certain spatial capture–recapture models can be formulated precisely in terms of this temporary emigration model, but with ϕ depending on individual i via its location relative to the trapping array.

To clarify this notion, consider a superpopulation of N individuals that are *ever* exposed to sampling. During a sampling event there are N_j individuals instantaneously exposed to sampling, where $N_j \sim \text{Bin}(N, \phi)$. However, we cannot generally observe N_j. The observed count of individuals, n_j, is a binomial outcome with sample size N_j: $n_j \sim \text{Bin}(N_j, p)$. This simple hierarchical model provides one interpretation/mechanism of what happens in spatial sampling problems. That is, at any point in time there are N_j individuals exposed to sampling from among the superpopulation of size N (individuals that are *ever* exposed to sampling). These N individuals occupy, not the actual sample unit, but a larger 'effective sample area' that includes the nominal sample unit as a subset.

The above development assumes that ϕ is constant for all samples and all individuals. In practice, when there is a spatial organization to the sampling apparatus, we expect ϕ to vary (at least) by individual. For example, consider the lizard data analyzed in the previous two chapters. Instead of summarizing the data in terms of capture frequencies, we associate with each individual capture the spatial location of capture, and we put these spatial capture histories on a map. For these data, there were 68 lizards captured during the survey, a total of 134 times. The 134 capture locations are shown in Figure 7.1. In this context, we might expect individuals captured near the border of the 9 ha plot to experience less exposure to capture because they may not be present on the 9 ha plot during some samples. This heuristic notion has led some investigators (e.g., Boulanger and McLellan, 2001) to consider using the mean location of capture as an individual covariate and to apply the methods described in Chapter 6. The problem is that mean location is an imperfect observation of the latent variable that is directly

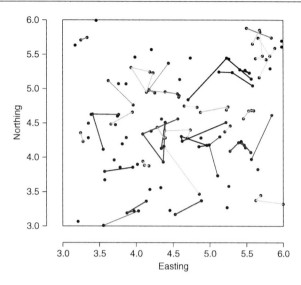

Figure 7.1. Locations of 68 flat-tailed horned lizards captured a total of 134 times on a 9 ha plot in southwestern Arizona. Captures of the same individual are connected by lines of the same color and thickness.

responsible for heterogeneity in exposure – that being, an individual's home range or territory. To formalize the usage of location of capture information, we require a hierarchical model that links encounter to observed location, and observed location to this notion of individual 'place'. We provide a formal analysis of this phenomenon in Section 7.3, where we define an individual's activity center and its relation to the individual's capture locations.

7.1 DISTANCE SAMPLING AS AN INDIVIDUAL COVARIATE MODEL

Distance sampling is an important class of methods and models for conducting inference about abundance. The basic idea is that we survey an area and record distances from the point of observation to any individuals that are encountered. The data are distance measurements to n individuals $x_1, x_2, \ldots, x_n$. In the standard situation the sample unit is a long transect and the distances are recorded by an observer traversing the center line of the transect. We suppose that distances are measured perfectly, and x is continuous.

In practice, distance sampling is often described as a procedure for estimating density, having more of a heuristic motivation than a model-based formulation.

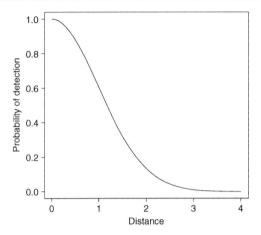

Figure 7.2. A half-normal detection function having $\sigma = 1$.

Here we formulate it as a hierarchical model and describe then how it relates to other models within the broader class of spatial capture–recapture models and, even more broadly, models containing individual effects. We note that distance sampling comes with a lot of methodological esoterica, which we avoid here because we are only concerned with its model-based origins. In addition, there are comprehensive books on distance sampling, including Buckland et al. (2001) and Buckland et al. (2004b), which the interested reader should consult.

A fundamental element of distance sampling methodology is choice of and inference about the *detection function*, which describes the probability of detection as a function of distance from point of observation to individual. Some analysts have strong preferences for particular shapes and features of detection functions (e.g., that they have a 'shoulder'), but the basic requirement is that the domain be $[0, 1]$. The detection function is related directly to the distribution of x for the individuals that were detected – the 'distance function'. The detection and distance functions are related by Bayes' rule, in the sense that $[y = 1|x]$ is the probability of detection given x whereas $[x|y = 1]$ is the conditional distribution of the distance measurement given that an individual was detected. We develop this idea shortly. A common distance function is the half-normal function given by

$$p(x; \sigma) = \Pr(y = 1|x) = \exp(-x^2/\sigma^2)$$

which is shown in Figure 7.2. We use this distance function in several analyses described subsequently.

7.1.1 Technical Formulation

Distance sampling can be viewed as a special case of the individual covariate models described in Chapter 6. In those models, we observe a covariate x on n individuals, and we have capture frequencies (or histories) based on J replicate samples of the population. Distance sampling is equivalent to an individual covariate model with a single replicate sample of the population, and the individual covariate is the distance from the point of observation to the individual. Thus, we observe the pair $(y, x) = (1, x_i)$, instead of (y_i, x_i) for y_i a general binomial count based on J samples. In the general case (i.e., Chapter 6), there is direct information about detectability from the replication, whereas in distance sampling there is no replication (i.e., normally $J = 1$). Information about detection probability stems from the *assumed* parametric relationship between detection probability and distance and certain other assumptions that we will address subsequently. In this regard conceptually, distance sampling is similar to the variable area sampling design described in Section 4.6.

While we considered several probability distributions for covariates in Chapter 6, it is customary in distance sampling to assume that

$$[x] = \mathrm{U}(0, B_x)$$

for some upper bound of distance measurement B_x (it can be the case theoretically that $B_x = \infty$). To link this more precisely with the class of spatial capture–recapture models, we note that uniformity of distances derives from the fundamental assumption of uniformity of *locations* of individuals, say $\mathbf{s}$. This assumption has been relaxed in certain specific contexts (Hedley et al., 1999; Royle et al., 2004), and also has given way to more distinctly design-based views that derive from random placement of the transect itself which ensure, regardless of $[x]$, that $\mathrm{E}[n] = Np$ for p constant. This is often used as the basis for claims of robustness in applications of the distance sampling method.

We previously gave the joint distribution of the observations, including n, for the individual covariate models in Chapter 6, which was:

$$[\mathbf{y}, \mathbf{x}, n | N, \sigma, \theta] = \left\{ \prod_{i=1}^{n} \mathrm{Bin}(y_i | J, p(x_i; \sigma)) \right\} \left\{ \prod_{i=1}^{n} g(x_i | \theta) \right\}$$
$$\times \frac{N!}{n!(N-n)!} (1 - \pi_{\mathrm{enc}})^{N-n}, \qquad (7.1.1)$$

where θ is the (possibly vector-valued) parameter of the covariate distribution g, $\mathbf{y} = (y_1, y_2, \ldots, y_n)$ (similarly for $\mathbf{x}$) and π_{enc} is the marginal probability of encounter, i.e.,

$$\pi_{\text{enc}} = \Pr(y = 1 | \sigma, B_x) = \int p(x; \sigma) g(x | B_x) \, dx = \int p(x; \sigma)/B_x \, dx.$$

We will analyze this general model subject to the assumptions of classical distance sampling to show that distance sampling is a special case of this individual covariate model, being that which arises under $g(x) \propto 1$ (uniform distribution of distance to individuals, or 'uniformity'), and when $J = 1$. The uniformity assumption yields:

$$[\mathbf{y}, \mathbf{x}, n | N, \sigma] = \left\{ \prod_{i=1}^{n} \text{Bin}(y_i | J, p(x_i; \sigma)) \right\} \left(\frac{1}{B_x} \right)^n$$

$$\times \frac{N!}{n!(N-n)!} (1 - \pi_{\text{enc}})^{N-n} \qquad (7.1.2)$$

and $J = 1$ produces (after combining terms):

$$[\mathbf{y}, \mathbf{x}, n | N, \sigma] = \left\{ \prod_{i=1}^{n} \frac{p(x_i; \sigma)}{B_x} \right\} \frac{N!}{n!(N-n)!} (1 - \pi_{\text{enc}})^{N-n} \qquad (7.1.3)$$

which can be maximized for different choices of the detection function $p(x)$. That is, we can base inference on the likelihood obtained from the joint distribution specified by Eq. (7.1.3).

From Eq. (7.1.3), we can multiply and divide by π_{enc}^n and combine terms, which produces

$$[\mathbf{y}, \mathbf{x}, n | N, \sigma] = \left\{ \prod_{i=1}^{n} \frac{p(x_i; \sigma)}{B_x \pi_{\text{enc}}} \right\} \frac{N!}{n!(N-n)!} \pi_{\text{enc}}^n (1 - \pi_{\text{enc}})^{N-n}. \qquad (7.1.4)$$

Finally, if we ignore the second term (that outside of curly braces), so that inference is based only on the partial likelihood, then we have precisely the conditional likelihood that is used to obtain the classical distance sampling estimator of N, which we derive in the next section.

7.1.2 Classical Derivation

The classical derivation of distance sampling methods (Burnham and Anderson, 1976; Buckland et al., 2001) is based on the conditional distribution of x given that the observation appeared in the sample, a concept analogous to the 'conditional likelihood' that we have encountered previously. We review this derivation here because we believe that it is useful and instructive to review the probability calculus and the precise manifestation of model assumptions.

As before, x is the covariate 'distance' and $y = 1$ is the event 'captured'. For inference, we require the joint pdf of the observed distance measurements. Bayes' rule yields

$$[x|y = 1, \sigma] = \frac{[y = 1|x, \sigma][x]}{[y = 1|\sigma]},$$

where $[y = 1|x, \sigma] \equiv p(x; \sigma)$ is the detection function, $[x] = 1/B_x$, and $[y = 1|\sigma]$ is thus the integral of the distance function – the average probability of detection. The quantity $[x|y = 1]$ is the distribution of distance to detected individuals, which is estimated by the empirical distribution of the n observed distances.

The distance sampling likelihood is derived as follows. Suppose $x_1, x_2, \ldots, x_n$ are observed distances. The likelihood is then:

$$
\begin{aligned}
\prod_{i=1}^{n} [x_i | y = 1] &= \prod_{i=1}^{n} \left\{ \frac{[y_i = 1|x_i, \sigma][x_i]}{[y_i = 1|\sigma]} \right\} \\
&= \prod_{i=1}^{n} \left\{ \frac{[y_i = 1|x_i, \sigma]}{[y_i = 1|\sigma]} \right\} \qquad [[x] = 1/B_x] \\
&= \prod_{i=1}^{n} \left\{ \frac{p(x_i; \sigma)/B_x}{\int_x p(x; \sigma)/B_x \, dx} \right\} \qquad [\text{def. of detection fn.}] \\
&= \prod_{i=1}^{n} \left\{ \frac{p(x_i; \sigma)}{B_x \pi_{\mathrm{enc}}(\sigma)} \right\}
\end{aligned}
$$

which is the same as the partial likelihood from Eq. (7.1.4). This is maximized to obtain $\hat{\sigma}$ or, equivalently, $\pi_{\mathrm{enc}}(\hat{\sigma})$ from which the classical *conditional estimator* of N is obtained:

$$\hat{N} = \frac{n}{\pi_{\mathrm{enc}}(\hat{\sigma})}.$$

This estimator of N is almost never presented in distance sampling analyses. Instead, it is customary use the estimator of density as given by:

$$\hat{D} = \frac{\hat{N}}{A} = \frac{n}{\pi_{\mathrm{enc}}(\hat{\sigma}) A},$$

where A is the total area sampled. This last expression can be found in, for example, Buckland et al. (2001, p. 38).

7.1.3 Distance Sampling Remarks

7.1.3.1 Conditional formulation vs. unconditional

The conditional formulation of estimators developed in the previous section is the standard practice in distance sampling, and it is very rare to see a fully model-based analysis, i.e., based on the joint distribution of the observations and n (Eq. (7.1.4)). We prefer the latter as it is more consistent with our view of hierarchical modeling. It preserves the N's in the model. In some cases there may be interest in describing additional model structure on spatially (or temporally) indexed N parameters. For example, modeling spatial variation in N in the form of a response to habitat or landscape covariates (e.g., Royle et al., 2004). We present a framework that enables this in Chapter 8.

7.1.3.2 Inference about N or inference about D?

Using the joint distribution of the observations – or the 'unconditional likelihood' approach – we obtain an estimate of $\hat{N}$ from which we compute density as $\hat{N}/A$. The conditional approach is based on a different estimator of N. The main distinction is that in the latter case, i.e., in standard distance sampling applications, $\hat{N}$ is almost never reported. There seems to be some debate, at least informally, over whether N or D is the natural parameter. But these parameters are statistically equivalent in the sense described by Sanathanan (1972). As such, we see no particular benefit to analysis based on the conditional formulation of the model, and we believe that it is generally less flexible, and also obscures the basic model-based derivation of distance sampling that renders it equivalent to an individual covariate model, with certain specific model assumptions.

7.1.4 Bayesian Analysis of Distance Sampling by Data Augmentation

Likelihood analysis of either the unconditional or conditional likelihoods described in the previous section can be achieved without difficulty. The main issue is that the probability of capture has to be computed by integration, which can be accomplished easily in **R**, and we have provided the basic tools to do this in previous chapters. Our purpose here is not to espouse the virtues of distance sampling nor even to encourage people to use it in practice. Instead, our aim is to link the method to the broader hierarchical modeling framework of individual effects models and to provide a technical and conceptual linkage between the method of distance sampling and others in this broad class.

Bayesian analysis can be accomplished by specifying prior distributions for N and σ, and by devising a method for sampling from the joint posterior distribution using MCMC. As with the individual covariate models of the previous chapter, we adopt an approach to Bayesian analysis based on data augmentation. We begin by assuming that $N \sim \text{Bin}(M, \psi)$ for some large M. We suppose also that ψ has a $\text{U}(0,1)$ prior. When ψ is removed by integration, the resulting marginal prior for N is discrete uniform on $(0, 1, 2, \ldots, M)$. Data augmentation can then be implemented by physically augmenting the observed vector of n observations (which happen to all be equal to 1 in distance sampling) with $M - n$ zeros, thus yielding the data vector $y_1, y_2, \ldots, y_n, y_{n+1}, \ldots, y_M$. We only observe the covariate distance on the first n values of y, and the remaining $M - n$ are regarded as missing values (as in the individual covariates models, see Section 6.5).

The hierarchical construction of the prior for N can be implemented by introducing a sequence of binary indicators w_i ($i = 1, 2, \ldots, M$) where $w_i \sim \text{Bern}(\psi)$. The zero-inflation parameter, ψ, takes the role of N in the model (recall that M is fixed a priori). In this chapter, we depart from the use of z for these latent indicator variables for clarity, because we use z in later sections to represent exposure to trapping due to movement of individuals.

The main benefit of adopting data augmentation is that it yields a hierarchical reparameterization for which a simple and efficient Bayesian analysis by conventional MCMC methods can be achieved. The hierarchical model for the augmented data consists of the following 3 model components:

- $w_i \sim \text{Bern}(\psi)$ for $i = 1, 2, \ldots, M$;
- $[y_i | p(x_i)] = \text{Bern}(w_i p(x_i))$ with $p_i = p(x_i; \sigma)$,
- $[x_i] = \text{U}(0, B_x)$.

With suitable priors for ψ and σ, estimation and inference are straightforward using conventional Markov chain Monte Carlo (MCMC) methods. For the analysis of the following section, we assumed $\psi \sim \text{U}(0,1)$ and, for the parameter of the half-normal distance function, a proper uniform prior, $\sigma \sim \text{U}(0, B_\sigma)$ as described below. While one can implement an MCMC algorithm for this problem in **R** without difficulty, we provide an implementation in **WinBUGS** in the following analysis.

7.1.5 Example: Analysis of Burnham's Impala Data

The data considered here are the impala data from Burnham et al. (1980, p. 63)[2], reporting on a line transect study to estimate density of ungulates in Africa. For these data, 73 animals were observed on a 60 km transect. Data recorded were sighting distance and angle from the transect, and sighting distances were truncated at 400 m. These were converted to perpendicular distances and scaled by dividing

---

[2]Burnham et al. acknowledge P. Hemingway.

Table 7.1. Results of fitting the distance sampling model to the impala data using the hierarchical formulation of the model under data augmentation, as implemented in **WinBUGS**.

parameter	mean	sd	2.5%	50%	97.5%
σ	1.870	0.170	1.570	1.850	2.250
N	179.900	22.950	140.000	178.000	229.000
D	3.742	0.472	2.917	3.708	4.75

the resulting perpendicular distances by 100 m. The **R** instructions for setting up the analysis are provided in the Web Supplement.

The **WinBUGS** model specification is given in Panel 7.1 using the half-normal detection function. Note that both N (size of the sampled population on the 60×0.8 km strip) and D (density) are derived parameters, the former being a function of the latent indicator variables, the latter being a function of model parameters through N. Note that the distance function parameter σ is given a uniform distribution on the interval $[0, 10]$ where the upper bound was chosen arbitrarily, but larger than 4 (the maximum distance class). A check of the posterior reveals no sensitivity to this bounded prior distribution.

The simulation was carried out for 20000 iterations after 2000 burn-in, thinned by 2, and using 3 chains initialized with random starting values. This yielded a total of 30000 posterior draws upon which estimates of posterior summaries were made (Table 7.1). These are consistent with previously reported estimates, as we might expect given the non-informative prior we assumed. The estimated posterior distributions of N and σ are shown in Figure 7.3.

7.1.6 Joint Distribution of the Augmented Data

We can use data augmentation to formally reparameterize the model. For simple models, including distance sampling, we can specify the model for the augmented data analytically and obtain MLEs of ψ and σ instead of N (or D) and σ. We provide that analysis here, and an **R** function for carrying out the analysis is provided in the Web Supplement. In Chapter 5 we provided a similar treatment of Model M_0.

The joint distribution of the observations, n, and N conditional on σ, M, and ψ is:

$$[\mathbf{x}|n, \sigma][n|N, \sigma][N|M, \psi].$$

Its precise form is

$$[n, \mathbf{x}, N | \psi, \sigma, M] = \left\{ \prod_{i=1}^{n} \frac{p(x_i; \sigma)/B_x}{\pi_{\mathrm{enc}}} \right\}$$
$$\times \frac{N!}{n!(N-n)!} \pi_{\mathrm{enc}}^n (1 - \pi_{\mathrm{enc}})^{N-n} \mathrm{Bin}(N | M, \psi), \quad (7.1.5)$$

where $\pi_{\mathrm{enc}} = 1 - f(0|\sigma)$ as before. As we noted in Section 3.7.2, this can also be maximized to obtain the MLE's of the 3 parameters N, σ and ψ. Alternatively, we can remove N from the likelihood, by summation, to obtain $[n|M, \psi, \sigma]$, thus formally replacing N as an unknown parameter by ψ. This is essentially the objective of data augmentation, as now the data set is of fixed size, having M elements (and some values of x that are unknown) instead of a variable dimension parameter space. So what does the joint distribution of the observations and augmented data look like, after N is removed? This result is obtained by the compounding of two binomial distributions (see the results described in Section 5.1),

```
model {
  sigma~dunif(0,10)
  psi~dunif(0,1)
  sigma2<-sigma*sigma

  for(i in 1:(nind+nz)){
    w[i]~dbern(psi)
    x[i]~dunif(0,Bx)      # Bx = strip width input as data
    logp[i]<-   -((x[i]*x[i])/sigma2)
    p[i]<-exp(logp[i])
    mu[i]<-w[i]*p[i]
    y[i]~dbern(mu[i])
  }
  N<-sum(w[1:(nind+nz)])
  D<- N/48   # conversion to ind/km*km
}
```

Panel 7.1. WinBUGS specification of distance sampling model for the impala data, with the half-normal detection function.

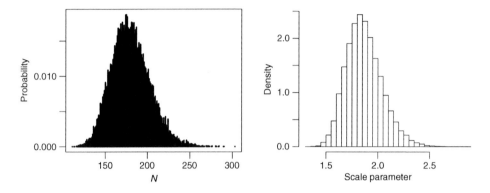

Figure 7.3. Posterior of N (left panel) and σ (right panel) for the impala data.

which produces

$$[n, \mathbf{x}|\psi, \sigma, M] = \left\{ \prod_{i=1}^{n} \frac{p(x_i; \sigma)/B_x}{\pi_{\text{enc}}} \right\}$$

$$\times \frac{M!}{n!(M-n)!} (\psi \pi_{\text{enc}})^n \left\{1 - \psi \pi_{\text{enc}}\right\}^{M-n} \qquad (7.1.6)$$

which is a function only of ψ and σ. This simplifies slightly to yield the joint likelihood:

$$L(\psi, \sigma | n, \mathbf{x}, M) \propto \left\{ \prod_{i=1}^{n} \psi p(x_i; \sigma) \right\} \frac{M!}{n!(M-n)!} \left\{1 - \psi \pi_{\text{enc}}\right\}^{M-n} . (7.1.7)$$

This can be maximized easily using standard numerical methods.

7.2 DISTANCE SAMPLING WITH MEASUREMENT ERROR

One assumption of classic distance sampling is that distances are measured without error. While it is debatable as to whether or not this can be achieved in practice, and whether an inability to do so is practically important, the analysis of the problem of measurement error using hierarchical models provides an interesting case study of their application. This situation is largely hypothetical as people seldom deal explicitly with measurement error. Conventional wisdom suggests that distance sampling estimators are relatively 'robust' to measurement error (and also, there is a claim that recording distance class instead of actual distance resolves the problem, and so practitioners of distance sampling try to do that). On the other hand, the basic design formulated here (based on replicate samples) is not so hypothetical as the design is now widely applied (i.e., 'mark recapture distance sampling,' Alpízar-Jara and Pollock (1996); Borchers et al. (1998)), but we haven't seen the problem of accounting for measurement error addressed analytically. While we think this is an interesting component of the problem, our motivation for considering the problem is to build a framework for analysis of the spatial capture–recapture model described in Section 7.3. Such models are closely related to the measurement error models we describe here. The relationship is somewhat more precise if we were to develop the problem in terms of measurements of the *locations* of individuals, instead of distance to individuals. However, it is not usually the case in practice that locations are measured, and the mathematics is slightly more tedious in that case.

To consider inference in the presence of distance measurement error, we introduce a departure from the conventional distance sampling design (having $J = 1$) and suppose instead that J replicate samples are obtained (e.g., using two observers, $J = 2$), yielding an encounter history on each individual and replicate observations of the distance to each individual observed. This modifies the observation model so that y_i, the total number of detections of individual i, is $\text{Bin}(J, p(x_i))$ where the true distance to individual i is x_i as before. The observation model could as well be multinomial (e.g., in a double observer situation in which the observers had different p parameters) but for clarity we assume that the replicate samples have the same p. We denote the replicate distance measurements as u_{ij} $(j = 1, 2, \ldots, J)$. We require a model for the observations u_{ij} conditional on the truth, $[u_{ij}|x_i]$. Here, we suppose that $u_{ij}|x_i \sim \text{N}(x_i, \delta^2)$ where δ^2 is the measurement error variance.

The main thing to note is that we now have a two-component observation model – $[y|x]$ and $[u|x]$, and, both are defined *conditional on* x. In particular, the encounter observations, y, are conditional on x, *not* on the observed locations u. We now have an interesting 3-stage hierarchical model consisting of components (dependence on parameters is suppressed) $[y|x]$, $[u|x]$ and $[x]$. The joint distribution of the observations and the latent distance variables (x) is a slight extension of the joint distribution that arose for the simpler individual covariate model, because we now have measurements (u) to be concerned with. The joint distribution of $i = 1, 2, \ldots, n$ observations is:

$$[\mathbf{y}_i, \mathbf{u}_i, \mathbf{x}_i, n | N, \sigma^2, \delta^2] = \left\{ \prod_{i=1}^{n} [y_i|x_i] \left(\prod_{j=1}^{y_i} [u_{ij}|x_i] \right) [x_i] \right\}$$
$$\times \frac{N!}{n!(N-n)!} (1 - \pi_{\text{enc}})^{N-n}. \qquad (7.2.1)$$

In this expression, $[y_i|x_i] = \text{Bin}(J, p(x_i))$. The limit of the second product is y_i, the number of detections of individual i (i.e., $y_i \in \{1, 2, \ldots, J\}$). The last part of this expression, the part with N in it, is similar to that in Eq. (7.1.1). In this case the probability of encounter, π_{enc}, is computed as under the ordinary distance sampling model, i.e.,

$$(1 - \pi_{\mathrm{enc}}) = \Pr(y = 0) = \int_x (1 - p(x; \sigma))^J (1/B_x) \, \mathrm{d}x.$$

The $[u|x]$ component of the model is irrelevant for uncaptured individuals.

We could do a Bayesian analysis of this model using data augmentation with only slight modifications to Panel 7.1, or by devising an MCMC algorithm based on Gibbs sampling. The full-conditional distributions for the model parameters have convenient forms (except for σ). Classical analysis based on likelihood is somewhat more difficult, but we pursue aspects of that analysis here in order to clarify the consequences of measurement error in this particular problem. We provide the conceptual outline of the approach here, absent mathematical details and some formal argument.

The motivation is that the true distances x are latent variables and, as it stands, the model (the detection function) is expressed in terms of x, whereas we observe u. We need to get x out of the likelihood. Recall that the detection function is

$$\Pr(y = 1|x) = \exp(-x^2/\sigma^2).$$

Thinking about this like any other latent variable problem, we can think about computing the marginal probabilities obtained by integrating x^2 out of this conditional-on-x detection function. That is, we need:

$$\Pr(y = 1) = \mathrm{E}_{x^2}\{\exp(-x^2/\sigma^2)\}.$$

This is conceptually straightforward, if we only had the distribution of x^2. Note that we observe some information about a particular x_i, in the form of (one or more) imperfect measurements, u_{ij}. This suggests that what we should do is take this expectation over the posterior distribution of x_i^2 given the observations of u_{ij}. We can motivate this more formally by noting that a typical strategy (Basu, 1977) in this situation is to condition on a sufficient statistic for x_i (the latent variable). That is, we need to express the model for y_i conditional on a sufficient statistic for x_i. The sufficient statistic for x_i is the mean of observations $\bar{u}_i$, provided that δ^2 is known. We will assume that δ^2 is known, but the added complexity of dealing with unknown δ^2 is not too great. So we need to derive the conditional distribution $x_i^2|\bar{u}_i$ which will allow us to evaluate

$$\Pr(y_i = 1|\bar{u}_i) = \mathrm{E}_{x^2|\bar{u}}\{\exp(-x_i^2/\sigma^2)\} = \int_0^\infty \exp(-x_i^2/\sigma^2)[x_i^2|\bar{u}_i]\mathrm{d}x_i^2.$$

To a close approximation $[x_i|\bar{u}_i]$ is normal with mean $\bar{u}_i$ and variance δ^2/J. Therefore, $(\sqrt{J}x_i/\delta)^2$ has a non-central chi-squared distribution with non-centrality parameter $\lambda = \bar{u}_i^2/(\delta^2/J)$. We recognize

$$\mathrm{E}_{x_i^2|\bar{u}_i}\{\exp(-x_i^2/\sigma^2)\} = \int_0^\infty \exp(-x_i^2/\sigma^2)[x_i^2|\bar{u}_i]\mathrm{d}x_i^2$$

as the moment-generating function (MGF) of a non-central chi-square distribution[3]. It transpires that

$$\Pr(y_i = 1|\bar{u}_i) = \left(1 + \frac{\delta^2}{\sigma^2}\right)^{-1/2} \exp\left\{-\frac{\bar{u}_i^2}{\sigma^2 + \delta^2}\right\}.$$

This expression yields the likelihood (i.e., using Eq. (7.1.2)) of y_i and n as a function of $\bar{u}_i$ (instead of x_i) and the parameters σ^2 and δ^2. It is worth pondering the form of this 'effective detection function' and its implications.

(1) When distance is measured with error, we can use the mean observed distance in the half-normal distance function. This supports the heuristic notion in the related trapping grid problem of using the mean location of capture as an individual covariate (Boulanger and McLellan, 2001).

(2) Note that the effective scale parameter is $\sigma^2(1 + \delta^2/\sigma^2)$ which tends to σ^2 as $\delta^2 \to 0$, as it should. Moreover, the 'intercept' of the implied distance function $(1 + \frac{\delta^2}{\sigma^2})^{1/2}$, is less than 1, but tends to 1 as $\delta^2 \to 0$.

(3) This intuitive form also supports the general phenomenon (in regression problems) that measurement error attenuates (flattens) the regression curve toward the mean. In this case, it flattens the detection function.

(4) That the intercept of the implied distance function is non-zero suggests that the phenomenon of measurement error might appear to be 'imperfect detection on the line'. We see that adjustment for the latter phenomenon will not necessarily resolve the bias inherent in the scale parameter that results from measurement error. That is, if we were to use two observers to account for imperfect detection on the line – we might not be addressing the correct problem – measurement error.

7.3 ESTIMATING DENSITY FROM LOCATION-OF-CAPTURE INFORMATION

In this section, we consider a design that is not widely used, but one that provides a nice transition from the distance sampling model toward what proves to be a

---

[3]You can look this one up on Wikipedia.

very broad and useful class of problems (namely, trapping arrays; Section 7.4). The situation we consider here is an 'area search' type of protocol wherein some prescribed area – a quadrat or other areal unit – is subjected to a uniform search intensity such that individuals contained in the unit have constant detection probability. The quadrat is searched multiple times (i.e., $J > 1$), generating a *spatial* capture-history on n individuals. The lizard data that were introduced in Chapter 5 are of this type. The data are shown in Figure 7.1. A 9 ha plot 300 m on a side was searched on 14 (almost consecutive) days (see Royle and Young, 2008, for details). While the basic protocol does not appear to be widely used, we suppose that it could be generally applicable to other reptile and amphibian species that are easily captured by hand.

The main issue in dealing with this problem, as we have noted previously, is that there is movement of individuals onto and off of the plot between samples. Thus, non-capture of an individual could be either because the individual was present on the quadrat but not encountered, or because the individual had moved off of the quadrat. The effect of this is to yield under-estimates of p, and the estimated N is an estimate of some superpopulation of individuals that is ever exposed to capture (Kendall et al., 1997). Equivalently, $\hat{N}$ applies to some effective sample area of the quadrat – the area over which individuals have been exposed to sampling, which is unknown. Spatial capture–recapture information allows us to resolve this issue. We can model the movement process explicitly and obtain estimates of absolute density that, in effect, are adjusted (by a model) for movement bias.

We present a spatially explicit, capture–recapture model for estimating density from area-search sampling under a conventional multiple sample capture–recapture study design. The development here follows Royle and Young (2008), who specified a hierarchical model in terms of individual activity centers (described subsequently) and a model of individual movement conditional on the activity centers. This movement model is, essentially, an explicit model of temporary emigration. We elaborate on this shortly. The hierarchical model is completed by specifying a model for the observations conditional on the location of individuals during each sample occasion. The modeling objective is to estimate the absolute density of individuals in the survey plot. Under the proposed model, this objective is accomplished by estimating the number of activity centers contained within the delineated sample unit. We emphasize Bayesian analysis by data augmentation to carry out analyses.

Before proceeding, we note that the basic hierarchical model structure of this 'area search' problem is similar to that considered in the distance sampling with measurement error problem. Conceptually, the distance sampling model was based on the joint specification $[y|x][u|x][x]$ for encounter data y, distance measurements u and actual distance x. In the present case the analogous model is $[y|u][u|x][x]$ for encounter data y, observed *location* u and 'activity center' x (we will depart from use of x to represent a two-dimensional vector but use it here to clarify the analogy between models). The main technical distinctions are that the latent variable here is bivariate (but related to distance by a transformation), and the model of encounter y is conditional on observations u and not x. The model has other structural similarities. For example, we adopt the use of a bivariate normal 'movement kernel' to relate u to x in this model, just as we did in the distance sampling model (where it was used as a model for the detection function). An important distinction is that, here, $[u|x]$ is a biological process related to movement, whereas in the distance sampling model it was purely an observation process (measurement error).

7.3.1 Model Formulation

We assume that N individuals are distributed over some region denoted by $\mathcal{S}$ and that each of these individuals has a fixed 'activity center', say $\mathbf{s}_i = (s_{1i}, s_{2i})$. An individual's movements are centered around $\mathbf{s}_i$ in a probabilistic way. Similar to concepts such as home range, territory, or utilization distribution, this activity center is not a precise biological concept. However, it is amenable to a precise mathematical definition that allows us to interpret repeated observations of individuals in space. We suppose that a region $\mathcal{X}$ is sampled J times generating encounter histories $\mathbf{y}_i = (y_{i1}, y_{i2}, \ldots, y_{iJ})$ on n individuals and auxiliary information on the individual location, $\mathbf{x}_{ij}$, the two-dimensional coordinate at which individual i exists during sample j. We regard $\mathbf{x}_{ij}$ as the outcome of a random variable that is observed whenever $y_{ij} = 1$, in which case $\mathbf{x}_{ij}$ is a capture location. The value of $\mathbf{x}_{ij}$ is missing whenever $y_{ij} = 0$. We regard $\mathbf{x}$ as the outcome of a movement process, and we imagine that $\Pr(y_{ij} = 1)$ should be dependent on $\mathbf{x}_{ij}$. We will formalize these notions shortly.

We suppose that $\mathcal{X}$ is a strict subset of $\mathcal{S}$, i.e., $\mathcal{X} \subset \mathcal{S}$. Thus, there may be individuals that can be encountered in the sample whose activity center is an element of $\mathcal{S}$ but not $\mathcal{X}$. We suppose that $\mathcal{X}$ is subjected to a uniform search intensity such that individuals located within $\mathcal{X}$ are susceptible to a constant probability of encounter. In the case of the lizard data, $\mathcal{X}$ is the 9 ha plot shown in Figure 7.1. The inference objective considered here is estimating the density of individuals – i.e., their activity centers – in any arbitrary polygon located within $\mathcal{S}$. It is convenient (but not necessary) to consider the polygon $\mathcal{X}$, which is the sampled area having size $A(\mathcal{X})$. We could as well consider a rectangle containing $\mathcal{X}$, the minimum convex hull, or even a subset of $\mathcal{X}$ itself. However, for our present purposes we define the absolute density of individuals in the survey plot $\mathcal{X}$ as follows:

$$D = \frac{1}{A(\mathcal{X})} \sum_{i=1}^{N} I(\mathbf{s}_i \in \mathcal{X}).$$

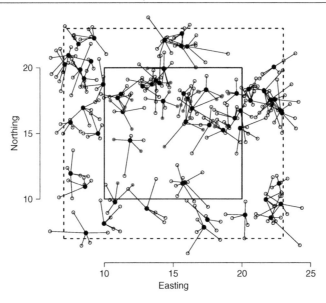

Figure 7.4. Simulated spatial capture–recapture data set for an 'area search' type of design. The surveyed area is the 10×10 sample plot nested within 16×16 quadrat, containing 60 individual activity centers (solid black circles) of which 24 are contained within the sample plot. All locations of each individual are marked with open black circles. Captures are indicated with red.

As an example of this system, consider Figure 7.4, which shows a surveyed quadrat (the smaller square) of dimension 10×10, which is $\mathcal{X}$, is nested within a larger quadrat $\mathcal{S}$, (the dashed polygon), a square of dimension 16×16. The activity centers of $N = 60$ individuals are indicated by the solid black circles. These were generated uniformly over $\mathcal{S}$ (the other symbols on the figure will be described shortly). We will provide an analysis of these simulated data below.

We require a model and a technical strategy to formalize the analysis of that model. To begin the technical development of a model for this system, we first outline our conceptual approach. As always, we require the joint distribution of the observations which we formulate here as a hierarchical model that includes the conditional distribution of the encounter data *given* the individual locations $[y|\mathbf{x}]$ and a model for the individual locations $[\mathbf{x}]$. We have a minor technical issue to confront, that being that there are only n observed individuals and some number, say $N - n$, which are unobserved (never captured). We pretend for a moment that we had observed all N individuals so that we can focus on writing down a model for a hypothetical complete data set, composed of a complete set of encounter histories.

7.3.1.1 The process model

Our reason for introducing the individual activity center $\mathbf{s}$ is because it simplifies the formulation of a model for the individual locations $\mathbf{x}$. That is, by thinking conditionally we are able to formulate apparently sensible models for $\mathbf{x}$ even though $\mathbf{s}$ is not observable. Following Royle and Young (2008), we assume that, conditional on $\mathbf{s}_i$, an individual moves about its activity center randomly according to some bivariate probability density function, $g(\mathbf{x}|\mathbf{s}, \theta)$. We assume here that the components of $\mathbf{x}$ are independent normal random variables, such that $\mathbf{x}_{ij}|\mathbf{s}_i, \sigma^2 \sim \mathrm{N}(\mathbf{s}_i, \sigma^2\mathbf{I})$. Absent the latent variable $\mathbf{s}$, it is not clear how to specify a model for $\mathbf{x}$ that would be clearly interpretable either in terms of movement or some other biological process.

There are a number of possibilities for extending this movement model, which might depend on the species and scale of the area. For example, we can allow for anisotropic movements by specifying

$$\boldsymbol{\Sigma} = \begin{pmatrix} \sigma_1^2 & 0 \\ 0 & \sigma_2^2 \end{pmatrix}$$

but we should probably have a priori hypotheses to support such structure. Also, in some instances a two-dimensional random walk might be appropriate. We think that in most capture–recapture surveys, it would be very difficult to estimate very complex models of movement due to data limitations.

Next, we require a model for the latent variables $\mathbf{s}_i$. We regard the N activity centers as the realization of a spatial point process. As such, we have considerable latitude in the conceptual formulation of models for $\mathbf{s}$. We think the development of general point process models in this context is an important research area. For now, we adopt the usual uniformity assumption (i.e., underlying distance sampling), wherein activity centers are uniformly distributed over $\mathcal{S}$:

$$\Pr(\mathbf{s}) = \frac{1}{A}; \quad \text{for } \mathbf{s} \in \mathcal{S},$$

where A is the area of $\mathcal{S}$.

Specification of $\mathcal{S}$ is necessary in this model because it represents, essentially, the prior distribution for $\mathbf{s}$. We will adopt an approach to analysis based on MCMC that requires simulation of the underlying point process, and we must provide the domain over which that point process is defined, $\mathcal{S}$. Also note that implicit in the nature of the sampling problem, there exists individuals outside of $\mathcal{X}$ that are exposed to sampling. They have to come from somewhere. Having said that, definition of $\mathcal{S}$ is not terribly important, provided that it is sufficiently large so that it includes the exposed population of individuals. This should be judged satisfactory if the posterior of model parameters is not truncated (in particular, the posterior of σ).

7.3.1.2 The observation model

To complete the model specification, we require an observation model that describes the manner by which the locations $\mathbf{x}_{ij}$ are observed, i.e., the conditional relationship between y_{ij} and $\mathbf{x}_{ij}$. The observation model is derived as follows. If $\mathbf{x}_{ij}$ is contained in $\mathcal{X}$ during sample j, then individual i is detected with probability p. Otherwise, the individual cannot be detected and $y_{ij} = 0$ with probability 1. That is, y_{ij} is a deterministic zero in this case. These two possibilities are manifest precisely in the following model for the observations.

$$
\begin{aligned}
y_{ij} &= 0 & \text{if } \mathbf{x}_{ij} \notin \mathcal{X} \\
y_{ij} &\sim \text{Bern}(p_{ij}) & \text{if } \mathbf{x}_{ij} \in \mathcal{X}.
\end{aligned}
\tag{7.3.1}
$$

We have the usual flexibility in modeling p_{ij} (for example, as a function of time, individual, or prescribed covariates).

7.3.1.3 Illustration: simulated data

Figure 7.4 shows an example of a realization of the process described by the model above, and the resulting pattern of observations. As noted previously, the sample plot is a square of dimension 10×10 units, and this plot is nested within a larger square of 16×16 units. We simulated $N = 60$ individuals (the large black dots) and subjected them to sampling on $J = 5$ occasions with $p = 0.25$. Movements were governed by the bivariate normal having $\sigma = 1$. Individual locations for each of the 5 samples are connected to the center of activity by lines.

In all, 24 of the 60 individuals had their center of activity within the sample quadrat. In this case, only 18 individuals were observed during sampling, of which 17 had their activity centers located within the plot. The 18 individuals were observed a total of 31 times. Simulating and fitting data of this sort provides great enjoyment, and so we provide the **R** instructions for simulating and plotting realizations of the process in the Web Supplement.

The statistical objective is to estimate the number of centers within the 10×10 sample quadrat when confronted only with the capture histories of the 18 individuals and the locations of the blue circles in Figure 7.4.

7.3.2 Bayesian Analysis by Data Augmentation

The model has a relatively simple hierarchical structure, being composed of a sequence of 3 probability distributions for observations y_{ij}, locations of individuals during each sample, $\mathbf{x}_{ij}$, and the activity centers $\mathbf{s}_i$. The joint distribution is the product of these 3 component models which is (omitting dependence on parameters) of the form $[y|x][x|s][s]$. Because the probability structure of the model (data, latent variables, and parameters) is well-defined, we can now proceed by applying probability calculus to this model, the formal basis for analysis and inference whether one adopts a Bayesian or classical likelihood approach. In particular, due to the simple hierarchical structure, it is easy to devise an MCMC algorithm to draw samples from the posterior distribution of the model parameters and the various latent variables of interest (primarily, the $s_1, s_2, \ldots, s_N$). This model is also easily analyzed using **WinBUGS** with the aid of data augmentation. We provide that implementation here and sketch out the major elements of a native **R** implementation shortly.

Data augmentation proceeds as in the individual covariate models described in Chapter 6. We augment the observed encounter histories with $M - n$ vectors of all-zero encounter histories: $y_i \equiv 0$; for $i = n + 1, \ldots, M$. For these individuals, we treat the x_{ij} locations as missing. In **R** we can pad the location information with NA to indicate missing values, and this gets treated by **WinBUGS** properly. The **WinBUGS** model specification is given in Panel 7.2. Note that, in the model specification, we check whether each **s** is included in the sampled quadrat using the step instruction in the BUGS language. This allows us to tally-up the number inside the box. This component of the **WinBUGS** model specification would be more difficult to analyze in **WinBUGS** for arbitrary polygons, but the model could easily be discretized to allow for that situation. Finally, note that the prior for the locations x_{ij}, a normal random effect, does not have compact support and so we have experienced problems with **WinBUGS** not choosing effective updating methods, as we described in Section 6.4.2. As such, we truncate the prior for movement by δ units below and above the boundaries of $\mathcal{S}$. In this case, $\delta = 5$ and so movements more than 5 units outside of $\mathcal{S}$ are not permitted.

7.3.3 Analysis of Simulated Data

The model was fit to the data set depicted in Figure 7.4. Recall that 24 of the 60 individuals had their center of activity within the sample quadrat, and, with $p = 0.25$, only 18 individuals were observed in the sample a total of 31 times. Using data augmentation, the model was fitted by MCMC in **WinBUGS** (see Panel 7.2).

Posterior summary statistics are shown in Table 7.2. The posterior mean (95 percent interval) of $N(\mathcal{X})$ was 23.45 (18, 33) which, while relatively diffuse as expected, works out to be centered close to the truth (in this case, 24). The estimates (posterior means) of detection probability and σ were 0.286 and 0.902, respectively.

```
model {

psi~dunif(0,1)
sigma~dunif(0,10)
tau<-1/(sigma*sigma)
p~dunif(0,1)
delta<-5
for(i in 1:(nind+nz)){
  mu1[i]~dunif(S1l,S1u)
  mu2[i]~dunif(S2l,S2u)
  w[i]~dbern(psi)

  for(j in 1:J){
     # determines if point is in box
     in1[i,j]<-step(S1[i,j]-(S1l+3))
     in2[i,j]<- step( (S1u-3) - S1[i,j])
     in3[i,j]<-step(S2[i,j] - (S2l+3))
     in4[i,j]<- step( (S2u-3)-S2[i,j])
     inplot[i,j]<-in1[i,j]*in2[i,j]*in3[i,j]*in4[i,j]

     muy[i,j]<-inplot[i,j]*w[i]*p
     S1[i,j]~dnorm(mu1[i],tau)I(S1l-delta,S1u+delta)
     S2[i,j]~dnorm(mu2[i],tau)I(S2l-delta,S2u+delta)
     Y[i,j]~dbern(muy[i,j])
  }
  }
for(i in 1:(nind+nz)){
  tmp1[i]<- step(mu1[i] - (S1l+delta))
  tmp2[i]<- step((S1u-delta) - mu1[i])
  tmp3[i]<- step(mu2[i] - (S2l+delta))
  tmp4[i]<- step((S2u-delta) - mu2[i])
  centerin[i]<-tmp1[i]*tmp2[i]*tmp3[i]*tmp4[i]
  realind[i]<- w[i]*centerin[i]
}
  NX<-sum(realind[1:(nind+nzeroes)])
  D<-NX/9
}
```

Panel 7.2. **WinBUGS** implementation of spatial model for the lizard data. The parameters of interest, $N(\mathcal{X}) = \sum_{i=1}^{N} I(\mathbf{s}_i \in \mathcal{X})$ and $D = N(\mathcal{X})/9$ are derived parameters, and are calculated by evaluating whether each $\mathbf{s}_i$ is located within $\mathcal{X}$.

Table 7.2. Parameter estimates for the simulated data in which $N = 60$, $\sigma = 1$, $p = 0.25$, and $N(\mathcal{X})$, the number of individuals within the surveyed plot, was 24.

parameter	mean	sd	2.5%	50%	97.5%
N	53.800	12.810	34.000	52.000	84.000
$N(\mathcal{X})$	23.450	4.063	18.000	23.000	33.000
ψ	0.456	0.116	0.268	0.443	0.717
p	0.286	0.060	0.174	0.284	0.408
σ	0.902	0.054	0.805	0.899	1.015

7.3.4 Analysis of the Lizard Data

We now consider the flat-tailed horned lizard data collected on a 9 ha plot of dimension 300 m × 300 m. The focus of the study was to evaluate methods of estimating density and occupancy using several analytic techniques (Young and Royle, 2006). For this plot, there were 14 capture occasions over 17 days (14 June to 1 July 2005). A total of 68 individuals were captured 134 times. The distribution of capture frequencies was (34, 16, 10, 4, 2, 2) for 1 to 6 captures respectively, and no individual was captured more than 6 times. The plot boundaries in a scaled coordinate system along with the capture locations are shown in Figure 7.1.

The model was fit in **WinBUGS** using the specification given in Panel 7.2. Two Markov chains were run for 20000 iterations, after discarding 1000, and using a thinning rate of 2. Posterior summaries of the net 20000 posterior samples are given in Table 7.3. In this table, $N(\mathcal{X})$ is the number of activity centers within the survey plot boundaries. We estimate that approximately 77 individuals have home range centers within the 9 ha study plot. The posterior mean density is $D = N(\mathcal{X})/9 = 8.62$ lizards/ha, with a 95 percent Bayesian confidence interval of $(7.556, 10.110)$. The posterior mean of the spatial movement parameter was 0.187. Recall that the plot was scaled to a 3×3 unit square, so $\sigma = 0.187$ represents about 6.25 percent of the dimension, or about 19 m relative to the original dimension of 300 m. The parameter ψ is the complement of the zero-inflation parameter described in Section 7.1.4. That is, it determines the inclusion probability for a member of the augmented list of size M. We note that its posterior mass is concentrated away from the boundary $\psi = 1$, and so we judge that sufficient pseudo-individuals were added to the data set.

It is useful to put the estimate of $N(\mathcal{X})$ into context with that obtained using typical closed population methods, which we applied to these data in previous chapters. The estimate of N under Model M_0 (Section 5.3.2) yielded $\hat{N} = 82.02$ and $\hat{p} = 0.117$ whereas the logit-normal heterogeneity model (Section 6.2.1) yielded $\hat{N} = 106.8$. These estimates are consistent with the expected influence of temporary emigration (that is, they should be estimating the size of some population that

Table 7.3. Parameter estimates for the lizard data. $N(\mathcal{X})$ is the number of home range centers located within the 9 ha study plot whereas N is the population size of individuals in $\mathcal{S}$ of which the surveyed area $\mathcal{X}$ is a formal subset.

parameter	mean	sd	2.5%	50%	97.5%
N	204.100	22.050	165.000	203.000	251.000
$N(\mathcal{X})$	77.600	5.890	68.000	77.000	91.000
D	8.622	0.654	7.556	8.556	10.110
ψ	0.641	0.074	0.505	0.637	0.794
p	0.125	0.013	0.100	0.125	0.152
σ	0.187	0.012	0.164	0.187	0.212

applies to an area larger than the nominal 9 ha unit). In the present case, we see that the estimate of $N(\mathcal{X}) = 77.6$ under the spatial model (Table 7.3) is somewhat lower than that obtained under the closed population model, and considerably lower than that obtained under the logit-normal version of Model M_h.

7.3.5 Effective Sample Area

There is a formal relationship between effective sample area A_e, abundance and density. We feel that at the present time many ecologists perceive methods as being distinct based on whether they provide an estimate of A_e, or density, or N, or some combination (or something altogether different). However, any conceptually coherent procedure should imply a clear relationship between these quantities via the specified model.

To motivate the calculation of effective sample area, suppose that $\mathcal{S}$ was discrete, being composed of a large number of pixels of unit area $s_k(k = 1, 2, \ldots)$. Suppose further that each pixel had associated with it a binary indicator variable z_k such that $\phi_k = \Pr(z_k = 1)$. Prior to sampling we flip a coin for each pixel, having success probabilities ϕ_k, and each pixel for which $z_k = 1$ is sampled; therefore, in this case, the sampled area (in units of pixels) is the outcome of a stochastic process, being the sum of the z_k indicators:

$$A_e = \sum_k z_k.$$

Consider a 10×10 grid, flip a coin 100 times, and all of the grid cells that come up 'heads' are included in the sample. The sampled area is the sum of the area associated with those grid cells. Naturally we might estimate this quantity by its expected value

$$\hat{A}_e = \sum_k \phi_k.$$

This is the expected or effective sample area. In the 10×10 grid example, if the 100 coins flips all have $\phi_k = 0.5$, then the effective sample area during each sample is 50. That is, you expect to count individuals from 50 grid cells during the survey. An alternative way to frame this concept of effective sample area is to suppose that $s_k; (k = 1, 2, \ldots)$ are the activity centers of hypothetical lizards and each lizard flips a coin with probability ϕ_k to determine whether it will be located within the sampled plot. Whether it is lizards that are flipping coins to decide their exposure to sampling, or biologists flipping coins to decide whether a pixel will be sampled, both situations yield the same conceptual formulation of the effective sample area. Note that $1 - \phi_k$ is interpretable as the temporary emigration probability (Kendall et al., 1997).

Now, suppose that the sampling will be conducted on more than one occasion, say $J = 2$ times. In this case, then a lizard will be exposed to sampling if $z_k = 1$ in either (or both) samples. Thus, the lizard is exposed with probability $\epsilon_k = 1 - (1 - \phi_k)^2$. In this case, the effective sample area is

$$\hat{A}_e = \sum_k \epsilon_k.$$

We see that the effective sample area is not just a function of ϕ_k, but also the number of times that the population is exposed to sampling. Effective sample area increases with J, the number of temporal samples.

This concept extends directly to the continuous space situation. That is, the effective sample area is the integral:

$$A_e = \int_s \epsilon(s) \mathrm{d}s,$$

where $\epsilon(s) = \mathrm{Pr}(\text{exposed to sampling}|s)$. As before, 'exposed to sampling' is the event that $x \in \mathcal{X}$ *at least once* during the J samples. Let $\phi(s) = \mathrm{Pr}(\mathbf{x} \in \mathcal{X})|\mathbf{s})$; then if the J samples are independent,

$$\epsilon(s) = 1 - (1 - \phi(s))^J.$$

We emphasize that the probability of exposure, and hence the effective sample area, both depend on the effort expended sampling – the number of samples, J. We do not believe that this is widely appreciated in spatial sampling problems. Note also that $\mathrm{Pr}(\mathbf{x} \in \mathcal{X}|\mathbf{s})$ is a function of the parameter σ, and so the effective sample area is also a function of the movement process of the species being sampled, as we would expect. An individual has more exposure to sampling as its movement distribution becomes concentrated in the surveyed region which is influenced by both σ and $\mathbf{s}$. For example, an individual having $\mathbf{s}$ right in the middle of the quadrat should have exposure nearly equal to 1.00, whereas an individual's exposure should decrease to 0 as **s** moves away from the sampled area. Exposure to sampling is also related to the total sampling effort (J), such that more effort will increase exposure. Thus, with $\sigma = 0.187$ we would not expect to capture an individual one-half unit from the sample plot in 1 or 2 samples, but with $J = 14$, we should have a reasonable chance of capturing the individual.

Figure 7.5 shows the effective sample area of the 9 ha quadrat under various values of σ. The color of a pixel (let **s** index the pixel) was obtained by placing a normal density centered at **s** having prescribed σ. Then, the normal pdf was evaluated at a fine grid of points and those points within the grid box were summed and divided by the overall total. The A_e shown in this figure is in units of ha, and so the estimate of A_e for $\sigma = 0.187$ is 12.84 ha. We note that it would be easy to obtain a characterization of uncertainty associated with A_e within a Bayesian framework, as A_e is just a function of parameters for which we obtain posterior samples.

The quantity $\phi(\mathbf{s})$ can be computed easily, as the solution to the integral:

$$\phi(\mathbf{s}; \sigma) = \int_x I(\mathbf{x} \in \mathcal{X})[\mathbf{x}|\mathbf{s}]\mathrm{d}\mathbf{x}. \qquad (7.3.2)$$

This is the overlap of a bivariate normal kernel with the region $\mathcal{X}$, which can be computed using `pnorm` in **R** (if $\mathcal{X}$ is a simple polygon). In general, we find it efficient to apply a discrete approximation of the normal integral, by evaluating the kernel on a large grid of points and them summing them up within the box, and in total.

7.3.6 Remarks

The relevance of the buffer $\mathcal{S}$ around the sampled region $\mathcal{X}$ is apparent from Eq. (7.3.2). In order for this integral to be unaffected by the size of $\mathcal{S}$, it must be large enough so that $\Pr(\mathbf{x} \in \mathcal{X}|\mathbf{s})$ is negligible for any $\mathbf{s} \notin \mathcal{S}$. As we have noted, $\mathcal{S}$ is, in effect, a prior distribution on **s** – the potential origin of captureable individuals. Thus $\mathcal{S}$ must be sufficiently large so as to contain all potentially captureable individuals. Prescribing a particular $\mathcal{S}$ can be thought of as truncating the prior distribution on **s**, which should not be too detrimental provided that $\mathcal{S}$ is large relative to σ. As $\mathcal{S}$ becomes large relative to σ, there will be no effect on estimates of $N(\mathcal{X})$ and other parameters due to this truncation of the prior. This concept also applies to the data augmentation idea, where we impose a uniform $[0, M]$ prior on N.

Under the spatial capture–recapture model described here, effective sample area A_e is a by-product of the model. That is, it is a function of the canonical model parameters. This is one nice feature of the model-based formulation – it produces

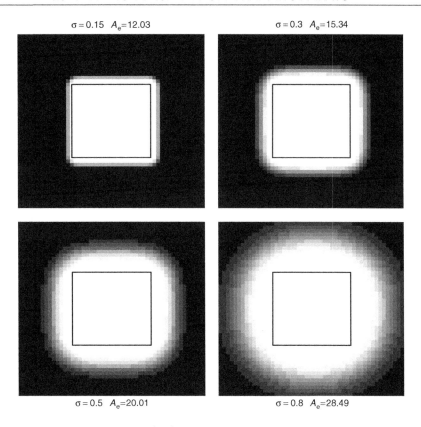

Figure 7.5. Effective sample area (A_e) of a 9 ha plot for various values of σ. Units of A_e are hectares (ha). The grayscale scheme in all cases is about 1.0 for the completely white and 0.0 for the completely black.

estimates of abundance, density, and effective sample area that are equivalent representations of the information in the data, under the prescribed model.

7.4 ESTIMATING DENSITY FROM TRAPPING ARRAYS

The use of trapping arrays for estimating abundance and density is widespread throughout animal ecology. Historically, trapping arrays were composed of baited traps in which individuals were captured, marked, and then subjected to repeated sampling in a conventional capture–recapture study. The classical example is found in small-mammal trapping where grids of live traps are used. A good example are the *Microtus* data which we introduced in Chapter 5. More recently, with the advent of DNA analysis technology, the use of hair-snares or marking obtained from scat sampling has become popular in many carnivore surveys, especially of bears. In addition to genetic marking, camera trapping is quite popular for certain carnivores such as tigers (Karanth, 1995; Karanth and Nichols, 1998; Karanth et al., 2006), ocelots (Trolle and Kéry, 2003; Trolle and Kéry, 2005), jaguars (Wallace et al., 2003; Maffei et al., 2004; Silver et al., 2004; Soisalo and Cavalcanti, 2006) and other species with unique markings. In such cases, the arrays may be irregular, but the basic problem of estimating density is relevant regardless of the method or geometry of the trapping grid.

The conventional approach to the analysis of density from these systems is to apply closed population models, as we did in Chapters 5 and 6, and then attempt to convert those estimates to densities using a wide array of heuristically motivated but essentially ad hoc methods (Dice, 1938; Hansson, 1969; Otis et al., 1978; Wilson and Anderson, 1985b,c; White and Shenk, 2001). The conventional view of the matter is expressed concisely by Trolle and Kéry (2003) who, in the analysis of camera trapping data on ocelots, wrote:

> To convert the estimate of population size into an estimate of density, we followed the procedure adopted by Karanth and Nichols (1998). We first calculated a core area as the minimum convex polygon defined by all trapping stations. This core area was unlikely to contain the entire home range of all trapped ocelots. Instead, it is likely that some ocelots had home ranges that extended beyond the core area. To account for that, we added a boundary strip to obtain the total area from which our animals were taken. Strip width was given by half the mean maximum distance moved by ocelots caught on more than 1 trap. This ad hoc approach has little theoretical justification, but it appeared to work well in simulation studies of Wilson and Anderson (1985), and it was the only available means of estimating boundary strip width in our situation. Density was then obtained by dividing the population size estimate by the estimated total area.

We agree that these procedures are ad hoc in the sense that the model or range of conditions for which they might work is poorly understood and difficult to characterize theoretically, and there is little basis for their extension.

In this section, we describe the basic trapping array problem and then its rendering as a hierarchical model. As with the previous situations considered in this chapter, there is a model for locations of individuals – their activity centers, and a model describing individual exposure to traps (in the previous section, this model for exposure was induced by an explicit movement model). When formally described by a hierarchical model, density estimation from trapping arrays is amenable to analysis either using Bayesian or classical likelihood methods. As was the case in the previous section, when properly specified in terms of spatial organization of individuals, their movements, and their observation, there is a formal relationship among the effective trap area, temporary emigration, abundance, and density. The material presented here comes from Royle et al. (2008).

7.4.1 Model Formulation

As in the previous section, we suppose that each individual in the population has a center of activity, $\mathbf{s}_i = (s_{1i}, s_{2i})$. We suppose that the population of N centers $\mathbf{s}_i; i = 1, 2, \ldots, N$ are distributed uniformly over some region, say $\mathcal{S}$. We will prescribe $\mathcal{S}$ (e.g., by specifying coordinates of some polygon that contains the sample locations). As an example, consider Figure 7.7. This figure shows a 10×10 array of traps with unit spacing (the black dots) within some hypothetical region bounded by a square of dimension 18×18 units, which is shown by the dashed boundary. This large square is $\mathcal{S}$.

Let $\mathcal{X}$ denote any arbitrary polygon within $\mathcal{S}$. One inference objective is to estimate the number of points contained within $\mathcal{X}$. For example, in practice, $\mathcal{X}$ is commonly defined as the convex hull containing the traps or it could be the minimum area rectangle, or a prescribed region of suitable habitat (Royle et al., 2008), etc. It could be $\mathcal{S}$, which is itself some arbitrarily large polygon that contains the trapping array.

Sampling is carried out by a network of K traps, having locations $\{\mathbf{x}_k; k = 1, 2, \ldots, K\}$. We suppose that the probability of an individual being captured is a function of the distance from the trap to its activity center, and one or more parameters θ.

The observations generated from a trapping array are the encounter histories $\mathbf{y}_i = (y_{i1}, y_{i2}, \ldots, y_{iJ})$ and the corresponding trap identities $\mathbf{h}_i = (h_{i1}, h_{i2}, \ldots, h_{iJ})$. For convenience, we will define $h_{ij} = 0$ whenever $y_{ij} = 0$. Formally, let h_{ij} be an integer $0, 1, 2, \ldots, K$ indicating the trap identity in which individual i was captured during sample j. Thus, $h_{ij} = 0$ indicates no capture of individual i at sampling occasion j, and non-zero values indicate the trap location at which the animal is caught at occasion j. As an illustration, suppose a population is sampled by a 10×10 array of traps numbered 1 to 100, then we could have encounter histories such as, for $J = 6$, $(0, 9, 10, 0, 0, 9)$ for an individual captured in trap 9 (during samples 2 and 6) and trap 10 (during sample 3). The observations h_{ij} aggregate the encounter history information (the binary y_{ij} observations) and the spatial location-of-capture information, in the form of trap identity.

The model can be formulated in terms of a conditional model for trap-of-capture given y_{ij}, say $[h_{ij}|y_{ij}]$ and then the marginal probability $[y_{ij}]$. Note that $\Pr(h_{ij} = 0|y_{ij} = 0) = 1$ and so it suffices to describe the conditional probability mass function of $h_{ij}|y_{ij} = 1$, which is the probability of capture in any trap k given capture, and the probability of capture $\Pr(y_{ij} = 1)$. Both of these should depend on the activity center of individual i, and we develop a specific form of dependence shortly.

The h_{ij} observations constitute a categorical random variable having probabilities $\boldsymbol{\pi}_i$ which are conditional on the latent variable $\mathbf{s}_i$. The vector $\boldsymbol{\pi}_i$ has length $K + 1$, the '0 cell' corresponding to the event 'not captured'. We will denote the observation model, conditional on $\mathbf{s}_i$, by:

$$h_{ij}|\mathbf{s}_i, \theta \sim \mathrm{Cat}(\boldsymbol{\pi}_i(s_i, \theta)). \tag{7.4.1}$$

A categorical random variable is equivalent to a multinomial trial – a multinomial observation based on a sample of size 1, having cell probabilities $\boldsymbol{\pi}_i$. To make this equivalence, we construct a vector of length $K + 1$ having a single '1' for element h_{ij}, with the remaining elements being zero. The vector constituted in this way is a multinomial trial. Then, for $j = 1, 2, \ldots, J$ replicate observations, the resulting trap frequencies are multinomial based on a sample size of J, and cell probabilities $\boldsymbol{\pi}_i$.

The multinomial cell probabilities can be factored according to

$$\pi_{ik} = \Pr(\text{captured in trap } k | \text{captured}, \mathbf{s}_i) \Pr(\text{captured}|\mathbf{s}_i)$$

which are verbal expressions of the two components $[h|y = 1]$ and $\Pr(y = 1)$ identified previously. In what follows, we describe these model components in terms of models for both 'process' and observation. The process component of the model is described in terms of individual exposure to trapping. This exposure process plays the role of the movement process in the area search problem considered in the previous section. The observation component of the model is described in terms of a Bernoulli detection component, conditional on exposure.

7.4.1.1 Modeling exposure to traps

We suppose that an individual's *exposure* to a trap k is described by some function $g(||\mathbf{s}_i - \mathbf{x}_k||; \sigma)$ which depends on the distance between the individual's center of activity $\mathbf{s}_i$ and the trap location $\mathbf{x}_k$. In the analyses presented below, we use the normal kernel centered about $\mathbf{s}_i$ and having scale parameter σ^2, so that

$$g(||\mathbf{x}_k - \mathbf{s}_i||; \sigma) = \exp\{-||\mathbf{x}_k - \mathbf{s}_i||^2/\sigma^2\}.$$

An individual's total exposure to the trap array is:

$$e_i(\sigma, \mathbf{s}_i) = \sum_k g(||\mathbf{x}_k - \mathbf{s}_i||; \sigma).$$

The net probability of exposure of an individual should be proportional to $e_i(\sigma, \mathbf{s}_i)$,

$$\phi_i = \Pr(\text{exposed to trapping}|\mathbf{s}_i) \propto e_i(\sigma, \mathbf{s}_i).$$

There is no unique way to parameterize this relationship. Royle et al. (2008) and Gardner et al. (2008) used slightly different formulations. Royle et al. (2008) defined this total exposure probability relative to that of some hypothetical maximum exposed individual. That is, given σ, there exists a location $\mathbf{s}_0$ that has the highest total exposure to trapping. (This could also be the single individual in the population of N individuals that has the highest exposure; it doesn't matter except in how the parameters of the model are interpreted.) The maximum exposure to the array, taken across all individuals in the population, will be denoted by $e_0(\sigma) = \max_i(e_i)$, which is a function of σ and the trapping array configuration (their number and array geometry). Then, define

$$\phi_i(\sigma, \mathbf{s}_i) = \frac{e_i(\sigma, \mathbf{s}_i)}{e_0(\sigma)}.$$

To illustrate how ϕ_i varies spatially relative to a trapping array, Figure 7.6 shows a 10×10 grid of traps having unit spacing. The exposure kernel is the normal with $\sigma = 2$, and the image plot shows the exposure of an individual as a function of $\mathbf{s}$ in two-dimensional space. We see that individuals within the array have very high total exposure to the array, and the total exposure decreases rapidly as distance from the outer band of traps increases. Integrating the surface shown in Figure 7.6 yields the effective trap area of the array during a single sample ($J = 1$). In this case, the effective area is 100.0014, compared with the nominal area of 81 which contains the traps.

7.4.1.2 The capture process

Conditional on an individual being exposed to trapping, we assume constant probability of capture,

$$\Pr(\text{capture}|\text{exposure}) = p_0,$$

where p_0 is a parameter to be estimated. Thus, the apparent probability of capture for an individual having activity center $\mathbf{s}_i$ is

$$p_i = p_0 \phi_i(\sigma, \mathbf{s}_i). \tag{7.4.2}$$

We see that heterogeneity in apparent capture probability is induced by heterogeneity in exposure through ϕ_i. The heterogeneity in capture probability induced by spatial location relative to the trapping array can thus be described as a scaling (by p_0) of Figure 7.6.

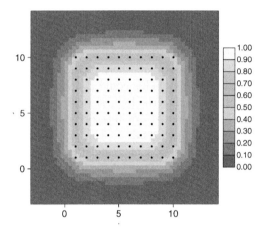

Figure 7.6. Probability of exposure of individuals to a 10 by 10 grid of traps as a function of their center of activity, s.

7.4.1.3 The observation model

Given Eq. (7.4.2), we can obtain two results. First, let y_i be the total number of times that individual i was captured,

$$y_i = \sum_{j=1}^{J} y_{ij}.$$

Then (under independence of captures), the total number of captures is a binomial random variable with success probability that varies by individual, depending on that individual's location relative to the trapping array, according to:

$$y_i \sim \text{Bin}(J, p_i(p_0, \sigma, \mathbf{s}_i)),$$

where $p_i(p_0, \sigma, \mathbf{s}_i)$ is the net probability of capture for individual i, from Eq. (7.4.2). Second, we can write

$$p_i = \frac{p_0}{e_0} \sum_{k} g(||x_k - s_i||; \sigma).$$

Therefore (i.e., from the Law of Total Probability), the multinomial cell probabilities from Eq. (7.4.1) are

$$\pi_{ik} = \frac{p_0 g(||x_k - s_i||; \sigma)}{e_0} \quad \text{for } k = 1, 2, \dots, K$$

$$\pi_{ik} = 1 - p_i \quad \text{for } k = K + 1$$

(recall, the last cell $K + 1$ corresponds to not captured). As such, the probability of capture in trap k conditional on capture, is

$$\gamma_{ik} = \frac{\pi_{ik}}{p_i} \quad \text{for } k = 1, 2, \ldots, K$$

$$= \frac{g(||x_k - s_i||; \sigma)}{e_i(\sigma, \mathbf{s}_i)}.$$

We emphasize that there are K of these cell probabilities, having lost 1 to conditioning.

In the Bayesian analysis described subsequently, either the unconditional multinomial (having $K + 1$ probabilities π_{ik}) or the conditional-on-capture multinomial (having K probabilities γ_{ik}) can be easily implemented. Programs in **R** as well as **WinBUGS** are available in the Web Supplement.

7.4.2 Example

Figure 7.7 shows an example of a realization of observations generated according to the considerations laid out previously. The trap array consists of 100 traps organized in a 10×10 array of unit spacing (traps are indicated by black dots). The array is nested within a larger square of dimension 18×18 over which $N = 200$ individual centers of activity were generated uniformly (the red dots). The inference problem is to estimate how many red dots are located within the 10×10 grid, noting that we can only observe the black dots, some of which correspond to red dots outside of the grid. To be precise, we will define $\mathcal{X}$ as the minimum-area-rectangle that contains the trapping grid. This is a square of dimension 9×9 units.

The detection kernel was the exponential having $\sigma = 2$ and $p_0 = 0.30$. For a simulated study with $J = 5$ periods, 63 unique individuals were captured a total of 95 times, the detection frequencies being $(38, 18, 7)$ individuals captured 1 to 3 times, respectively (no individuals were caught >3 times). In Figure 7.7 all captured individuals are connected to the traps in which they were captured by lines. There were 49 individuals located within the 10×10 trapping grid and 37 of those were captured. In this chapter, we have emphasized that spatial juxtaposition of the sampling apparatus with individual home ranges and movements induces heterogeneity in encounter probability. In this example, the summary statistics for p_i for the 200 individuals were $(0.001, 0.025, 0.095, 0.192)$ for first quartile, median, mean, and third quartile, respectively. Evidently, this detection kernel yields fairly extreme heterogeneity and low individual detection probabilities.

Fitting the standard closed-population model (M_0) to the observed sample of 63 individuals, yields an estimate of $\hat{N} = 90.5$, and the estimated p is about 0.21.

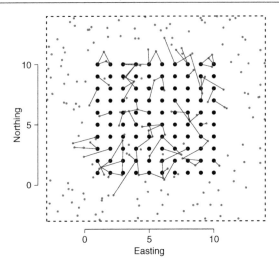

Figure 7.7. Simulated realization of trapping grid with captured individuals. Simulated captures of individuals (red dots) were made by 10×10 grid of traps (black dots). The trap(s) in which each individual was captured are indicated with blue lines.

The problem is that we don't know the effective trap area of the grid and so this estimate of $\hat{N}$ is, ostensibly, little more than an index. The naive density based on $\hat{N}$ and the physical area of the trapping array is $90.5/81 = 1.117$, whereas the actual known density is $200/(17^2) = 0.692$.

7.4.3 Analysis of the Model

As with all previous applications of closed population models in this book, the model is amenable to a relatively simple analysis by MCMC and implementation in **WinBUGS** using data augmentation. This is straightforward except that calculation of e_0 is something of a computational burden. We can avoid this by finding an upper bound for e_0, say e^*, perhaps by trial and error. Then, we pass **WinBUGS** this value as data which then precludes having to compute e_0 every time σ is updated. This merely induces a reparameterization so that the detection probability of the most exposed individual is $p_0(e_0/e^*)$, not p_0 as described in the development of the model. This reparameterization may be necessitated by practicality in problems with many individuals or large trapping grids. A **WinBUGS** implementation of this model is shown in Panel 7.3 and a similar, but more efficient formulation was used in the analysis of Gardner et al.

(2008). In this model specification, `emax` is the value e^* referred to above, and the indicator variables induced by data augmentation are given by w_i for individuals $i = 1, 2, \ldots, M$.

To obtain estimates of exposed population, the number of individuals within any arbitrary polygon $\mathcal{X}$, or density, we must tally-up the relevant subset of points $\mathbf{s}$ at each step of the MCMC algorithm. In Panel 7.3, the line

```
N<-sum(w[1:(nind+nzeroes)])
```

computes the total number of points within $\mathcal{S}$. To define a polygon that is a subset of $\mathcal{S}$ (which is a square in this case) can be difficult or at least tedious in **WinBUGS** unless it is a regular polygon. In the analysis of the tiger data, we also estimated the number of activity centers within the minimum area rectangle containing the trapping array, say $N(\mathcal{X})$. Whether or not each point $\mathbf{s}_i$ is in that polygon can easily be accomplished with standard point-in-polygon algorithms.

Royle et al. (2008) have devised a discrete-space implementation of this model in which $\mathcal{S}$ is approximated by an arbitrarily fine set of points. In that case, a reasonably efficient MCMC algorithm can be devised in **R**, based on data augmentation (which we expect to be available in the Web Supplement). We augment the data set with $M - n$ pseudo-individuals, and associated latent indicator variables w_i. Then, the basic structure of the algorithm is very similar to that outlined in Section 7.3. The main component is that we have to update each $\mathbf{s}_i$. There is a distinction between those corresponding to $w_i = 1$ and $w_i = 0$. In the former case, draws of $\mathbf{s}$ can be made easily using a Metropolis step (the full conditional is proportional to a multinomial likelihood kernel). In the latter case, we draw $\mathbf{s}$ from the uniform prior. The remaining steps of this algorithm are similar to previous models.

7.4.4 Analysis of the Tiger Camera Trapping Data

Camera trapping is now widely used to study carnivore populations that can be distinguished by stripe or spotting patterns (O'Brien et al., 2003; Trolle and Kéry, 2003; Wallace et al., 2003; Karanth et al., 2004; Kawanishi and Sunquist, 2004; Silver et al., 2004; Wegge et al., 2004; Trolle and Kéry, 2005; Soisalo and Cavalcanti, 2006). We provide an analysis of the tiger camera trapping data that was introduced in Chapter 5. The analysis used here follows closely to that presented in Royle et al. (2008). These data are from surveys of tigers in the Nagarahole reserve in the state of Karnataka, southwestern India. Tiger stripe patterns are unique, and individuals are readily identified from photographs. This population has been studied via camera trap methods by Karanth and associates since 1991 (e.g., Karanth (1995); Karanth and Nichols (1998); Karanth et al. (2006)). The data used here were collected from 24 January to 16 March 2006 from sampling at 120 trap stations (see Figure 7.8). In this figure, the minimum area rectangle that encloses the array is shown. We will denote this polygon by $\mathcal{X}$ and provide an estimate of $N(\mathcal{X})$ and the density, D over $\mathcal{X}$.

We provide here an analysis of the model using the native **R** implementation (Royle et al., 2008) based on approximating $\mathcal{S}$ with a fine grid of approximately $10\,000$ potential activity centers having spacing of approximately one-third of a km. The analysis by data augmentation used 800 augmented encounter histories, and thus $M = 855$. The minimum area rectangle containing the 120 camera trap locations was 679.4 km^2. For comparison, the convex hull around the trap array has area approximately 462.5 km^2. The estimates are summarized in Table 7.4.

```
model {
    psi~dunif(0,1)
    sigma~dunif(0,5)
    p0~dunif(0,1)

    for(i in 1:(nind+nz)){
        x0g[i]~dunif(Xl,Xu)
        y0g[i]~dunif(Yl,Yu)
        w[i]~dbern(psi)
        capprob[i]<- p0*total.exposure[i]/emax
        mu[i]<-w[i]*capprob[i]
        y[i]~dbin(mu[i],J)

        for(j in 1:ngrid){
            dist2[i,j]<- ( pow(x0g[i]-grid[j,1],2)  +  pow(y0g[i]-grid[j,2],2) )
            exposure0[1,j]<- exp(-dist2[i,j]/sigma)
            condlp[i,j]<- exposure0[i,j]/total.exposure[i]
        }
        total.exposure[i]<-sum(exposure0[i,1:ngrid])
    }
    for(i in 1:nind){
        for(t in 1:y[i]){
            h[i,t] ~ dcat(condlp[i,1:ngrid])
        }
    }
    N<-sum(w[1:(nind+nz)])
}
```

Panel 7.3. WinBUGS model specification for the trapping grid model fit to the tiger data. The model is based on a conditional factorization of the multinomial cell probabilities in which the number of detections for each individual is a binomial random variable, and the trap of capture is a conditional multinomial as described in the text. In this specification, emax is supplied as data and must be greater than the maximum total exposure at any location in $\mathcal{S}$.

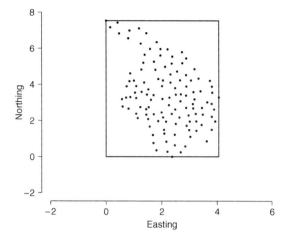

Figure 7.8. Tiger camera trapping array, composed of 120 traps in the Nagarahole reserve in the state of Karnataka, southwestern India. A unit of distance on this map is 5 km. The minimum area rectangle is shown enclosing the trap array.

Table 7.4. Posterior summaries of model parameters for the tiger camera trapping data. Here, A_e is the area exposed to trapping (in $J = 12$ samples) – or the effective sample area of the 120 trap array. It is a derived parameter under the model. $N(\mathcal{X})$ is the number of activity centers located within the minimum area rectangle having area 679.4 km^2, and D is the density per 100 km^2. The number of unique individuals observed was 45. ψ is the zero-inflation parameter introduced by data augmentation.

parameter	mean	sd	2.5%	median	97.5%
ψ	0.263	0.051	0.176	0.258	0.377
A_e	704.847	26.934	655.124	704.159	759.319
$N(\mathcal{X})$	91.852	15.928	66.000	90.000	129.000
D	13.511	2.345	9.715	13.247	18.988
σ	0.586	0.071	0.467	0.579	0.743
p	0.123	0.024	0.081	0.122	0.172

As we have noted previously, one benefit of a well-defined probability model for spatial capture–recapture is that it produces a cohesive summary of abundance and effective sample area. The latter is a derived parameter, being equal to the probability that an individual at **s** is exposed to sampling during the J samples, integrated over **s**. This is calculated at each iteration of the MCMC algorithm by evaluating the bivariate normal pdf describing movements at every point on the grid, and then calculating the exposure of each 'grid cell' to the trap array, and then summing over **s**. For the tiger data, the probability of exposure as a function

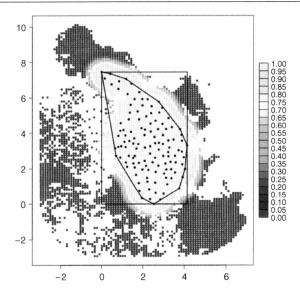

Figure 7.9. Probability of exposure to trapping in a $J = 12$ sample camera trapping study. The color of each pixel is the $\Pr(\text{exposure to sampling})$ of a tiger with activity center at that pixel. Only pixels judged to be suitable habitat are included. The sum of all pixels is the effective sample area of the trapping array. Reproduced from Royle et al. (2008).

of location is shown in Figure 7.9 (taken from Royle et al. (2008)). The sum of these exposure probabilities is 704.847.

The posterior mean of σ is 0.586 – the typical distance moved, from center of range, between samples, is about 2.5 km and 95 percent of the movements are less than about 5 km from the center of activity. The estimated density is about 13.5 tigers per 100 km². For comparison, if we use the estimate of N from model M_h reported in Chapter 6, which was 111.7, along with the estimated effective area, the computation yields a density estimate of 10.04 tigers per 100 km². As we noted previously (in the lizard example), the relationship between $\hat{N}$ under the heterogeneity model and effective area is only heuristic, and not precise mathematically. Thus, we should not expect equivalence of the two estimates.

7.5 SUMMARY

In this chapter, we considered spatial capture–recapture models. These are classical capture–recapture situations in which the spatial organization of the sampling apparatus yields, in addition to simple binary indicator of capture for each individual, the spatial location of capture. These methods include distance sampling, as well as a wide variety of trapping array or trapping grid situations. We discussed also a conceptual intermediate type of sampling based on an area search, wherein individuals within some prescribed area are subject to a uniform search intensity. Hierarchical models provide something of a conceptual, technical, and practical unification of this broad class of spatial capture–recapture models. These models are variations of individual covariate models (Chapter 6). We also unified the analysis of these models, which is achieved naturally by data augmentation.

A fundamental component of the hierarchical model is the underlying point process model that governs the distribution of individuals over the region from which individuals are exposed to sampling. While the models of this chapter were presented using the assumption that individuals are distributed uniformly in space – the fundamental distance sampling hypothesis – the explicit introduction of this point process yields great potential for extending the model to accommodate biologically relevant features of the distribution of organisms in space. We believe that this is the main practical benefit that the hierarchical formulation of the problem (potentially) yields. For example, it is natural to consider point process models that exhibit interaction (e.g., inhibition models, Markov point processes, and processes that allow for clustering, etc.), which can be used to model the interaction of individuals defending territories, or sex differences, or even interactions among species. A second potential area of extension of these models is in the modeling of spatial variation in the intensity function (i.e. local density) due to variations in habitat quality and suitability. Finally, point process models can be extended to spatio-temporal settings (Rathbun and Cressie, 1994) to allow for the explicit consideration of demographic processes. Such models can exploit multi-year capture history data on individuals. The filling-up of available habitat by individuals (especially of rare species) is an ecological process of some importance in conservation biology and management. Spatial capture–recapture models allow us to make explicit inferences about the underlying point process. To extend the framework described in this chapter, we only need to be able to simulate from these processes to do inference under such models within a Markov chain Monte Carlo framework. Fortunately much is known about simulation from point process models as contemporary inference under these models is largely simulation-based (Møller and Waagepetersen, 2004).

8

METAPOPULATION MODELS OF ABUNDANCE

Surveys of natural populations often include some form of spatial replication wherein a large number of small, but spatially distinct, sample units are selected from the population. In surveys of bird populations, for example, 'point counts' are often conducted at a set of randomly selected locations within a park, refuge, or forest. In such cases the spatial domain of sampling provides a somewhat operational definition of the biological population being surveyed.

As described in previous chapters, animals that are available to be sampled in these surveys often go undetected; thus, the number of animals observed in a sample provides a negatively biased index of abundance. The bias may be ignorable if it is reasonable to assume that detection or capture rate is identical at all sample locations and at all sample times, or if covariates responsible for variation in detectability can be identified and their effects modeled indirectly (Link and Sauer, 2002); however, such is often not the case, even when 'standardized' methods of data collection are employed.

A common remedy to detection bias is to employ a sampling protocol that allows both abundance *and* detection to be estimated simultaneously. Such protocols include distance sampling, sampling with multiple observers, capture–recapture, and removal sampling (Nichols et al., 2000; Borchers et al., 2002; Farnsworth et al., 2002; Williams et al., 2002). The class of statistical models developed for each of these protocols represents, in some sense, a mechanism for 'transforming' the observed animal counts into an index of abundance that is unbiased with respect to variation in detectability of animals. While these models and protocols have proven to be enormously useful, they can be difficult to apply in spatially replicated surveys. For example, in surveys of terrestrial salamanders, animals are often detected beneath coverboards or natural cover (logs, sticks, and rocks) within fixed-width transects or plots (Bailey et al., 2004). The number of salamanders that are vulnerable to sampling at each location may be small (even zero); consequently, the observed counts are generally sparse and provide little information about 'local abundance' (i.e., at the scale of the sample unit). A common solution to this problem is to aggregate counts among sample units and to develop elaborate models of heterogeneity in detection that ignore the spatial indexing of the sample units – that is, to practice what we refer to as an 'observation-driven view' (see Chapter 1). The capture–recapture analysis of White (2005) is a good example. In using this approach, focus shifts from modeling variation in abundance of animals to modeling variation in their detectability, and the estimator of total abundance (i.e., among all sample units) essentially amounts to an adjustment of the observed aggregate count that conditions on estimated nuisance parameters of the detection model. This solution may be appropriate for highly mobile species with large home ranges (e.g., large carnivores; Nichols and Karanth, 2002); however, other approaches are needed for species that are unlikely to move among sample locations in the time required to complete the survey.

Although the problem of low counts in spatial replicates provides ample motivation for new approaches to data analysis, inferential problems can also occur when the location-specific counts provide reasonably precise estimates of abundance and detection. For example, suppose the objective of the survey is to estimate the level of association between site-specific abundance and one or more potential measures of habitat. The conventional approach to this inference problem is to build regression models wherein the estimates of site-specific abundance are treated as though they were data (Boulinier et al., 2001; Cam et al., 2002b; Doherty et al., 2003). This approach amounts to computing 'statistics on statistics' and does not properly account for the uncertainty in the estimation of abundances. There is clearly a need to develop an inferential framework that combines the observation model, which includes a detectability component, with the ecological model of habitat association.

In this chapter we describe an alternative modeling framework wherein the model of detectability is augmented with a model that specifies variation in local abundance of animals. We adopt a hierarchical approach where prior distributions may be used to specify stochastic variation in abundance or detection among sample units. This approach also allows spatial covariates that are thought to be informative of detection or abundance to be included; thus, systematic sources of variability (e.g., habitat features) may be explored and quantified by specifying models with or without such covariates. The hierarchical framework also allows abundance to be estimated or predicted for individual locations or groups of locations (regions). Such estimates are often used to assess the consequences of location-specific management actions. Another conceptual benefit of the hierarchical framework is its unified treatment of inference problems for both animal abundance and occurrence. These two quantities are often viewed independently, leading some investigators to estimate occurrence using models of binary (presence/absence) data formed by quantizing the counts of individual animals (e.g., logistic-regression or site-occupancy models (MacKenzie et al.,

2002, 2006; Manly et al., 2002; Sargeant et al., 2005)).   In contrast, in our modeling framework, estimators of occurrence are derived quite naturally from the fundamental equivalence between occurrence and the event that at least one animal is present (i.e., abundance $N > 0$).

8.1 A HIERARCHICAL VIEW OF THE POPULATION

Consider a survey wherein a particular sampling protocol is applied at various locations judged to be representative of the entire population.   In this survey the sample includes a set of spatially-indexed counts observed by sampling a corresponding set of spatially distinct local populations.   Conceptually, the population being sampled may be viewed as a *metapopulation* (Hanski and Simberloff, 1997) of relatively isolated and spatially distinct local populations connected demographically by occasional dispersal of animals. Strict adherence to the metapopulation view is not essential, but it does provide a natural hierarchy for building statistical models of animal abundance and occurrence. For example, at the bottom of this hierarchy, counts are observed at individual sample locations (Figure 8.1).   The particular sampling protocol determines, in large part, an appropriate model of the counts.   For example, multinomial models are induced by capture–recapture sampling or by removal sampling; however, the precise form of the multinomial cell probabilities depends on which of these sampling protocols is used (see Section 8.2). In both cases the model of counts observed at a particular sample location depends on the unknown abundance of animals at that location and on the unknown probability of detection (per individual) at that location. Therefore, these unknowns (abundance and detection) are intermediate-level parameters in the hierarchy and are defined *locally* (i.e., for an individual sample location). The top of the hierarchy is reserved for parameters that apply across the entire population of animals.   These *metapopulation-level* parameters specify the extent to which abundance and detection differs among local populations.   Such differences may include *systematic* sources of variation (e.g., effects of habitat) based on measurable covariates and *stochastic* sources of variation for surveys where location-specific covariates are unavailable.

The hierarchy illustrated in Figure 8.1 is an example of a 'mixed model', a term widely used to describe statistical models that include both 'fixed effects' (metapopulation parameters that are fixed among sample units) and 'random effects' (local parameters that vary among sample units). Scientific interest may be focused on either set of parameters, depending on the context of the study. The hierarchical framework provides a formal connection between these two sets of parameters and allows both to be estimated. This chapter includes a variety of

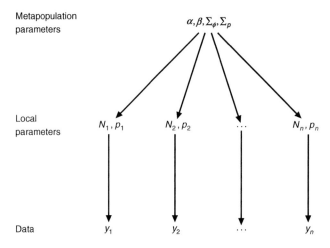

Metapopulation
parameters

$\alpha, \beta, \Sigma_\phi, \Sigma_p$

Local
parameters N_1, p_1 N_2, p_2 $\cdots$ N_n, p_n

Data y_1 y_2 $\cdots$ y_n

Figure 8.1. A hierarchy induced by sampling a metapopulation of animals at n spatially distinct locations.

examples chosen to illustrate the flexibility of this framework and the pros and cons of alternative modes of inference (classical vs. Bayesian).

8.2 MODELING COUNTS, ABUNDANCE, AND DETECTABILITY

In this section we summarize a general framework for building hierarchical models of animal abundance and occurrence from spatially replicated counts. Much of our summary is drawn from Royle and Dorazio (2006), who describe a class of models that fits within this framework. Here we provide a less-technical treatment, focusing on the conceptual issues related to sampling, model building, and parameter estimation.

8.2.1 Sampling Protocols and Observation Models

The observed counts, which lie at the base of the hierarchy in Figure 8.1, provide a natural starting point for modeling. Let $\boldsymbol{y}_i$ ($= y_{i1}, y_{i2}, \ldots, y_{iJ}$) denote the counts observed while sampling animals at the ith location ($i = 1, \ldots, n$) with a particular protocol. We assume that the number of animals N_i at this location remains fixed in the time required to complete sampling. In practice, one never knows whether this assumption is strictly satisfied; however, it is likely to hold at least approximately in surveys where the timing and spatial extent of sampling may be chosen to ensure that births, deaths, and movements of animals into and out of a sample location are unlikely events. The type of model specified for the counts $\boldsymbol{y}_i$ depends primarily on the sampling protocol. For example, in some surveys it may only be possible to make $J > 1$ independent point counts of the local population because individual animals cannot be uniquely identified. Point counts may be taken by a single observer in a sequence of visits (e.g., days) or by multiple observers sampling at different times of the same day. In either case one might reasonably model the counts as J independent outcomes of binomial sampling with common index N_i and success probability p_i:

$$[\boldsymbol{y}_i \mid N_i, p_i] = \prod_{j=1}^{J} \mathrm{Bin}(y_{ij} \mid N_i, p_i). \tag{8.2.1}$$

In other words, in each of the J sampling events, each of the N_i animals is assumed to have been 'sampled' independently with probability p_i. The joint pmf in Eq. (8.2.1) is sometimes referred to as a product-binomial model.

Another protocol that is often applied when individual animals cannot be identified uniquely is removal sampling. In this case animals are physically removed from the local population during each of $J > 1$ successive sampling events, in an attempt to decrease the local population's size. This protocol induces a dependence among the J counts so that their joint distribution may be modeled using a multinomial distribution:

$$\boldsymbol{y}_i \mid N_i, \boldsymbol{\pi}_i \sim \mathrm{Multin}(N_i, \pi_{i1}, \pi_{i2}, \ldots, \pi_{iJ}), \tag{8.2.2}$$

where $\pi_{ij} = p_i(1 - p_i)^{j-1}$ denotes the probability that animals in the ith local population are captured and removed during the jth removal, given that they had not been removed previously.

The multinomial distribution is also useful for modeling counts of animals that can be distinguished uniquely, say using natural or artificial marks. Such counts arise in a variety of sampling protocols, including capture–recapture, double-observer, and distance sampling. For example, suppose individual animals in the ith local population are observed during each of 2 capture occasions. This yields 3 kinds of observations: animals detected in both occasions, animals detected in the first but not the second occasion, and animals detected in the second but not the first occasion. Let the observed counts $\boldsymbol{y}_i = (y_{i1}, y_{i2}, y_{i3})$ correspond to the respective number of animals in these 3 categories. If p_{i1} and p_{i2} are used to denote the probabilities of capturing an animal in sampling occasions 1 and 2, respectively, then the multinomial probabilities associated with the 3 kinds of observations are

$$\pi_{i1} = p_{i1}p_{i2}$$
$$\pi_{i2} = p_{i1}(1 - p_{i2})$$
$$\pi_{i3} = (1 - p_{i1})p_{i2}.$$

Therefore, the joint distribution of counts obtained by sampling the ith local population is specified conditionally in terms of the unknown number of animals N_i and the unknown detection probabilities p_{i1} and p_{i2}.

The above examples illustrate a common feature of the hierarchical framework. The model of the observed counts at each sample location is specified only in terms of the number and detectability of animals *at that sample location*. This model can reasonably be assumed to have a simple structure because the differences in detectability or capture of animals in a local population are likely to be much lower than those that may exist among individuals in the entire metapopulation.

8.2.2 Modeling Abundance and Detection

The parameters of the count models described in Section 8.2.1 correspond to unknown states of the local populations. Additional modeling, which includes the introduction of metapopulation-level parameters, is needed to specify how these local-level parameters are interrelated.

8.2.2.1 Abundance

A common choice for modeling local abundances is the Poisson distribution. Thus, we might assume $N_i|\lambda_i \sim \text{Po}(\lambda_i)$, where λ_i denotes the mean abundance of animals that are available to be detected at the ith sample location. Given this assumption, the simplest model that could be considered is that all local populations share the same mean abundance (i.e., $\lambda_i = \lambda$). Of course, this model is unlikely to approximate the heterogeneity in abundance in many natural populations because the spatial distribution of animals is often highly variable. In practice, the sources of variation may be unknown or they may be known and modeled as functions of measurable, location-specific covariates (Royle and Dorazio, 2006). In fact, the primary objective of some surveys is to estimate the level of association between abundance and one or more spatially varying covariates for the purposes of characterizing or quantifying habitat. In such surveys one might use a Poisson-regression formulation of local mean abundances, such as $\log(\lambda_i) = \beta_0 + \beta_1 x_i$, to estimate the regression slope parameter β_1, which quantifies the association between λ_i and habitat covariate x_i measured at the ith sample location. In this example, the regression coefficients, β_0 and β_1, represent metapopulation-level parameters of the hierarchical model. Scientific interest may be focused solely on computing inferences for these parameters; alternatively, one also may be interested in using estimates of these parameters to make *predictions* of mean abundance at unsampled locations where measures of the abundance covariate x are available (say, using geographic information systems). We note that the prediction of site-specific *occurrence* is a trivial calculation given a model of local abundance. For example, $\Pr(N > 0) = 1 - \exp(-\lambda)$ under a Poisson model of local abundance N; therefore, given a model-based prediction of λ at unsampled location k, say $\hat{\lambda}_k$, the mean occurrence at that location is predicted using $1 - \exp(-\hat{\lambda}_k)$. Such predictions can be used to compute range maps of the spatial distribution of occurrence of a species.

How do we construct models of local abundance if the sources of variation in abundance are unknown or if inclusion of location-specific covariates fails to account for enough of the variation? Answers to this question invariably involve making distributional assumptions about the local mean abundances, which are treated as random outcomes of one or more stochastic processes. For example, one might assume that the variation in λ among sample locations can be modeled exchangeably using a gamma distribution, say $\lambda_i | a, b \sim \text{Gamma}(a, b)$. This assumption implies that local abundances N_i have a negative binomial distribution, which has a long history of use in ecology (e.g., see He and Gaston (2000a)). However, a variety of alternatives exist for modeling extra-Poisson variation (i.e., overdispersion) in the local abundance of animals. For example, one could model exchangeable variation in λ_i by assuming $\log(\lambda_i) | \beta, \sigma \sim \text{N}(\beta, \sigma^2)$, which implies a log-normal distribution for λ_i (see Section 4.4.2). The log-normal and gamma distributions have the advantage of adding only two metapopulation-level parameters to the hierarchical model. This frequently simplifies parameter estimation, especially in surveys where the number of sample locations n is relatively low. A disadvantage of these models is that their relatively simple form (e.g., unimodality) may not adequately approximate the actual pattern of variation in mean abundances. For example, animals may be spatially clustered due to patchiness in habitat, but the covariates associated with this clustering may not be observed. In this case more sophisticated models of heterogeneity in abundance (e.g., Dirichlet process priors (Dorazio et al., 2008) or geostatistical models of spatial correlation (Royle et al., 2007b)) may be formulated.

For now, it suffices to note that additional assumptions and metapopulation-level parameters can be used to construct realistic models of animal abundance and that these models often include both stochastic and systematic components of variation.

8.2.2.2 Detection

The same approach may be used to model heterogeneity in detection among local populations. For example, known sources of heterogeneity may be specified by modeling associations between location-specific covariates and p_i (on the logit-scale as in logistic-regression). This formulation ensures that estimates of p_i are confined to the $(0, 1)$ interval. In addition, stochastic sources of variation in detection may be modeled using distributional assumptions, such as $p_i | a, b \sim \text{Be}(a, b)$ or $\text{logit}(p_i) | \alpha, \sigma \sim \text{N}(\alpha, \sigma^2)$.

Specific recommendations for building models of animal abundance and detection depend to a great extent on the nature and amount of available data. In Section 8.3 we provide a variety of examples to illustrate the construction and use of hierarchical models in the analysis of real data sets.

8.3 ESTIMATING MODEL PARAMETERS FROM DATA

In our hierarchical view of the population (Figure 8.1), the total number of local parameters (N_i and p_i) in a model increases with the sample size n; therefore, to fit the model and estimate its metapopulation-level parameters, we must reduce the number of local parameters to be estimated. In some cases this may be achieved with simplifying assumptions, such as $\lambda_i = \lambda$ (for all i) or $\log(\lambda_i) = \beta_0 + \beta_1 x_i$. Alternatively, the local parameters can be eliminated by integration (when feasible) to form a model of the counts that depends only on the metapopulation-level parameters. Elimination of local-level parameters by integration is routinely used in classical statistics to compute maximum-likelihood estimates (Berger et al., 1999); however, this technique can also be used in Bayesian analyses to simplify or reduce calculations. The following sections contain examples that illustrate the estimation of both local and metapopulation-level parameters using classical and Bayesian methods.

8.3.1 Binomial Mixture Models

As noted earlier a binomial observation model is induced by a relatively simple, but widely used, sampling protocol wherein repeated observations of the same local population are obtained independently. When this observation model is combined with a model of heterogeneity in local abundances (i.e., heterogeneity among sample locations), we obtain a mixture of binomials. For example, assuming $N_i | \lambda \sim \text{Po}(\lambda)$ produces a binomial-Poisson mixture model. This model was developed originally by Royle (2004c) to estimate the average abundance of bird populations. Subsequently it was extended to quantify associations between abundance and environmental

Table 8.1. A subset of ovenbird point counts observed at several sample locations (rows) and days (columns).

				Day of sampling					
1	2	3	4	5	6	7	8	9	10
0	0	0	0	0	0	0	0	0	0
0	0	0	0	0	0	0	0	1	0
0	0	0	0	0	0	0	0	0	0
0	0	0	1	1	0	0	0	0	1
2	1	1	2	2	2	1	2	2	1
0	0	1	1	0	0	0	0	0	0
1	0	1	1	0	0	0	0	0	0
0	0	0	0	0	0	0	0	0	0
0	1	1	1	0	0	0	0	0	0
0	0	0	0	2	1	2	2	2	2
0	1	0	1	0	1	0	1	2	0

gradients (Dodd and Dorazio, 2004; Kéry et al., 2005) and to estimate maps of animal occurrence using spatial covariates of abundance (Royle et al., 2005). The following subsections include two examples, one involving counts of individual birds and the other involving counts of bird territories. In each example, we develop binomial mixture models and fit them using both classical (likelihood-based) and Bayesian methods for purposes of comparison.

8.3.1.1 Example: ovenbird point counts

Point counts of ovenbirds (*Seiurus aurocapillus*) were taken at each of 50 sites located along a sample route included in the North American Breeding Bird Survey (Link and Barker, 1994). The sites were evenly spaced (about one-half mile apart) and were visited on 10 separate days of the breeding season. On each day the number of birds detected by sight or song was recorded at each sample location. This protocol resulted in a 50×10 matrix of counts. A portion of this matrix is shown in Table 8.1 to illustrate the sparseness in the data. Such low counts are typical in avian point-count surveys.

Let's consider a fairly simple model of the ovenbird counts. Let y_{ij} denote the number of birds detected at the ith sample location and jth day. To specify the model, we make two distributional assumptions:

$$y_{ij}|N_i, p \sim \text{Bin}(N_i, p)$$
$$N_i|\lambda \sim \text{Po}(\lambda).$$

This model contains 52 parameters: λ, the average abundance of ovenbirds among sample locations; p, the (fixed) probability of detection per bird; and $\{N_i\}$, the set of local abundance parameters.

8.3.1.2 Classical analysis

To fit this model by classical methods, we reduce the number of parameters to be estimated by forming an integrated likelihood function. This is accomplished by integrating the joint probability of $\boldsymbol{y}_i = (y_{i1}, y_{i2}, \ldots, y_{i10})$ and N_i over the admissible values of N_i to obtain the marginal probability of the counts $\boldsymbol{y}_i$ given only the metapopulation-level parameters λ and p:

$$q(\boldsymbol{y}_i | \lambda, p) = \sum_{N_i = \max(\boldsymbol{y}_i)}^{\infty} \left(\prod_j f(y_{ij} | N_i, p) \right) g(N_i | \lambda). \qquad (8.3.1)$$

In this equation f denotes the pmf of the binomial distribution and g denotes the pmf of the Poisson distribution. The marginal probability $q(\boldsymbol{y}_i | \lambda, p)$ cannot be expressed in closed form; therefore, in practice this probability is computed numerically by substituting a sufficiently large number for the upper index of summation in Eq. (8.3.1).

To compute maximum likelihood estimates of λ and p, we assume independence among sample locations and find the value of these parameters that maximizes the likelihood function $L(\lambda, p) = \prod_{i=1}^{n} q(\boldsymbol{y}_i | \lambda, p)$. Panel 8.1 contains **R** code for computing the logarithm of this likelihood function. When unconstrained optimizers, such as **R**'s built-in functions `nlm` or `optim`, are used, a 1-to-1 transformation of the model's parameters may be needed to ensure that the estimates are confined to their admissible range of values. For example, one normally would compute the MLE of $\log(\lambda)$ and then obtain the MLE of λ by inverting the transformation. In this way, the MLE of λ is guaranteed to be strictly positive (as required in the model), whereas the MLE of $\log(\lambda)$ may be positive or negative. Transformation is also useful when estimating detection probabilities. For example, one normally would compute the MLE of $\text{logit}(p)$ to ensure that an estimate of p is confined to the (0,1) interval. Such transformations were used in the analysis of ovenbird point counts; this yielded the following MLEs (and 95 percent confidence limits): $\hat{\lambda} = 2.12$ (1.66, 2.69), $\hat{p} = 0.32$ (0.27, 0.37). Based on these estimates, we conclude that in any given sampling occasion only 1 ovenbird is detected for every 3 that are present. In addition, we estimate that the average abundance of ovenbirds is 2.12 birds per sample location.

8.3.1.3 Bayesian analysis

Now suppose we wish to fit the model using Bayesian methods. First, we must complete the model by specifying prior distributions for λ and p. A common practice for specifying prior indifference in the magnitude of these parameters would be to assume the following pair of mutually independent priors: $p \sim \mathrm{U}(0, 1)$ and $\log(\lambda) \sim \mathrm{N}(0, 100^2)$. In other words, we assume equal or nearly equal probability density over the entire range of each parameter's value. The **WinBUGS** code for implementing this model is fairly simple (Panel 8.2) and amounts to little more than pseudo-code describing the distributional assumptions of the model. Fitting the model to the ovenbird point counts yields the following parameter estimates

```
LogLikelihood = function(yMat, nrepls, lambda, p, Ninfty=5000) {
  nsites = length(nrepls)
  logLike = rep(NA, nsites)
  for (i in 1:nsites) {
    y = yMat[i, 1:nrepls[i]]
    Nmin = max(y)

    # compute marginal likelihood for site
    siteSum = 0
    if (Nmin==0) {
      siteSum = siteSum + exp(-lambda)
      Nmin=1
    }
    N = Nmin:Ninfty
    logSum =  dpois(N, lambda=lambda, log=T)
    for (j in 1:nrepls[i]) {
      logSum = logSum + dbinom(y[j], size=N, prob=p, log=T)
    }
    siteSum = siteSum + sum(exp(logSum))

    logLike[i] = log(siteSum)
  }

  sum(logLike)
}
```

Panel 8.1. R code for the likelihood of the model of ovenbird point counts. This function allows the number of replicate counts per site to differ among sites.

(provided as posterior means with 95 percent credible limits): $\hat{\lambda} = 2.15 \ (1.66, 2.71)$, $\hat{p} = 0.31 \ (0.26, 0.37)$. Therefore, the classical and Bayesian estimates of the model's parameters appear to be in close agreement for this set of data, as often occurs in Bayesian analyses with non-informative priors and reasonably large samples. We note that from a practical standpoint, the code required to implement the Bayesian analysis was much simpler to develop than that required to fit the model by classical (non-Bayesian) methods.

8.3.1.4 Example: Swiss breeding bird survey

In this example we consider another avian point count survey. In contrast to the previous example, we consider species-specific counts of bird territories detected visually or aurally by observers walking along prescribed routes in Switzerland. Our focus here is an analysis of the number of willow tit territories observed in repeated visits to each route. The willow tit (*Parus montanus*) is one of the species that we considered earlier in the estimation of population site occupancy (Chapters 3 and 4). In that analysis the territory counts were quantized to apparent presence/absence data that were used to illustrate the development and application of site-occupancy models with heterogeneity in probabilities of detection and occurrence modeled as functions of observed covariates. Here, we use the same idea and covariates, except that we analyze the counts directly without reducing them to presence/absence data.

```
model {

  beta ~ dnorm(0, 1.0E-4)
  lambda <- exp(beta)
  p ~ dunif(0,1)

  for (i in 1:n) {
     N[i] ~ dpois(lambda)

     for (j in 1:J) {
     y[i,j] ~ dbin(p, N[i])
     }
  }
}
```

Panel 8.2. **WinBUGS** model specification for the model of ovenbird point counts.

Let y_{ij} denote the number of willow tit territories observed during the jth visit to the ith route (=sample unit). Each of 237 routes was visited either 2 or 3 times during the breeding season. We assume a binomial model of the counts

$$y_{ij}|N_i, p_{ij} \sim \text{Bin}(N_i, p_{ij}),$$

where N_i denotes the unknown number of territories that are available to be detected along the ith route and p_{ij} denotes the probability of detection during the jth visit to that route.

We assume that the detectability of willow tits may differ among routes and among visits due to varying sampling effort and to the phenology of breeding behavior. In particular, two covariates, intensity and date, are used to model these potential sources of variation in detectability. The first, intensity, is a measure of sampling intensity; the second, date, is used to capture changes in breeding behavior (e.g. calling frequency of birds) during the season. The most complex model that we consider is

$$\text{logit}(p_{ij}) = \alpha_0 + \alpha_1\, \texttt{intensity}_{ij} + \alpha_2\, \texttt{date}_{ij} + \alpha_3\, \texttt{date}_{ij}^2$$

which allows the probability of detection to increase to some maximum level and then decline over the course of the breeding season.

We also assume that the number of willow tit territories N_i may differ among routes, due to differences in habitat characteristics and to differences in the lengths of routes sampled. Two covariates, elev and forest, are used to model differences in habitat associated with differences in elevation and forest cover. A single covariate, length, specifies the length of route sampled. In the most complex model that we consider, the route-level abundance of willow tit territories is specified as follows:

$$N_i|\lambda_i \sim \text{Po}(\lambda_i)$$
$$\log(\lambda_i) = \beta_0 + \beta_1\, \texttt{elev}_i + \beta_2\, \texttt{elev}_i^2 + \beta_3\, \texttt{forest}_i + \beta_4\, \texttt{length}_i.$$

The quadratic term, $\texttt{elev}^2$, allows us to specify improvements in willow tit habitat with increases in elevation and then a decline in habitat at the highest elevations.

8.3.1.5 Classical analysis

The most complex model that we fit includes 246 parameters: 9 metapopulation-level parameters $(\boldsymbol{\alpha}, \boldsymbol{\beta})$ and 237 local parameters $(\{N_i\})$. To fit this model by classical methods, we reduce the number of parameters to be estimated by forming an integrated likelihood function. The marginal probability of the counts observed at the ith route, $q(\boldsymbol{y}_i|\boldsymbol{\alpha}, \boldsymbol{\beta})$, is essentially the same as that used in the analysis of ovenbird counts (Section 8.3.1.1), except that λ and p vary among routes as functions of $\boldsymbol{\beta}$ and $\boldsymbol{\alpha}$, respectively:

$$q(\boldsymbol{y}_i|\boldsymbol{\alpha},\boldsymbol{\beta}) = \sum_{N_i=\max(\boldsymbol{y}_i)}^{\infty} \left(\prod_j f(y_{ij}|N_i,p_{ij}) \right) g(N_i|\lambda_i).$$

To compute maximum likelihood estimates of $\boldsymbol{\alpha}$ and $\boldsymbol{\beta}$, we assume independence among routes and find the value of these parameters that maximizes the likelihood function $L(\boldsymbol{\alpha},\boldsymbol{\beta}) = \prod_{i=1}^{n} q(\boldsymbol{y}_i|\boldsymbol{\alpha},\boldsymbol{\beta})$. The **R** code for evaluating $L(\boldsymbol{\alpha},\boldsymbol{\beta})$ is almost identical to that used in the analysis of ovenbird counts, the difference being that `lambda` is a vector and `p` is a matrix (see the Web Supplement for the complete **R** code). The values of `lambda` and `p` depend on the covariates and on the values assigned to $\boldsymbol{\alpha}$ and $\boldsymbol{\beta}$ during numerical optimization.

We fit a variety of plausible models to the willow tit territory counts (Table 8.2). The model with the lowest value of AIC contained all of the covariates of detection and abundance that we considered. This result is perhaps not so surprising, as it is consistent with the results of fitting the Royle and Nichols (2003) model of site occupancy to presence/absence of willow tit territories. The site-occupancy analysis is described in detail in Chapter 4; we summarize the results of that analysis in Table 8.2 to facilitate comparisons with the analysis of territory counts.

The MLEs corresponding to the best fitting model (model 1) are similar, but not identical, when computed from counts vs. presence/absence of territories. Some of the differences are no doubt due to errors in estimation because there is a loss of information that occurs when territory counts are quantized to presence/absence data. For example, the analysis of presence/absence data yields

$$\mathrm{SE}(\hat{\boldsymbol{\beta}}) = (0.324, 0.268, 0.224, 0.131, 0.122);$$

in contrast, the analysis of territory counts yields less uncertainty in $\hat{\boldsymbol{\beta}}$:

$$\mathrm{SE}(\hat{\boldsymbol{\beta}}) = (0.137, 0.192, 0.138, 0.066, 0.065).$$

We used the MLEs of $\boldsymbol{\beta}$ to compute route-specific estimates of mean abundance as follows:

$$\log(\hat{\lambda}_i) = \hat{\beta}_0 + \hat{\beta}_1\,\mathtt{elev}_i + \hat{\beta}_2\,\mathtt{elev}_i^2 + \hat{\beta}_3\,\mathtt{forest}_i + \hat{\beta}_4\,\mathtt{length}_i.$$

MLEs from the two sources of data generally produced similar estimates of abundance (Figure 8.2).

Table 8.2. Comparison of models fitted to territory counts of willow tits and to apparent presence/absence of willow tits. Maximum likelihood estimates are tabulated for parameters associated with the probability of detection per territory (α_0, `intensity`, `date`, `date`^{2}) and with territory abundance (β_0, `elev`, `elev`^{2}, `forest`, `length`).

	Analysis of point counts									
	Detection parameters				Abundance parameters					
Model	α_0	intensity	date	date^2	β_0	elev	elev^2	forest	length	AIC
1	−0.230	0.495	−0.320	0.191	−0.004	2.383	−0.896	0.661	0.309	1211.8
2	−0.063	0.427	−0.307	0.158	−0.437	1.449		0.940	0.219	1269.7
3	0.196	0.425	0.079	0.042	0.231			0.671	0.091	1644.0
4	0.507	0.257			0.395					1783.2
5	0.644				0.374					1784.6
6	0.525	0.252	−0.021		0.392					1785.2
7	0.449	0.306	−0.036	0.045	0.396				0.105	1785.4
8	0.513	0.228	−0.067	0.057	0.385					1786.6

	Analysis of presence/absence data									
	Detection parameters				Abundance parameters					
Model	α_0	intensity	date	date^2	β_0	elev	elev^2	forest	length	AIC
1	−0.471	0.490	−0.083	0.012	0.237	1.911	−0.947	0.588	0.278	414.4
2	−1.160	0.419	−0.064	−0.038	0.465	1.167		0.922	0.147	437.4
3	−0.936	0.247	0.203	−0.051	0.863			0.587	−0.132	502.8
4	−0.856				1.080					526.0
5	−0.842	0.278			0.949					526.6
6	−0.843	0.239	0.150		0.928					528.0
7	−0.842	0.253	0.163	−0.077	0.994					529.7
8	−0.851	0.218	0.164	−0.078	1.016				−0.092	531.0

8.3.1.6 Bayesian analysis

The covariates included in our analysis of willow tit territories appear to have accounted for a substantial portion of the variation in route-level abundances. But suppose we suspect that additional sources of variation exist among routes. How do we model these extra-Poisson sources of heterogeneity in abundance?

One approach is to expand the model of route-level abundances by adding a parameter b_i that specifies random variation in the mean number of territories as follows:

$$N_i|\lambda_i \sim \mathrm{Po}(\lambda_i)$$
$$\log \lambda_i = \beta_0 + \beta_1\,\texttt{elev}_i + \beta_2\,\texttt{elev}_i^2 + \beta_3\,\texttt{forest}_i + \beta_4\,\texttt{length}_i + b_i$$
$$b_i|\sigma \sim \mathrm{N}(0,\sigma^2),$$

where σ parameterizes the extent of this random variation. A classical statistician might refer to the parameter b_i as a log-normal random effect.

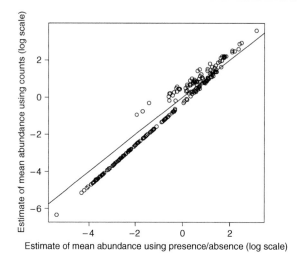

Figure 8.2. Comparison of route-level abundances of willow tit territories estimated from point counts (ordinate) or presence/absence data (abscissa). A 45 degree line is superimposed.

To fit this expanded model by the method of maximum likelihood, we would need to integrate both N_i and b_i from the model to obtain a likelihood function based on the metapopulation-level parameters α, β and σ. This approach was used in Chapter 4 to estimate the parameters of an abundance-based occupancy model (see Panel 4.1). Instead, here we opt to conduct a Bayesian analysis of the territory counts. This requires us to complete the model by specifying a set of mutually independent priors for the metapopulation-level parameters; thus we assume $\beta \sim N(\mathbf{0}, 10^2 \mathbf{I})$, $\alpha_0 \sim N(0, 1.6^2)$, $(\alpha_1, \alpha_2, \alpha_3) \sim N(\mathbf{0}, 10^2 \mathbf{I})$, and $\sigma \sim U(0, 100)$, where $\mathbf{I}$ denotes an identity matrix of appropriate size. The prior assumed for α_0 approximates a $U(0, 1)$ prior for $\text{expit}(\alpha_0)$. The Web Supplement contains **R** and **WinBUGS** code needed to fit this model to the data.

Table 8.3 contains summary statistics (posterior means and credible intervals) obtained by fitting this model to the willow tit territory counts. For comparison with the likelihood analysis, we also fit the model without assuming extra-Poisson variation in route-level abundances. The posterior means associated with this simpler model are consistent with the MLEs reported in Table 8.2. However, the parameter estimates obtained by fitting the overdispersed model suggest that the abundance covariates account for some, but not all, of the heterogeneity among routes. For example, the 95 percent credible interval for the overdispersion parameter σ clearly exceeds zero, and the posterior of σ has negligible probability density near zero.

Table 8.3. Bayesian analysis of willow tit territory counts. Variation in route-level territory abundances was modeled with and without extra-Poisson variation as parameterized by the overdispersion parameter σ.

	Poisson model		Overdispersed model	
Parameter	Posterior mean	95% credible interval	Posterior mean	95% credible interval
α_0	-0.220	$(-0.558, 0.118)$	-1.065	$(-1.822, -0.308)$
intensity	0.485	$(0.220, 0.751)$	0.256	$(-0.125, 0.638)$
date	-0.324	$(-0.520, -0.127)$	-0.211	$(-0.379, -0.043)$
date^2	0.194	$(0.059, 0.330)$	0.088	$(-0.022, 0.197)$
β_0	-0.014	$(-0.286, 0.258)$	-0.228	$(-0.935, 0.479)$
elev	2.406	$(2.026, 2.786)$	2.409	$(1.854, 2.963)$
elev^2	-0.912	$(-1.183, -0.640)$	-1.051	$(-1.548, -0.554)$
forest	0.660	$(0.531, 0.790)$	0.922	$(0.541, 1.303)$
length	0.307	$(0.179, 0.434)$	0.390	$(0.033, 0.748)$
σ			1.444	$(1.086, 1.803)$

8.3.2 Multinomial Mixture Models

In Section 8.2.1 we noted that a multinomial observation model is induced by a variety of different sampling protocols (e.g., mark-recapture, removal sampling, and double-observer sampling) and that the counts that arise by applying a protocol at a particular site depend on the same multinomial index parameter (say, N_i for the ith site). This latent parameter is, in effect, the local abundance of animals at that site. When the multinomial observation model is combined with a model of heterogeneity in local abundances (i.e., heterogeneity among sample locations), we obtain a mixture of multinomials. For example, assuming $N_i|\lambda \sim \text{Po}(\lambda)$ produces a multinomial-Poisson mixture model.

As with the binomial mixture models, multinomial mixture models have been developed and used to solve a variety of inference problems. For example, Royle et al. (2004) used counts observed in spatially replicated distance samples to estimate spatial variation in local abundance. Similarly, hierarchical models of multinomial counts have been used to improve site-specific estimates of abundance in removal surveys (Dorazio et al., 2005, 2008) and to produce maps of the spatial distribution of abundance and occurrence from spatially-indexed, capture–recapture data (Royle et al., 2007b). The following subsections include two examples, one involving double-observer counts of manatees detected in an aerial survey and another involving removal counts of stream salamanders. In each example, we develop multinomial mixture models and fit them using classical (likelihood-based) and Bayesian methods.

Figure 8.3. The West Indian manatee.

8.3.2.1 Example: double-observer counts of manatees

The West Indian manatee (*Trichechus manatus*) is a marine mammal that lives in coastal waters of the southeastern United States, the Caribbean islands, eastern Mexico and Central America, and northern South America (Figure 8.3). In Florida manatees occupy both freshwater and saltwater habitats and undergo extensive, but seasonal, migrations in response to changes in water temperature and food availability. However, most of the time manatees are relatively sedentary, being slow swimmers.

Aerial surveys are often used to monitor manatees in coastal waters of Florida because these animals are easily observed from low-flying aircraft. In a recent survey groups of manatees were observed in 30 transects that contained 3 spatially-distinct habitat zones: inshore, transitional, and offshore. The transitional zone is located between inshore and offshore zones. Transect length and width in each zone were fixed by design. The purpose of the survey was to estimate the average abundance of manatee groups in each habitat zone. To account for imperfect detection of these animals, two observers (say, A and B) were used to enumerate groups of manatees independently. Data were recorded so that at the end of each transect it was possible to determine the number of groups seen by both observers (y_1), the number of groups seen by observer A but not B (y_2), and the number of groups seen by observer B but not A (y_3). Therefore, estimating the number of groups in a transect amounted to estimating the number of groups *not seen* by either observer (y_0), while accounting for the imperfect detectability of animals by each observer.

The observed counts, which we denote by $\boldsymbol{y}_i = (y_{i1}, y_{i2}, y_{i3})$ for the ith transect, contain many zeros (Table 8.4). To account for the dependence of counts

Table 8.4. A subset of the transect-level, double-observer counts of manatees detected in 3 habitat zones (inshore I, offshore O, and transitional T).

Habitat zone	y_1	y_2	y_3
O	0	0	0
T	0	0	0
I	1	0	0
O	0	0	0
O	0	0	0
T	0	0	0
I	0	0	0
O	0	3	1
O	2	2	1
O	0	0	0
T	0	0	0
O	0	0	0

within a transect induced by double-observer sampling, we assume a multinomial observation model

$$\boldsymbol{y}_i | N_i, p_A, p_B \sim \text{Multin}(N_i, \pi_{i1}, \pi_{i2}, \pi_{i3})$$
$$\pi_{i1} = p_A p_B$$
$$\pi_{i2} = p_A(1 - p_B)$$
$$\pi_{i3} = (1 - p_A)p_B,$$

where p_A and p_B denote the detection probabilities of observers A and B, respectively, and N_i denotes the unknown number of manatee groups in the ith transect. To specify the effects of habitat on N_i, we further assume

$$N_i | \lambda_i \sim \text{Po}(\lambda_i)$$
$$\log(\lambda_i) = \boldsymbol{x}_i \boldsymbol{\beta},$$

where $\boldsymbol{x}_i = (x_{i1}, x_{i2}, x_{i3})$ denotes a row vector of dummy variables used to codify the habitat zone of the ith transect. For example, $\boldsymbol{x} = (1, 0, 0)$ indicates habitat zone 1 (inshore), $\boldsymbol{x} = (0, 1, 0)$ indicates habitat zone 2 (offshore), and $\boldsymbol{x} = (0, 0, 1)$ indicates habitat zone 3 (transitional). This coding gives the elements of metapopulation-level parameter $\boldsymbol{\beta}$ a simple interpretation: $\exp(\beta_j)$ = the average number of manatee groups per transect in habitat zone j. In addition, this coding yields a *hierarchically centered* parameterization that produces lower posterior correlations among parameters and faster convergence when using MCMC methods of analysis (Gelfand et al., 1995; Gilks and Roberts, 1995).

8.3.2.2 Classical analysis

To compute maximum likelihood estimates of the metapopulation-level parameters, the local abundance parameters must be removed by integration. We could compute the marginal probability of the counts by summing over the admissible values of N_i, as we did in the analysis of ovenbird counts (cf. Section 8.3.1.1):

$$q(\boldsymbol{y}_i|\boldsymbol{\beta}, p_A, p_B) = \sum_{N_i=y_i.}^{\infty} f(\boldsymbol{y}_i|N_i, \boldsymbol{\pi}_i)g(N_i|\lambda_i), \qquad (8.3.2)$$

where $y_{i.} = \sum_{j=1}^{3} y_{ij}$ is the sum of the counts observed in transect i, f denotes the pmf of a multinomial distribution with index N_i and parameters $\boldsymbol{\pi}_i = (\pi_{i1}, \pi_{i2}, \pi_{i3})$, and g denotes the pmf of a Poisson distribution with mean $\lambda_i = \exp(\boldsymbol{x}_i\boldsymbol{\beta})$. However, this brute-force calculation is really unnecessary because the integration can be done analytically, as we now demonstrate. First, we expand the right-hand side of Eq. (8.3.2) while omitting the location subscript i for convenience:

$$q(\boldsymbol{y}|\beta, p_A, p_B) = \sum_{N=y.}^{\infty} \frac{N!}{y_1!y_2!y_3!(N-y.)!} \pi_1^{y_1} \pi_2^{y_2} \pi_3^{y_3} (1-\pi.)^{(N-y.)} \frac{\exp(-\lambda)\lambda^N}{N!},$$

where $\pi. = \sum_{j=1}^{3} \pi_j$ and $\beta = \log(\lambda)$. Next, we simplify and bring terms that don't involve N outside the summation:

$$q(\boldsymbol{y}|\beta, p_A, p_B) = \frac{\pi_1^{y_1} \pi_2^{y_2} \pi_3^{y_3} \exp(-\lambda)}{y_1!y_2!y_3!} \sum_{N=y.}^{\infty} \frac{(1-\pi.)^{(N-y.)}\lambda^N}{(N-y.)!}.$$

Now, we define a substitution, $k = N - y.$, which requires a change in the lower limit of summation:

$$q(\boldsymbol{y}|\beta, p_A, p_B) = \frac{\pi_1^{y_1} \pi_2^{y_2} \pi_3^{y_3} \exp(-\lambda)}{y_1!y_2!y_3!} \sum_{k=0}^{\infty} \frac{(1-\pi.)^k \lambda^{k+y.}}{k!}$$

$$= \frac{(\lambda\pi_1)^{y_1}(\lambda\pi_2)^{y_2}(\lambda\pi_3)^{y_3}\exp(-\lambda)}{y_1!y_2!y_3!} \sum_{k=0}^{\infty} \frac{\{\lambda(1-\pi.)\}^k}{k!}. \quad (8.3.3)$$

Note that the sum in Eq. (8.3.3) is a kernel for a $\text{Po}(\lambda(1-\pi.))$ distribution; therefore,

$$q(\boldsymbol{y}|\beta, p_A, p_B) = \frac{(\lambda\pi_1)^{y_1}(\lambda\pi_2)^{y_2}(\lambda\pi_3)^{y_3}\exp(-\lambda)\exp(\lambda(1-\pi.))}{y_1!y_2!y_3!}$$

$$= \frac{(\lambda\pi_1)^{y_1}(\lambda\pi_2)^{y_2}(\lambda\pi_3)^{y_3}\exp(-\lambda\pi.))}{y_1!y_2!y_3!}$$

$$= \prod_{j=1}^{3} \frac{(\lambda\pi_j)^{y_j}\exp(-\lambda\pi_j)}{y_j!}. \qquad (8.3.4)$$

In other words, we have proven in Eq. (8.3.4) that the marginal distribution of counts obtained by integrating a multinomial-Poisson mixture is mathematically equivalent to a joint distribution of conditionally independent Poisson distributions. Therefore, we are effectively specifying a kind of Poisson regression model that allows Eq. (8.3.2) to be replaced with the following equivalent expression:

$$q(\boldsymbol{y}_i|\boldsymbol{\beta}, p_A, p_B) = \prod_{j=1}^{3} \mathrm{Po}(y_{ij}|\lambda_i \pi_{ij}). \qquad (8.3.5)$$

The practical significance of this theoretical result is that the marginal probability $q(\boldsymbol{y}_i|\boldsymbol{\beta}, p_A, p_B)$ can be evaluated with *much less* computational effort by using Eq. (8.3.5) instead of Eq. (8.3.2). This computational savings can dramatically reduce the time needed to fit such models because many evaluations of $q(\boldsymbol{y}_i|\boldsymbol{\beta}, p_A, p_B)$ may be required to compute the MLE of a model's parameters by numerical optimization. Multinomial-Poisson mixtures arise in a variety of problems (e.g., see Section 8.3.2.5), so it is wise to exploit analytic marginalization when possible.

The **R** code required to compute MLEs for the model of double-observer counts is given in Panel 8.3. The matrix of dummy variables used to codify the habitat zones is indicated by the function argument xMat. This matrix can be computed with a single assignment statement in **R**:

```
xMat = model.matrix(~hab-1)
```

where hab represents a categorical vector (i.e., an **R** factor variable) of habitat zones associated with the transects that were surveyed (first column of data in Table 8.4).

The MLEs of the model parameters indicate that the two observers had nearly equal detection probabilities ($\hat{p}_A = 0.58$, $\hat{p}_B = 0.54$). In addition, the estimated number of manatee groups per transect differed significantly among habitat zones (inshore = 0.14, offshore = 0.58, transitional = 0.04), as indicated by a likelihood-ratio comparison of models with and without habitat effects ($p = 0.0007$). Evidently, there were significantly more manatees in the offshore zone than in the other 2 zones.

8.3.2.3 Bayesian analysis

This conclusion is supported by a Bayesian analysis of the manatee counts. To perform this analysis, we assume a set of mutually independent priors for the metapopulation-level parameters: $p_A \sim \mathrm{U}(0,1)$, $p_B \sim \mathrm{U}(0,1)$, and $\boldsymbol{\beta} \sim \mathrm{N}(\mathbf{0}, 100^2 \boldsymbol{I})$, where $\boldsymbol{I}$ is an identity matrix. In other words, we assume equal or nearly equal probability density over the entire range of each parameter's value. The **WinBUGS** code for fitting this model (Panel 8.4) amounts to little more than pseudo-code for the distributional assumptions of the model.

As in the likelihood-based analysis, the Bayesian analysis reveals substantial differences in abundance of manatee groups in different habitat zones (Figure 8.4). The posteriors associated with mean abundance of manatees in inshore and transitional zones overlap somewhat; in contrast, the posterior distribution of mean abundance associated with the offshore habitat zone clearly supports higher values. Therefore, we may conclude that the offshore zone contained more manatee groups than the other two habitat zones.

8.3.2.4 Estimating local abundances removed by marginalization

In the previous section we used analytic marginalization to remove the local abundance parameters (i.e., $\{N_i\}$) from the multinomial-Poisson mixture model. Suppose we are interested in estimating these parameters for scientific reasons (e.g., to compute an estimate of the total abundance of animals at all sample locations).

```
Model.Fit = function(yMat, xMat) {

  objFunc = function(param) {
    nsites = dim(yMat)[1]
    pA = rep(expit(param[1]), nsites)
    pB = rep(expit(param[2]), nsites)
    beta =  param[-(1:2)]
    phi - xMat %*% beta
    lambda = as.numeric(exp(phi))
    piMat = cbind(pA*pB, pA*(1-pB), (1-pA)*pB)
    logLike = dpois(yMat, lambda*piMat, log=T)
    -sum(logLike)
  }

  ySum = apply(yMat, 1, sum)
  pGuess = 0.5
  lambdaGuess = mean(ySum)/pGuess
  paramGuess = c(rep(logit(pGuess), 2), rep(log(lambdaGuess), dim(xMat)[2]))

  fit = optim(par=paramGuess, fn=objFunc, method='BFGS', hessian=T)
  list(LogLikelihood=-fit$value, mle=fit$par, covMat=chol2inv(chol(fit$hessian)))
}
```

Panel 8.3. R code for fitting the model of double-observer counts of manatee groups by the method of maximum likelihood.

How can N_i be estimated once it is removed as a formal parameter of the model? The answer is provided by a single application of Bayes' rule:

$$p(N_i|\boldsymbol{y}_i, \lambda_i, \boldsymbol{\pi}_i) = \frac{f(\boldsymbol{y}_i|N_i, \boldsymbol{\pi}_i)g(N_i|\lambda_i)}{q(\boldsymbol{y}_i|\lambda_i, \boldsymbol{\pi}_i)}, \qquad (8.3.6)$$

where $q(\boldsymbol{y}_i|\lambda_i, \boldsymbol{\pi}_i)$ denotes a general expression for the marginal probability of the counts $\boldsymbol{y}_i$ given the Poisson mean λ_i and multinomial probabilities $\boldsymbol{\pi}_i$. (A specific example involving double-observer counts is given in Eq. (8.3.2).) In fact, it is easily shown that Eq. (8.3.6) simplifies to

```
model {

for (k in 1:3) {
    beta[k] ~ dnorm(0.0, 1.0E-4)
    lambda.mean[k] <- exp(beta[k])
}

pA ~ dunif(0,1)
pB ~ dunif(0,1)

pi.1 <- pA*pB
pi.2 <- pA*(1-pB)
pi.3 <- (1-pA)*pB

for (i in 1:n) {

    log(lambda[i]) <- inprod(x[i,], beta[])
    mu.1[i] <- lambda[i]*pi.1
    mu.2[i] <- lambda[i]*pi.2
    mu.3[i] <- lambda[i]*pi.3

    y[i,1] ~ dpois(mu.1[i])
    y[i,2] ~ dpois(mu.2[i])
    y[i,3] ~ dpois(mu.3[i])
}
}
```

Panel 8.4. WinBUGS code for fitting the model of double-observer counts of manatee groups.

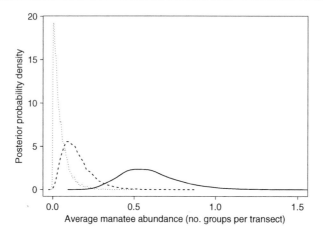

Figure 8.4. Comparison of posterior distributions of average manatee abundance in 3 habitat zones: inshore (dashed), offshore (solid), and transitional (dotted).

$$p(N_i|\boldsymbol{y}_i, \lambda_i, \boldsymbol{\pi}_i) = \frac{1}{(N_i - y_{i\cdot})!} \exp\left(-\lambda_i(1 - \pi_{i\cdot})\right) \left(\lambda_i(1 - \pi_{i\cdot})\right)^{N_i - y_{i\cdot}},$$

$$(8.3.7)$$

where $\pi_{i\cdot} = \sum_j \pi_{ij}$ and $N_i \geq y_{i\cdot}$. Note that the right-hand side of this equation is the pmf of a Poisson distribution with mean $\lambda_i(1 - \pi_{i\cdot})$. In this case the random variable is the number of animals not detected, which we denote by $y_{i0} = N_i - y_{i\cdot}$. Therefore, Eq. (8.3.7) implies that the conditional distribution of y_{i0}, given the observed the counts $\boldsymbol{y}_i$, is $\text{Po}(\lambda_i(1 - \pi_{i\cdot}))$. This result provides the basis for the so-called *best unbiased predictor* of N_i (Laird and Ware, 1982; Robinson, 1991),

$$\begin{aligned}
\text{E}(N_i|\boldsymbol{y}_i) &= \text{E}((y_{i\cdot} + y_{i0}) \mid \boldsymbol{y}_i) \\
&= y_{i\cdot} + \text{E}(y_{i0}|\boldsymbol{y}_i) \\
&= y_{i\cdot} + \hat{\lambda}_i(1 - \hat{\pi}_{i\cdot}),
\end{aligned}$$

whose estimation depends on the MLEs of λ_i and π_i. We described best unbiased predictors in other contexts in Chapter 2.

Eq. (8.3.7) is also useful in Bayesian analyses because it represents the conditional posterior probability of local abundance N_i given the counts $\boldsymbol{y}_i$ and the model parameters λ_i and $\boldsymbol{\pi}_i$. Thus, one could use an arbitrarily large posterior sample of these parameters to compute a sample of the marginal posterior of N_i by the method of composition (Tanner, 1996); alternatively, one could estimate any function of the marginal posterior distribution of N_i, such as an expectation or quantile, using

Monte Carlo integration (Robert and Casella, 2004). For example, a Bayesian estimate of the posterior mean of N_i could be computed as follows:

$$\mathrm{E}(N_i|\boldsymbol{y}_i) = y_i. + \frac{1}{R} \sum_{r=1}^{R} \lambda_i^{(r)} (1 - \pi_{i.}^{(r)}), \qquad (8.3.8)$$

where $\lambda_i^{(r)}$ and $\pi_{i.}^{(r)}$ denote random draws from a posterior sample of size R. An advantage of the Bayesian estimator of $\mathrm{E}(N_i|\boldsymbol{y}_i)$ is that it averages over the posterior uncertainty associated with estimating λ_i and $\pi_{i.}$. In contrast, the best unbiased predictor of N_i ignores this source of uncertainty, relying only on the point estimates, $\hat{\lambda}_i$ and $\hat{\pi}_{i.}$. The best unbiased predictor of N_i can be adjusted for errors in these point estimates; however, such adjustments may involve a substantial amount of additional calculation (Laird and Louis, 1987).

8.3.2.5 Example: removal counts of salamanders

During the summer months of 2001–2004, Jung et al. (2005) collected removal samples of salamanders from various streams of Shenandoah National Park, Virginia. The purpose of the survey was to assess the feasibility of removal sampling for long-term monitoring of stream salamanders. Species of two genera, *Desmognathus* and *Eurycea*, were numerically dominant in the samples; however, salamander counts of individual species were aggregated to the genus level because individual species were usually too difficult to distinguish in the field.

Jung et al. (2005) attempted to estimate abundances of *Desmognathus* and *Eurycea* at each sample location using conventional methods (Zippin, 1956). However, estimates could not be calculated for nearly half of the sample locations either because no salamanders were detected or because the number of salamanders captured in successive removals was too low to indicate that sampling had successfully depleted the local population.

Motivated by these difficulties, we fit a hierarchical model to the removal counts of *Desmognathus* and *Eurycea* observed at 18 locations that were sampled each year during 2001–2004. Our analysis excludes data from 7 locations that were sampled in 2002 but not in other years; in doing so, our analysis provides a fair comparison of annual estimates of salamander abundance. A subset of the removal counts included in our analysis is provided in Table 8.5. Note that the number of removal passes differs among sample sites, but it is limited to either 2 or 3 passes at each site.

We developed a hierarchical model of each year's removal counts in which salamander abundances and capture probabilities are assumed to vary among sample locations. Specifically, let y_{ij} denote the number of salamanders captured and removed in the jth pass of the ith sample location. Removal sampling induces a multinomial dependence among the counts observed at the same location (as

Table 8.5. A subset of removal counts of two salamander species, *Desmognathus* and *Eurycea*. Pass indicates the order of removal at each sample site.

Site	Pass	*Desmognathus*	*Eurycea*
17	1	0	0
17	2	0	0
18	1	0	0
18	2	0	0
13	1	3	15
13	2	0	2
13	3	1	18
14	1	5	9
14	2	2	11
14	3	6	6
7	1	13	5
7	2	5	2
7	3	5	2
8	1	5	3
8	2	2	4
8	3	0	6

noted in Eq. (8.2.2)) wherein $\pi_{ij} = p_i(1 - p_i)^{j-1}$ denotes the probability that salamanders at the ith location were captured and removed during the jth pass, given that they had not been removed previously. We assume a Poisson model for the local abundance of salamanders at the ith sample location (i.e., $N_i|\lambda_i \sim \text{Po}(\lambda_i)$). Thus, we have specified a multinomial-Poisson mixture, which, as described in Section 8.3.2.1, implies that the removal count y_{ij} may be modeled as a conditionally independent Poisson outcome with mean $\lambda_i p_i(1 - p_i)^{j-1}$.

Jung et al. (2005) did not measure any spatially varying covariates of salamander abundance or detectability; however, it seems prudent to assume that some level of heterogeneity in abundance and detection was present based on their preliminary findings. Therefore, we modeled extra-Poisson variation in abundance among sample locations and extra-binomial variation in detection probability as follows:

$$\log(\lambda_i)|\beta, \sigma \sim \text{N}(\beta, \sigma^2)$$
$$\text{logit}(p_i)|\alpha, \tau \sim \text{N}(\alpha, \tau^2).$$

To compute MLEs of the metapopulation-level parameters $(\alpha, \beta, \sigma, \tau)$, the local-level parameters $(\{\lambda_i\}, \{p_i\})$ must be eliminated by numerical integration. We can do these calculations quite accurately because the integrals are of low dimension; however, writing **R** code for such calculations can be time consuming. As an alternative, we implement the model in **WinBUGS** (Panel 8.5) to conduct a Bayesian analysis of the removal counts. This requires us to specify priors for the metapopulation-level parameters; therefore, we assumed equal or nearly equal probability density over the entire range of each parameter's value (e.g., $\text{expit}(\alpha) \sim U(0,1)$).

The results of our analysis are consistent with some, but not all, of the conclusions reported by Jung et al. (2005). For example, local abundances of both salamander species varied significantly among sample locations in every year. Posteriors of σ had negligible probability density near zero. A similar pattern was evident in local capture probabilities, as posteriors of τ had negligible probability density near

```
model {
# priors
 pMean ~ dunif(0,1)
 etaMean <- log(pMean) - log(1-pMean)
 etaSigma ~ dunif(0,10)
 etaTau <- pow(etaSigma, -2)

 beta ~ dnorm(0., 0.01)
 lambdaMean <- exp(beta)
 sigma ~ dunif(0,10)
 tau <- pow(sigma, -2)

# model of heterogeneity
 for (i in 1:nsites)
{
    eta[i] ~ dnorm(etaMean, etaTau)
    logit(p[i]) <- eta[i]
    b[i] ~ dnorm(beta, tau)
}

# marginal model of counts
 for (i in 1:n) {
    log(lambda[i]) <- b[site[i]]
    pi[i] <- p[site[i]] * pow(1-p[site[i]], pass[i]-1)
    mu[i] <- lambda[i] * pi[i]
    y[i] ~ dpois(mu[i])
} }
```

Panel 8.5. WinBUGS code for fitting the model of removal counts of salamanders. This specification allows the number of removal passes to vary among sample sites.

Table 8.6. Summary statistics for posterior distributions of mean abundance and mean probability of capture of salamanders collected at 18 locations by removal sampling. The posterior median and 95% credible intervals are tabulated for each parameter.

Year	*Desmognathus*				*Eurycea*			
	Abundance		Capture probability		Abundance		Capture probability	
2001	12.8	(3.6–37.8)	0.51	(0.14–0.84)	26.3	(10.5–57.7)	0.33	(0.12–0.58)
2002	19.5	(4.7–42.6)	0.19	(0.04–0.53)	69.6	(23.5–196.7)	0.10	(0.02–0.31)
2003	2.9	(0.6–18.9)	0.38	(0.05–0.68)	15.8	(5.4–42.0)	0.17	(0.04–0.43)
2004	5.7	(2.9–10.9)	0.55	(0.26–0.68)	14.9	(6.3–26.0)	0.31	(0.11–0.55)

zero. To determine whether mean salamander abundance and capture probability differed among years, we examined posteriors of $\exp(\beta)$ and $\mathrm{expit}(\alpha)$, respectively. Although mean abundances of *Desmognathus* and *Eurycea* appear to be lower in 2003 and 2004 compared to the two previous years (Table 8.6), there is considerable uncertainty in the estimates. Given the overlap in credible intervals, we cannot conclude that mean abundances differed among years. Similarly, although annual estimates of mean capture probabilities ranged from 0.10 to 0.55, we cannot conclude that capture rates differed among years, given the uncertainty in the estimates.

Based on this analysis, we conclude that removal sampling provides a satisfactory protocol for long-term monitoring of stream salamanders in Shenandoah National Park; however, we recommend an increase in the number and representativeness of sample locations to reduce the uncertainty in annual estimates of abundance.

8.4 SUMMARY

In this chapter, we have described a hierarchical view of the population and a modeling framework that can be used in the analysis of surveys that produce spatially-indexed counts of animals. This hierarchical modeling framework may be used to specify sources of variation in animal abundance and detection among sample locations and therefore admits a formal decomposition of the spatial variation in counts into these two components.

The hierarchical view may be applied to a variety of commonly used sampling protocols and provides a unified treatment of ecological inference problems. For example, hierarchical models of repeated point counts have been used to estimate population abundances (Royle, 2004c), to quantify associations between abundance and environmental gradients (Dodd and Dorazio, 2004; Kéry et al., 2005), and to estimate maps of animal occurrence using spatial covariates of abundance (Royle et al., 2005; Kéry et al., 2005). Similarly, hierarchical models of multinomial counts have been used to improve site-specific estimates of abundance in removal surveys (Dorazio et al., 2005), to model spatial variation in local abundances from distance sampling surveys (Royle et al., 2004), to develop geostatistical models of abundance and occurrence using spatially-indexed, capture–recapture data (Royle et al., 2007b), and to estimate abundance indices from categories of calling intensity (Royle, 2004b; Royle and Link, 2005).

We anticipate other applications of hierarchical models in the context of modeling and estimation of abundance. For example, some surveys may use multiple sampling protocols, such as a combination of point counts and double-observer counts, to improve the representativeness of the sample. A combination of sampling protocols would seem to be attractive if one of the protocols can be completed with less time or effort. Another potential application of the hierarchical view is to account for the effects of sampling design (e.g., stratification or clustering) in the model. In complex designs it is essential that the model include variables that are relevant to the data-collection process. Failure to do so can produce misleading inferences (e.g., see Chapter 7 of Gelman et al., 2004). Fortunately, the hierarchical modeling approach is sufficiently flexible to accommodate these and many other applications.

9

OCCUPANCY DYNAMICS

Species distribution and patch occupancy are fundamental concepts in biogeography, landscape ecology, and metapopulation biology. They are also of considerable interest in the conservation and management of animal and plant species. Moreover, studies of factors that influence occurrence and distribution, such as habitat or landscape structure, are the focus of many ecological investigations. In Chapter 3 we discussed simple models of occupancy and occurrence probability for closed systems – that is, when the occupancy status of patches or sample units is constant under repeated sampling. Such systems are said to be 'closed' with respect to occupancy status. The modeling framework that we developed in that chapter was based on a simple logistic regression model for the occupancy state variable, with an additional model component to allow for observation error.

In this chapter, we consider the situation in which the occupancy state variable evolves over time. That is, the occupancy status of a patch or site is allowed to change. Sites that are occupied at time t may become unoccupied (i.e. experience local extinction), and sites that are unoccupied at time t may become occupied (i.e., they are colonized). These systems are said to be 'open' to the demographic processes of extinction and colonization. Considering these open systems requires extensions of the state model to accommodate temporal variation in occupancy state and, potentially, spatio-temporal interactions. The observation model is unchanged from that considered in Chapter 3.

There is intrinsic theoretical interest in these temporally dynamic processes of extinction and colonization. This interest is often embodied in the 'metapopulation postulates.' Hanski (1999, pp. 224–225) identifies 4 essential conditions for a species to exist as a metapopulation:

(1) habitat patches support local breeding populations;

(2) no single population is large enough to ensure long-term persistence;

(3) the patches are not too isolated to preclude recolonization; and

(4) local dynamics are sufficiently asynchronous to make simultaneous extinction of all local populations unlikely.

Central to these metapopulation postulates are notions of temporal variation and spatial interaction among local populations. As such, spatial and temporal dynamics are important elements of the metapopulation conceptual framework, especially as dynamic attributes (extinction/colonization) relate to patch characteristics (e.g., patch area, configuration). For example, an important metapopulation concept is the 'connectivity' of patches, which is usually expressed as a function of distances between patches. We might expect that colonization and survival probabilities increase with connectivity of a network of patches, and this has obvious conservation and management relevance.

Naturally, in this chapter we emphasize the hierarchical formulation of dynamic models for occupancy. The model for the observed detection/non-detection data is expressed as the product of two component models: first a submodel for the observations *conditional* on the latent (unobserved or partially observed) process, and, secondly, a submodel for the latent occupancy process. Classical likelihood-based inference under these models (MacKenzie et al., 2003) is based on removing the latent state process from the model by marginalization. In this chapter, we follow closely the hierarchical formulation of these models described in Royle and Kéry (2007). Under the hierarchical modeling approach, we retain the state variables in the model, as we believe that this representation naturally leads to several important extensions (e.g., the inclusion of site-level effects, including random effects). A special case of these models with site-level effects are models of spatial dynamics (Section 9.6). This is analogous to the relationship between models with individual effects in capture–recapture (Chapter 6) and the corresponding spatial capture–recapture models (Chapter 7), where we think the hierarchical formulation of the model is also preferred. The hierarchical representation of dynamic occupancy models is simple and, we believe, is a more natural way to express the models because the state model is Markovian and can be expressed as a sequence of simple conditional probability models. The hierarchical formulation of occupancy models first described in Chapter 3 extends directly to temporally dynamic systems, spatial systems, and spatio-temporal systems. We emphasize Bayesian analysis of the hierarchical parameterization of dynamic occupancy models because likelihood analysis of complex spatial models is not practically feasible. Bayesian analysis of such models can be achieved directly using **WinBUGS**.

We begin by developing models of occupancy dynamics under the assumption that the state variable can be observed perfectly (that is, assuming that $p = 1$). This case allows us to focus on understanding the fundamental structure of the state process model. Then we incorporate imperfect observation of the state variable in later sections.

9.0.1 Background

Modeling and inference in metapopulation models has received considerable attention over the last 10 or 15 years. Much of the work has been on devising models of extinction and colonization, assuming the occupancy state was perfectly observable. The Markovian state model (absent spatial dynamics) was developed by Clark and Rosenzweig (1994). Hanski (1994), Day and Possingham (1995) and others have addressed spatial dynamics. Formalization of inference procedures has been addressed by Moilanen (1999); and O'Hara et al. (2002) and Ter Braak and Etienne (2003) provide a Bayesian treatment of the inference problem for occupancy models with temporal dynamics. These papers focus on the state process model and inference under that model supposing that the state-variable could be observed perfectly.

Several papers have considered the situation in which the state variable is subject to imperfect observation. Erwin et al. (1998) recognized the connection between these models and models for populations of individuals experiencing mortality and recruitment. The equivalence arises by equating a site or patch (or colony) with an individual. Thus, when detection is perfect, open population capture–recapture models can be used to obtain estimates of survival and recruitment. This work led to Barbraud et al. (2003), who continued to develop the analogy with capture–recapture, employing a 'multi-state' model under the 'robust design' (Pollock, 1982) to estimate colony survival and colonization in the presence of imperfect detection. The fundamental problem addressed by Barbraud et al. (2003) was, quoting directly from them (Barbraud et al., 2003, p. 116):

> However, in studies of colony dynamics, interior 0's in a colony site detection history can indicate either that breeding individuals were present but not detected, or that breeding individuals were not present (locally extinct), yet recolonized at a later time. In this respect, the modeling of colony site detection history data is similar to the modeling of species detection history data in community studies (Nichols et al. 1998), and is similar also to the modeling of capture–recapture data in the presence of temporary emigration. The robust design (Pollock 1982) provides the information needed to estimate quantities of interest in the presence of temporary emigration (Kendall et al. 1997), and also provides a basis for estimating colony dynamics parameters from detection history data.

Previously, Moilanen (2002) also considered non-detection or observation error in the form of 'false zeros' as a result of non-detection. MacKenzie et al. (2003) gave a comprehensive treatment of the problem with formal inference based on conventional likelihood methods. Royle and Kéry (2007) provided a fully-hierarchical formulation of the model that integrates the Markovian process model with the formal observation model allowing for non-detection.

9.1 OCCUPANCY STATE MODEL

We consider data obtained from repeated sampling of $i = 1, 2, \ldots, M$ spatial units (patches or 'sites'). The sites are sampled in $t = 1, 2, \ldots, T$ periods (which we refer to as primary periods) across which the occupancy state of sites may change due to local extinction and colonization. These primary periods are defined relative to the biology of the species in question and for that reason MacKenzie et al. (2003) referred to them as 'seasons'. In the case of many birds, amphibians, and other vertebrates in temperate climates, these primary periods or seasons might naturally correspond to an annual period, such as a breeding season.

We begin by developing the basic structure of the state model for open systems. Let $z(i, t)$ denote the true occupancy status of site i during primary period t, having possible states 'occupied' ($z = 1$) or 'not occupied' ($z = 0$). One parameter of interest is the probability of site occupancy (or the probability of occurrence) for period t, $\psi_t = \Pr(z(i, t) = 1)$. Changes in occupancy over time can be parameterized explicitly in terms of local extinction and colonization processes, analogous to population demographic processes of mortality and recruitment. Let ϕ_t be the probability that an occupied site 'survives' (i.e., remains occupied) from period t to $t + 1$, i.e., $\phi_t = \Pr(z(i, t+1) = 1 | z(i, t) = 1)$. Local extinction probability (ϵ_t) in the parameterization used by MacKenzie et al. (2003), is the complement of ϕ_t, i.e., $\epsilon_t = 1 - \phi_t$. In metapopulation systems, *local colonization* is the analog of the recruitment process in a classical population (i.e., of individuals). Let γ_t be the local colonization probability from period t to $t+1$, i.e., $\gamma_t = \Pr(z(i, t+1) = 1 | z(i, t) = 0)$.

The state model has a simple formulation in terms of initial occupancy probability, i.e., at $t = 1$, which we will designate ψ_1, local survival probabilities, $(\phi_1, \ldots, \phi_{T-1})$, and the recruitment (colonization) probabilities $(\gamma_1, \ldots, \gamma_{T-1})$. The initial occupancy states, i.e., for $t = 1$, are assumed to be *iid* Bernoulli random variables,

$$z(i, 1) \sim \text{Bern}(\psi_1) \quad \text{for } i = 1, 2, \ldots, M, \tag{9.1.1}$$

whereas, in subsequent periods,

$$z(i, t) | z(i, t-1) \sim \text{Bern}(\pi(i, t)) \quad \text{for } t = 2, 3, \ldots, T, \tag{9.1.2}$$

where

$$\pi(i, t) = z(i, t-1)\phi_{t-1} + [1 - z(i, t-1)]\gamma_{t-1}. \tag{9.1.3}$$

Thus, for a site that is occupied at $t-1$ (i.e., $z(i, t-1) = 1$), the survival component in Eq. (9.1.2) determines the subsequent state and $z(i, t)$ is a Bernoulli outcome with probability ϕ_{t-1}. Conversely, if a site is not occupied at time $t - 1$, then the recruitment component in Eq. (9.1.2) determines the subsequent state, and $z(i, t)$

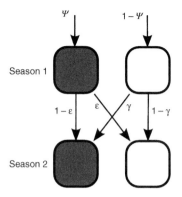

Figure 9.1. Occupancy state model, with 2 states. The parameters $\phi = 1 - \epsilon$ and γ are local survival and colonization probabilities, respectively and ψ is the initial occupancy probability. Graphic courtesy of Ian Fiske.

is a Bernoulli outcome with parameter γ_{t-1}. The expressions in Eqs. (9.1.1) and (9.1.2) define the state process model. Generalizations, where ϕ and γ may be structured spatially or temporally for instance, are described later. The dynamics of this simple system in which the state process has only 2 possible states is depicted graphically in Figure 9.1. Conceptually, these models extend directly to systems in which the state variable possesses >2 states (see Figure 9.2), and there has been some recent interest focused on such models (Royle, 2004b; Royle and Link, 2005; Nichols et al., 2007). Evidently, the two-state occupancy model described here is a particular case of a multi-state model (having 2 states) with a single unobservable state (Kendall and Nichols, 2002).

9.1.1 Metapopulation Summaries

The canonical parameters of the dynamic occupancy model are the initial occupancy probability, ψ_1, the survival probabilities $\{\phi_t\}_{t=1}^{T-1}$ and the colonization probabilities $\{\gamma_t\}_{t=1}^{T-1}$. In addition, a number of derived parameters are of interest. First, the occupancy probability at t can be computed recursively according to

$$\psi_t = \psi_{t-1}\phi_{t-1} + (1 - \psi_{t-1})\gamma_{t-1} \tag{9.1.4}$$

for $t = 2, \ldots, T$. MacKenzie et al. (2003) defined the growth rate as

$$\lambda_t = \frac{\psi_{t+1}}{\psi_t}.$$

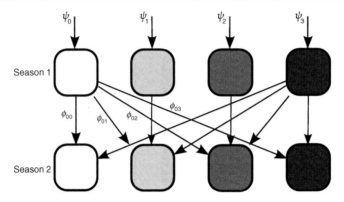

Figure 9.2. Multi-state occupancy model structure for a four-state process. The parameters ϕ are the state-transition probabilities governing local survival and colonization and ψ is the initial occupancy probability. Graphic courtesy of Ian Fiske.

Nichols et al. (1998a) defined *turnover* as *the probability that an occupied quadrat is a newly occupied one.* That is, turnover is the probability $\Pr(z(t-1)=0|z(t)=1)$. Bayes' Rule yields an expression for this in terms of previously defined model parameters:

$$\tau_t = \frac{\gamma_{t-1}(1-\psi_{t-1})}{\gamma_{t-1}(1-\psi_{t-1})+\phi_{t-1}\psi_{t-1}}, \tag{9.1.5}$$

for $t = 2, \ldots, T$. The denominator here is equal to ψ_t, by Eq. (9.1.4).

One useful summary of the dynamical system is the equilibrium occupancy probability (Hanski, 1994; MacKenzie et al., 2006, Chapter 7), i.e., the stable-state (occupancy) distribution. This is related to local survival and colonization according to:

$$\psi_t^{(eq)} = \frac{\gamma_t}{\gamma_t + (1-\phi_t)}.$$

This is the leading element of the dominant eigenvector of the 2×2 state transition matrix, and it is usually called the incidence function (Hanski, 1999, p. 85).

As we discussed in Section 3.7, there might be some interest in estimating finite-sample quantities, such as the number of presently occupied sites, the number of newly occupied sites, and similar quantities. These are functions of *realized* values of the occupancy state variables, not attributes of some hypothetical population. We may conduct inference about such quantities within a framework for Bayesian analysis of the hierarchical model without difficulty (Royle and Kéry, 2007).

9.1.2 Bayesian Analysis

The simple conditional specification of the hierarchical representation yields easily to Bayesian analysis using conventional methods of Markov chain Monte Carlo (MCMC). For the dynamic occupancy models, the Gibbs sampler is remarkably simple when conjugate prior distributions are used (see, Royle and Kéry, 2007, Appendix). But otherwise, more general methods can be employed. In practice, these models can be implemented without difficulty in **WinBUGS**. We provide an illustration in the following section.

9.1.3 Swiss Bird Data: European Crossbill Dynamics

We consider data from the Swiss Survey of Common Breeding Birds (Monitoring Häufige Brutvögel; MHB) over 4 years (2001 to 2004) for the European Crossbill (*Loxia curvirostra*) (Schmid et al., 1998, 2004; Kéry and Schmid, 2006). The data consist of replicate samples of 267 1 km^2 quadrats across Switzerland, sampled 2 or 3 times each during the breeding season. Other aspects of the survey have been described in Chapter 3. The crossbill is a medium-sized pine-seed-eating finch that is widespread in Switzerland. Its abundance and occurrence depend greatly on the cone set of conifers. In mast years, crossbills appear irruptively in many regions where otherwise they do not occur or are scarce. This species may have the greatest dynamics in site occupancy among all regular Swiss breeding birds (Royle and Kéry, 2007).

We assume a model in which local survival and colonization are year-specific. Further, we have reduced the data by taking only the second sample of the 3 replicates in each year. We will suppose that, for that survey, the species was detected perfectly. The **WinBUGS** specification of the model described in the previous section is given in Panel 9.1. While this model represents a fairly substantial extension over the ordinary logistic regression model (Chapter 3), its implementation in **WinBUGS** requires little or no additional technical attention. The **R** script to organize the crossbill data and estimate model parameters using **WinBUGS** is provided with the Web Supplement.

Posterior summaries of model parameters are given in Table 9.1. The population is characterized by low occupancy rates and reasonably constant survival over time. Conversely, colonization probabilities are relatively low but annual turnover is substantial, as a result of the low basic occupancy rate. Shortly we will compare these results to those obtained under a model where we allow for the possibility of imperfect observation of the occupancy state variable.

9.2 A GENERALIZED COLONIZATION MODEL: MODELING INVASIVE SPREAD

Under the model described in the previous section, site colonization is 'random' and does not depend on whether a site has previously been colonized and then suffered an extinction. All sites that are unoccupied at time t have the same probability of being colonized by $t+1$. There are many biological situations where we might wish to consider recolonization as being distinct from 'first colonization.' For example, in modeling the spread of invasive species or disease, distinct mechanisms should influence these two processes. We thus extend the model to distinguish initial colonization from recolonization.

Define the auto-covariate $A(i, t) = 1$ ('available' for initial colonization) if a (site) has never been colonized prior to t, and $A(i, t) = 0$ if the site has previously been colonized. Set $A(i, 1) \equiv 1; i = 1, 2, \ldots, M$. Formally, we can express $A(i, t)$ as the indicator function $A(i, t) = \prod_{k=1}^{t-1}(1 - z(i, k))$. Then, sites that are presently unoccupied ($z(i, t) = 0$) have different colonization rates depending on whether $A(i, t) = 1$ or $A(i, t) = 0$. An expression for the more general state model is

$$z(i, t + 1) \sim \text{Bern}(\pi(i, t + 1))$$

where

$$\pi(i, t + 1) = \phi_t z(i, t) + \gamma_{t+1}(1 - z(i, t))A(i, t) + \theta_{t+1}(1 - z(i, t))(1 - A(i, t)).$$

```
model {
  # prior distributions
  psi~dunif(0,1)
  for(t in 1:(T-1)){
     gamma[t]~dunif(0,1)
     phi[t]~dunif(0,1)
  }
  # state-model specification
  for(i in 1:M){
     z[i,1]~dbern(psi)
     for(t in 2:T){
         muZ[i,t]<- z[i,t-1]*phi[t-1] + (1-z[i,t-1])*gamma[t-1]
         z[i,t]~dbern(muZ[i,t])
     }
  }
}
```

Panel 9.1. **WinBUGS** model specification for a dynamic two-state occupancy model, assuming that the occupancy state can be observed perfectly.

Table 9.1. Posterior summaries of model parameters and certain derived parameters (below the horizontal line) for a dynamic occupancy model fitted to European Crossbill data from 2001–2004. Model assumes perfect observation of the state variable. q_x is the $100 \times x$th percentile of the posterior distribution.

parameter	mean	sd	$q_{0.025}$	$q_{0.50}$	$q_{0.975}$
ϕ_1	0.571	0.085	0.403	0.573	0.733
ϕ_2	0.506	0.068	0.372	0.506	0.638
ϕ_3	0.540	0.060	0.422	0.540	0.656
γ_1	0.151	0.023	0.108	0.150	0.200
γ_2	0.194	0.027	0.144	0.193	0.250
γ_3	0.111	0.022	0.071	0.110	0.158
ψ_1	0.122	0.020	0.085	0.121	0.164
ψ_2	0.202	0.025	0.156	0.201	0.252
ψ_3	0.257	0.027	0.207	0.257	0.311
ψ_4	0.221	0.025	0.174	0.220	0.273
λ_1	1.694	0.295	1.206	1.663	2.356
λ_2	1.288	0.178	0.978	1.275	1.675
λ_3	0.865	0.101	0.681	0.860	1.076
τ_1	0.656	0.064	0.525	0.657	0.775
τ_2	0.603	0.059	0.485	0.604	0.716
τ_3	0.372	0.062	0.254	0.371	0.498

This model becomes relevant in the context of modeling animal population dynamics (Chapter 10), as the Jolly–Seber state process model corresponds to the restriction $\theta_t = 0 \; \forall t$. That is, if we equate sites (a metapopulation unit) to individuals (a population unit), then the population dynamic model for a population of individuals is equivalent to a dynamic occupancy model wherein recolonization cannot happen (that being equivalent to raising the dead, let's say).

9.3 IMPERFECT OBSERVATION OF THE STATE VARIABLE

In the previous sections, we provided a brief analysis of the crossbill data under a model in which we assumed that detection of the species was perfect (that is, that the 'data' are equivalent to realizations of $z(i,t)$). As with the simple occupancy models for closed systems considered in Chapter 3, our ability (or inability) to observe the state variable of interest affects our inference about parameters that describe occupancy dynamics. In this section, we consider an observation process in which non-detection in the form of false negative errors is possible. As such, we introduce an additional parameter $p = \Pr(y = 1|z = 1)$. Its complement is the probability of a false absence, $1 - p = \Pr(y = 0|z = 1)$. As in Chapter 3, we extend the model by combining this observation component of the problem with the dynamic state model described in Section 9.1.

The consequence of ignoring the observation component of the model is that turnover appears higher than it actually is and ϕ appears lower. To illustrate, consider a situation with $T = 2$ primary periods. In this case, there are 4 possible occupancy sequences $(z(i,1), z(i,2))$ that describe a site's potential occupancy states: $\{(1,1),(0,1),(1,0),(0,0)\}$. In the presence of imperfect detection, some of the occupied states, i.e., $(1,1)$, $(1,0)$, and $(0,1)$, appear to be in the unoccupied state, $(0,0)$. The *apparent* fraction surviving from $t = 1$ to $t = 2$ is

$$n_{11}/(n_{11} + n_{10}),$$

where n_h is the frequency of sites having encounter history h. For constant p, ϕ and γ, this ratio has expectation

$$\frac{\psi p \phi p}{\psi p \phi p + [\psi p (1 - \phi) + \psi p \phi (1 - p)]} = p\phi.$$

Thus failure to account for observation error, in the form of imperfect detection, renders naive sample estimates of survival rate (and other parameters) difficult to interpret and, some would argue, meaningless with respect to the ecological system that generated the data. What does a local survival estimate mean if its expectation is $p\phi$, and we don't know p? We can affect our estimate of ϕ simply by doing a poor job sampling. As such, we believe that detection bias should be a fundamental consideration when modeling occupancy dynamics.

To resolve this problem, we need to bring additional information to bear on the problem. In Chapter 3, we extended the design to include replicate observations, and we adopt that approach here also. Suppose each site is sampled J times within each of T primary periods and that each site is closed with respect to its occupancy status within but not across primary periods. The idealized data structure is depicted in Table 9.2 for the case where there are $T = 3$ primary periods (i.e., years) and $J = 3$ replicates per primary period. A typical situation from which such data would arise is when sampling is repeated several times both within the breeding season of a species and over several years. This 'design' is usually referred to as the 'robust design' (Pollock, 1982; Kendall et al., 1997; Williams et al., 2002, Chapter 19), a term coined in conventional capture–recapture analyses.

9.3.1 Hierarchical Formulation

Denote the *observed* occupancy status of site i for survey j within primary period t as $y_j(i,t)$. We assumed that $y_j(i,t), j = 1, 2, \ldots, J$ are independent and

Table 9.2. Data structure and state variable matrix under the robust design with $T = 3$ primary periods (e.g., years) and $J = 3$ replicate samples per primary period. In practice, the state matrix is only partially observable.

site	\multicolumn{9}{c}{DATA MATRIX}			\multicolumn{3}{c}{STATE MATRIX}
	$t=1$			$t=2$			$t=3$			$t=1$	$t=2$	$t=3$
	$j=1$	$j=2$	$j=3$	$j=1$	$j=2$	$j=3$	$j=1$	$j=2$	$j=3$			
1	0	0	0	0	0	0	0	0	0	0	0	0
2	0	0	0	0	0	0	0	0	0	0	0	0
3	0	0	0	0	0	0	0	0	0	0	0	0
4	0	0	0	0	0	0	0	0	0	0	0	0
5	0	0	0	0	0	0	0	0	0	0	0	0
6	0	0	0	1	1	0	0	0	0	0	1	0
7	1	0	1	0	0	0	1	1	0	1	0	1
8	0	0	0	0	0	0	0	0	0	0	0	0
9	0	0	0	0	0	0	0	0	0	0	0	0
10	0	0	0	1	1	1	1	0	0	0	1	1
11	0	0	0	0	0	0	0	0	0	0	0	0
12	0	0	0	0	0	1	0	1	1	0	1	1
13	1	1	1	0	0	0	0	0	0	1	1	1
14	0	0	0	0	0	0	0	0	0	0	0	0
15	1	0	1	0	0	1	0	0	0	1	1	1
16	1	1	0	1	0	0	0	1	0	1	1	1
17	1	1	0	0	1	0	0	0	0	1	1	1

identically distributed for each site (i) and primary period (t). Let $\mathbf{y}(i,t) = (y_1(i,t), y_2(i,t), \ldots, y_J(i,t))$ be the vector of all observations for site i and primary period t. For convenience, we assume there are no missing values, but these are easily handled within the likelihood framework (MacKenzie et al., 2003), and basically irrelevant when the model is given a Bayesian analysis, such as implemented in **WinBUGS**. In such cases, missing values are simulated from their posterior distribution using standard MCMC methods.

This model is naturally formulated as a state-space or hierarchical model, in which we express the model by its two component processes: a submodel for the observations conditional on the unobserved state process, i.e., $\mathbf{y}(i,t)|z(i,t)$ and, secondly, a submodel for the unobserved or partially observed state process $\{z(i,t); i = 1, 2, \ldots, t = 1, 2, \ldots\}$. The classical likelihood formulation of this problem (MacKenzie et al., 2003) exploits an identical model structure, but removes the latent indicators of occupancy from the likelihood by integration. In contrast, in adopting a Bayesian framework for analysis and inference, retention of the latent variables in the model does not pose any difficulty. This has some advantages for certain model extensions, as we will see later.

9.3.1.1 Observation model

The observation model, specified conditional on the latent process $z(i,t)$, is given by

$$y_j(i,t)|z(i,t) \sim \text{Bern}(z(i,t)\,p_t).$$

Thus, if a site is occupied at time t ($z(i,t) = 1$), the data are Bernoulli trials with parameter p_t. If a site is unoccupied at time t ($z(i,t) = 0$), then the data are Bernoulli trials with $\Pr(y_j(i,t) = 1) = 0$. Obvious generalizations to accommodate structure in p can be obtained directly. For example, in many applications, sampling covariates are measured during each survey such as search effort, weather, or other environmental conditions. In this case, the observation model is Bernoulli with parameter $z(i,t)p_{it}$ and covariate effects are modeled on a suitable transformation of p_{it}. For example,

$$\text{logit}(p_{it}) = \alpha_0 + \alpha_1 x(i,t),$$

where $x(i,t)$ is the value of some covariate for site i and primary period t.

9.3.2 Swiss Bird Data: European Crossbill Dynamics

We return to the crossbill data and provide an analysis under the extended model in which imperfect observation of the state variable is considered. The **WinBUGS** specification of this more general model is given in Panel 9.2 and posterior summaries are tabulated in Table 9.3. We note that some derived parameters are provided in this table, and the **WinBUGS** model specification to compute those summaries is provided in the Web Supplement, as is the **R** script for organizing the data and executing **WinBUGS**.

There were large interannual changes in crossbill occupancy in Switzerland with increases between 2001 to 2002 and 2002 to 2003, followed by a decline in occupancy between 2003–04. The high turnover rates confirm the highly erratic occupancy pattern of the crossbill. Detection probability, p, was moderately high but did not change much across years. Despite this, the basic summary of the system is quite different than under the model with $p = 1$ (see Table 9.1). In particular, when we assume that $p = 1$, the apparent turnover (posterior means) were 0.656, 0.603 and 0.372 for the 3 annual intervals. Conversely, allowing for $p < 1$ yields posterior means of 0.499, 0.259, and 0.113. Thus, the system appears much more dynamic under the restricted model in which $p = 1$.

9.3.3 Likelihood Analysis of the Model

Despite our preference for the hierarchical formulation of models of occupancy dynamics, likelihood analysis of the model with simple temporal dynamics is

```
model {

# prior distributions
psi~dunif(0,1)
p[nyear]~dunif(0,1)
for(i in 1:(nyear-1)){
  gamma[i]~dunif(0,1)
  phi[i]~dunif(0,1)
  p[i]~dunif(0,1)
}

# process model specification
for(i in 1:nsite){
  z[i,1]~dbern(psi)
    for(t in 2:nyear){
      muZ[i,t]<- z[i,t-1]*phi[t-1] + (1-z[i,t-1])*gamma[t-1]
      z[i,t]~dbern(muZ[i,t])
    }
}

# observation model
for (t in 1:nyear){
    for(i in 1:nsite){
      for(j in 1:nrep){
        Py[i,j,t]<- z[i,t]*p[t]
        y[i,j,t] ~ dbern(Py[i,j,t])
      }
    }
  }
}
```

Panel 9.2. WinBUGS code for the hierarchical formulation of a dynamic occupancy model. Additional examples are in Royle and Kéry (2007) and the Web Supplement which also contains the **R** instructions for fitting this model.

Table 9.3. Posterior summaries of model and derived parameters (below horizontal line) from the dynamic occupancy model fitted to the Swiss Breeding Bird Survey occupancy data on the European Crossbill (*Loxia curvirostra*) from 2001 to 2004. q_x is the $100 \times x$th percentile of the posterior distribution. Reproduced from Royle and Kéry (2007).

parameter	mean	sd	$q_{0.025}$	$q_{0.500}$	$q_{0.975}$
p_1	0.584	0.044	0.493	0.584	0.666
p_2	0.493	0.037	0.422	0.493	0.564
p_3	0.566	0.033	0.504	0.566	0.629
p_4	0.574	0.037	0.499	0.574	0.643
ϕ_1	0.806	0.069	0.656	0.812	0.931
ϕ_2	0.855	0.046	0.758	0.858	0.938
ϕ_3	0.682	0.053	0.576	0.682	0.791
γ_1	0.259	0.037	0.189	0.257	0.334
γ_2	0.190	0.041	0.114	0.189	0.273
γ_3	0.071	0.029	0.022	0.068	0.133
ψ_1	0.242	0.029	0.190	0.241	0.300
ψ_2	0.391	0.035	0.323	0.390	0.461
ψ_3	0.450	0.034	0.386	0.450	0.517
ψ_4	0.346	0.032	0.286	0.345	0.409
λ_1	1.635	0.201	1.284	1.619	2.085
λ_2	1.157	0.099	0.978	1.153	1.371
λ_3	0.770	0.065	0.651	0.768	0.908
τ_1	0.499	0.058	0.384	0.500	0.610
τ_2	0.259	0.056	0.151	0.258	0.371
τ_3	0.113	0.046	0.032	0.110	0.213

straightforward. MacKenzie et al. (2003) provided a general characterization of open models in the presence of imperfect detection, and they described a likelihood-based framework for inference about model parameters. They exploited the similarity between this Markovian model for occupancy and classical multi-state capture–recapture models (in particular, models of temporary emigration of Kendall (1999)). The likelihood formulation is thus consistent with this broader class of capture–recapture models under the robust design. The details here are taken from MacKenzie et al. (2003) with some changes in notation.

As before, let $y_j(i, t)$ denote the *observed* occupancy status of site i for survey j within primary period t. And, let $\mathbf{y}_i$ be a column vector of all observations for site i organized in chronological order. Let $\phi_0 = (\psi_1, 1 - \psi_1)$ denote the occupancy state distribution at time $t = 1$. The model for the state variable $z(i, t)$ is a two-state Markov process having transition matrix

$$\Phi_t = \begin{pmatrix} \phi_t & 1 - \phi_t \\ \gamma_t & 1 - \gamma_t \end{pmatrix}, \tag{9.3.1}$$

where elements ϕ_t and γ_t are the local survival and colonization probabilities defined previously in Section 9.1.

Secondly, we need the probabilities of observing a particular $\mathbf{y}_i$ given the state vector $\mathbf{z}_i$, which we will denote by $\boldsymbol{\theta}_{y,t} = \Pr(\mathbf{y}|\mathbf{z})$. This is a function of parameters p_{ijt}. For example, given the encounter history $(1,1)$ within some primary period then

$$\boldsymbol{\theta}_{11} = \begin{pmatrix} \Pr(\mathbf{y} = (1,1)|z = 1) \\ \Pr(\mathbf{y} = (1,1)|z = 0) \end{pmatrix} = \begin{pmatrix} p_1 p_2 \\ 0 \end{pmatrix}.$$

Let $D(\boldsymbol{\theta}_{y,t})$ be the matrix with diagonal $\boldsymbol{\theta}_{y,t}$ and zeros off-diagonal. For example, if we have $J = 2$ secondary samples and given the encounter history $(1,1)$, then

$$D(\boldsymbol{\theta})_{y,t} = \begin{pmatrix} p_1 p_2 & 0 \\ 0 & 0 \end{pmatrix},$$

whereas, for the history corresponding to undetected, $(0,0)$,

$$D(\boldsymbol{\theta})_{y,t} = \begin{pmatrix} (1 - p_1)(1 - p_2) & 0 \\ 0 & 0 \end{pmatrix}.$$

Then, the probability of encounter history $\mathbf{y}_i$ is

$$\Pr(\mathbf{y}_i) = \phi_0 \left\{ \prod_{t=1}^{T-1} D(\boldsymbol{\theta}_{y,t}) \boldsymbol{\Phi}_t \right\} \boldsymbol{\theta}_y^T.$$

The likelihood is the product of the M such terms, according to:

$$L(\phi_0, \phi, \gamma, \mathbf{p}|\mathbf{y}_1, \ldots, \mathbf{y}_t) = \prod_{i=1}^{M} \Pr(\mathbf{y}_i).$$

This can be maximized in **R** easily. We provide a sample function that illustrates this with the Web Supplement. Software packages MARK (White and Burnham, 1999) and PRESENCE (MacKenzie et al., 2006) also implement a likelihood approach to inference for this model.

9.4 AUTO-LOGISTIC REPRESENTATION

It is evident from their construction (i.e., Eq. (9.1.2)) that dynamic occupancy models provide a means of parameterizing temporal auto-correlation in the binary occupancy state, similar to conventional auto-logistic models of correlated binary processes (e.g., Besag (1972)). However, our definition of the state model given by

Eq. (9.1.2) *appears* distinct from conventional specifications of such models which are linear functions of the state variable on the logit-scale. We note here that there is an exact equivalence between the representation given by Eq. (9.1.2) and more classical auto-logistic specifications of such models.

First, note that Eq. (9.1.2) can be rearranged to yield

$$z(i,t)|z(i,t-1) \sim \text{Bern}(\gamma_{t-1} + (\phi_{t-1} - \gamma_{t-1})\, z(i,t-1)) \qquad (9.4.1)$$

which resembles the regression-like structure of auto-logistic models with auto-regression parameter $b_{t-1} = \phi_{t-1} - \gamma_{t-1}$ except that, here, the Bernoulli success probability is specified on the probability-scale, instead of on the logit-linear scale. Note that there are some natural restrictions on these parameters, γ_{t-1} and b_{t-1} of Eq. (9.4.1), since $\gamma_{t-1} + b_{t-1} \leq 1$ and $\gamma_{t-1} + b_{t-1} \geq 0$, it must be that $-\gamma_{t-1} \leq b_{t-1} \leq 1 - \gamma_{t-1}$ (or vice versa). Alternatively, defining the Bernoulli success probability as $\pi_{it} = \text{Pr}(z(i,t) = 1|z(i,t-1))$, and considering models of the form

$$\text{logit}(\pi_{it}) = a_t + b_t z(i,t-1) \qquad (9.4.2)$$

does not require any restrictions on the parameters a_t and b_t of this reparameterized model. This reparameterization is precisely a conventional auto-logistic model in a time-series context. The new parameters a_t and b_t have simple relationships to the original parameters $(\gamma_{t-1}, \phi_{t-1})$. Namely:

$$\gamma_{t-1} - \text{logit}^{-1}(a_t)$$

and

$$\phi_{t-1} = \text{logit}^{-1}(a_t + b_t).$$

Note that we could transpose these relationships merely by substituting $1 - z(i,t-1)$ for $z(i,t-1)$ in Eq. (9.4.2).

We believe the auto-logistic parameterization will be more convenient in a number of situations, such as when covariates are thought to influence occupancy (rather than its components ϕ and γ), or modeling site-specific variation in occupancy, or spatial dynamics (i.e., classical auto-logistic models).

We have interacted with ecologists interested in modeling covariate effects on a 'net' occupancy rate in these dynamic occupancy models. Suppose that a single covariate is available, say x_i for site i, then it would be natural to include it as an additive effect according to

$$\text{logit}(\pi_{it}) = a_t + b_t z(i, t-1) + \beta x_i. \qquad (9.4.3)$$

To include separate effects of x for both colonization and survival probability, we need to extend the model to have an interaction of x with $z(i, t-1)$. That is,

$$\text{logit}(\pi_{it}) = a_t + b_t z(i, t-1) + \beta_1 x_i + \beta_2 x_i z(i, t-1).$$

This implies an effect of x on both survival and colonization probability. In particular, the effect on the logit of colonization probability is simply β_1 and the effect on the logit of survival probability is $\beta_1 + \beta_2$. The probability of occurrence at time t is defined recursively (as in Section 9.1.1). Under the auto-logistic parameterization with *no* covariate effect, ψ_t is

$$\psi_t = \text{expit}(a_t)\psi_{t-1} + \text{expit}(a_t + b_t)(1 - \psi_{t-1}).$$

While there is not a direct interpretation on occupancy of covariates effects on survival and colonization probability, the reduced model, on the boundary $\beta_2 = 0$ is more appealing and, importantly, the auto-logistic formulation appears to yield a more efficient Bayesian implementation, e.g., in **WinBUGS** because presumably, the parameters are more nearly orthogonal in the posterior. Under the auto-logistic parameterization, we find that Markov chains typically exhibit very low autocorrelation. It is worth emphasizing that parameterization is critically important in any statistical modeling activity (e.g., see Gilks and Roberts, 1995; Gelman, 2006) and see Section 8.3.2.1.

9.4.1 Covariate Models

We next consider a model for the crossbill data in which occupancy dynamics are parameterized to depend on covariates that describe variation in landscape structure. Specifically, we consider elevation and forest cover as in previous analyses of the Swiss BBS data. The latest Swiss breeding bird Atlas (Schmid et al., 1998), shows a maximum abundance at about 1400–1800 m. The crossbill feeds on pine seeds, and we therefore might expect an increasing response of abundance to forest cover (which is not classified by stand type in the data). Insofar as elevation and forest cover influence abundance, it should be the case that more forested and mid-elevation sites have higher survival probabilities and higher recolonization probabilities because the surrounding landscape has more spare birds to colonize a site.

We used the auto-logistic parameterization of the model to parameterize these effects and fit the model in **WinBUGS**. Specification of the model requires little additional detail beyond extending the linear predictor for $\Pr(z(i,t) = 1)$. We provide the **WinBUGS** model script with the Web Supplement. The main element of the model specification is the logistic regression structure:

```
for(i in 1:M){
 z[i,1]~dbern(psi)
   for(t in 2:T){
    logit(muZ[i,t])<- a[t-1] + b1[t-1]*z[i,t-1] + b2*forest[i]
       + b3*elev[i] + b4*(elev[i]*elev[i]) + b5*forest[i]*z[i,t-1]
       + b6*elev[i]*z[i,t-1] + b7*elev[i]*elev[i]*z[i,t-1]
    z[i,t]~dbern(muZ[i,t])
    }
 }
```

In this case, there are effects of each covariate on both colonization and survival (note the interaction of the covariates with the $z(i, t-1)$). Thus, for example, the effect of forest cover on survival probability is `b2 + b5` whereas the effect of forest cover on colonization probability is `b2`.

The results of fitting this model in **WinBUGS** are presented in Table 9.4. For clarity, the covariate effects are labeled by the name of the covariate. The derived parameters – the direct effects on ϕ and γ – are given in Table 9.5. We see that the effect of forest cover is most pronounced on survival probability, indicating higher survival probability with increasing forest cover. Further, while there is a concave response to elevation, the effect is largely to increase survival probability with increasing elevation (Figure 9.3, solid line). These effects are consistent with the expected pattern in abundance of the species and thus suggest a pattern of metapopulation dynamics that is inherited from variation in abundance. Conversely, there is little effect of forest cover on colonization probability, and a strong concave response of colonization to elevation (Figure 9.3, dashed line). These results suggest that more desirable habitats (higher forest cover and higher elevation, dominated by coniferous forests) have high occupancy rates and there is high turnover in more marginal habitats. This pattern would be fully consistent with current notions of site-dependent population regulation (Rodenhouse et al., 1997; Sergio and Newton, 2003).

9.5 SPATIAL AUTO-LOGISTIC MODELS

The occupancy modeling framework generalizes to accommodate explicit spatial structure, such as that representing contagion or similarity in occupancy status among sites. Such models are commonly used for image analysis applications and are also widely applied in other disciplines, including many ecological modeling problems (Heikkinen and Hogmander, 1994; Augustin et al., 1996; Hoeting et al.,

Table 9.4. Posterior summaries of model parameters and certain derived parameters for a dynamic occupancy model fitted to European Crossbill data from 2001–2004. Model has effects of elevation and forest cover on both local survival and colonization. q_x is the $100 \times x$th percentile of the posterior distribution. The model is parameterized in terms of a logit-linear response of occurrence probability with year-specific intercepts a and year-specific auto-logistic regression parameters $b1$. Covariate effects are labeled accordingly. See Section 9.4 for further details.

parameter	mean	sd	$q_{0.025}$	$q_{0.50}$	$q_{0.75}$
p_1	0.579	0.045	0.487	0.580	0.664
p_2	0.524	0.036	0.453	0.523	0.594
p_3	0.571	0.031	0.510	0.571	0.632
p_4	0.563	0.036	0.490	0.563	0.633
$a[1]$	0.167	0.283	−0.382	0.164	0.726
$a[2]$	−0.001	0.304	−0.606	0.000	0.594
$a[3]$	−1.372	0.544	−2.579	−1.320	−0.465
$b1[1]$	0.276	0.553	−0.766	0.267	1.383
$b1[2]$	2.007	0.618	0.858	1.984	3.259
$b1[3]$	2.204	0.689	0.994	2.159	3.651
forest	0.083	0.160	−0.234	0.085	0.397
elev	0.207	0.181	−0.148	0.208	0.559
elev^2	−1.441	0.265	−1.998	−1.431	−0.953
$\text{forest} * z(t-1)$	1.152	0.379	0.451	1.137	1.934
$\text{elev} * z(t-1)$	1.034	0.377	0.320	1.023	1.791
$\text{elev}^2 * z(t-1)$	0.953	0.467	0.067	0.947	1.878

2000; Gumpertz et al., 2000; Sargeant et al., 2005). In a recent paper, Sargeant et al. (2005), consider the development of spatial models for occupancy in the presence of imperfect observation of the state variable. Their model represents an extension of the classic auto-logistic model (Besag, 1972; Augustin et al., 1996) to include non-detection or false-negative errors. An alternative formulation of a model allowing for false-negative errors was described by Hoeting et al. (2000). We begin with a brief introduction to the classic auto-logistic model, followed by its extension to allow for imperfect detection of occupancy state.

9.5.1 The Auto-logistic Model

We suppose that spatial patches or units form a lattice of G discrete, non-overlapping units indexed by $i = 1, 2, \ldots, G$. Let $\mathcal{N}_i$ denote the collection of units that are neighbors of i. If the lattice is a regular grid, a common choice of neighborhood structure in many problems is a so-called first order neighborhood in which neighbors of i are defined as units north, south, east and west of i (a

Table 9.5. Posterior summaries of *derived* parameters for a dynamic occupancy model fitted to European Crossbill data, using the auto-logistic parameterization. The first 6 rows are the mean (logit-scale) colonization and survival probabilities, and the remaining parameters are the covariate effects on the logit of survival and colonization probability. For example, $for(\phi)$ is the coefficient on forest cover in the logit of survival probability.

parameter	mean	sd	$q_{0.025}$	$q_{0.50}$	$q_{0.75}$
$col[1]$	0.167	0.283	−0.382	0.164	0.726
$col[2]$	−0.001	0.304	−0.606	0.000	0.594
$col[3]$	−1.372	0.544	−2.579	−1.320	−0.465
$surv[1]$	0.443	0.473	−0.430	0.427	1.417
$surv[2]$	2.006	0.522	1.086	1.982	3.117
$surv[3]$	0.832	0.389	0.122	0.813	1.665
$for(\phi)$	1.235	0.323	0.648	1.216	1.921
$elev(\phi)$	1.240	0.312	0.663	1.227	1.892
$elev^2(\phi)$	−0.487	0.360	−1.186	−0.490	0.230
$for(\gamma)$	0.083	0.160	−0.234	0.085	0.397
$elev(\gamma)$	0.207	0.181	−0.148	0.208	0.559
$elev^2(\gamma)$	−1.441	0.265	−1.998	−1.431	−0.953

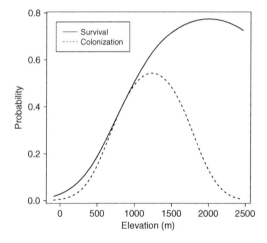

Figure 9.3. Estimated response of survival and colonization probability to elevation for the European Crossbill estimated from the Swiss BBS data.

'rooks' neighborhood'). Sometimes the diagonal cells are included. For an irregular lattice (e.g., counties or other administrative units), it is common to define the neighborhood to be units that share a boundary. Sargeant et al. (2005) apply their model to distribution modeling of kit foxes, in which the lattice consists of townships. Let n_i be the cardinality of $\mathcal{N}_i$, and define the spatial *auto-covariate*

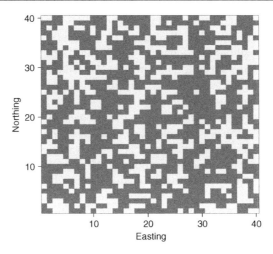

Figure 9.4. Binary image with auto-logistic parameters $\alpha = -0.5$ and $\beta = 1.5$ yielding an occupancy process with $\psi \approx 0.60$. The white cells are unoccupied and the colored cells are occupied.

$$x_i = \frac{1}{n_i} \left(\sum_{j \in \mathcal{N}_i} z_j \right).$$

Thus, x_i is the number of occupied neighbors of site i. The auto-logistic model is specified by the conditional distributions $[z_i | \mathbf{z}_{-i}]$ where $\mathbf{z}_{-i}$ denotes the vector of all latent state variables *except* i. In particular, $[z_i | \mathbf{z}_{-i}] \equiv \text{Bern}(\psi_i)$ where

$$\text{logit}(\psi_i) = \beta_0 + \beta_1 x_i.$$

Other (extrinsic) covariates, e.g., describing landscape or habitat could also be considered (and usually are). Accessible formulations of these models can be found in Hoeting et al. (2000) and Wintle and Bardos (2006).

We illustrate the analysis of these models with simulated data on a rectangular lattice of square sample units (quadrats) and for the special case where the state variable is observed perfectly. Even for this case, the likelihood for this model is intractable and analysis is carried out by approximate methods (e.g., 'pseudo-likelihood') or by MCMC under a fully-Bayesian treatment of the model. Bayesian analysis of the auto-logistic model was considered in the context of distribution modeling by Heikkinen and Hogmander (1994), Wu and Huffer (1997), Huffer and Wu (1998) and, using **WinBUGS**, Wintle and Bardos (2006).

We suppose the neighbors are the quadrats immediately to the north, south, east and west, so a site has a maximum of 4 neighbors. Let G be the size of the grid (number of grid points), `numnn[i]` is the number of neighbors each quadrat has, and `NN` is a matrix having dimension $G \times \max(\textbf{numnn})$ where NN[i,j] is the integer identity of neighbor j to quadrat i. This auto-logistic model under perfect observation of the state variable is shown in Panel 9.3. In this case, where z is observed, and the inference problem is only to estimate the parameters β_0 and β_1. In the Web Supplement we provide an **R** script for simulating data and fitting the model in **R**. It would also be possible to employ this model for the case where not all G quadrats were sampled, in which case a component of the inference problem might be to predict the missing values of z_i.

A realization from this process is shown in Figure 9.4, which has $\alpha = -0.5$ and $\beta = 1.5$. This generates a mean occupancy of about 0.60. The realization shown here was obtained by Gibbs sampling. Note that, historically, image analysis applications inspired the development of Gibbs sampling algorithms (Geman and Geman, 1984).

9.5.2 Imperfect Observation of the State Variable

An important technical consideration of these models is that the auto-covariate is unobserved. It is unobserved because we may observe the state variable imperfectly, and so we don't know whether any particular neighbor is occupied, even if the

```
model{

alpha ~ dnorm(0,.01)
beta  ~ dnorm(0,.01)

for(i in 1:nG){
    x[i,1]<-0
    for(j in 1:numnn[i]){
        x[i,j+1]<-x[i,j]+z[NN[i,j]]
    }
    logit(psi[i])<- alpha + beta*(x[i,numnn[i]+1]/numnn[i])
    z[i]~dbern(psi[i])
  }
}
```

Panel 9.3. WinBUGS model specification for an auto-logistic model with perfect observation of the binary state variable z.

neighboring unit was sampled. Also, we may not sample all G spatial units. However, the model is amenable to a Bayesian analysis by MCMC as missing or partially observed variables are treated no differently in the manner by which they are simulated. Almost all contemporary applications of auto-logistic type models are carried out using simulation-based methods.

In the context of binary maps, imperfect detection will cause the appearance of reduced spatial dependence than exists in the process itself. Since, presumably, spatial dependence is primarily the result of ecological processes of relevance (dispersal, migration, diffusive spread), understatement of spatial dependence should be detrimental to inference about such processes.

A hierarchical extension of the model is achieved with a formalization of the observation process. For example, consider the design in which J replicate samples are made of each spatial unit. Then, the observation model for the total number of detections for sample unit i is, for the case where detection probability is constant,

$$y_i \sim \text{Bin}(J, p\, z_i).$$

In general, p may vary among sites based on search intensity and other factors and so additional model structure on p might be desired. Note that this observation model is precisely equivalent to that considered in previous models of occurrence in which imperfect detection was considered, e.g., the models of Chapter 3. Sargeant et al. (2005) used a slightly different observation model. In their study, they sampled for the presence of kit fox tracks, and sampling was conducted until the presence of foxes was confirmed, or until 3 samples were conducted. This is a classic 'removal' design (e.g., see MacKenzie et al., 2006, p. 102).

Under this form of observation error, the model described previously (and implemented in Panel 9.3) is unchanged as a description of the *state process*. Naturally, since we have seen this observation model so many times before, extending the **WinBUGS** model specification requires little additional model description. This is shown in Panel 9.4. As before, the **R** code to simulate data and execute **WinBUGS** is provided in the Web Supplement. Many modifications to this observation model are possible, as have been elaborated on elsewhere in this book.

9.5.2.1 Remarks on the auto-logistic model

While the model, as formulated here, is ideally suited for spatial processes that are defined on a discrete lattice, any region could be rendered into a lattice arbitrarily, with points assigned to grid cells within which they are located. Alternatively, the model can be given a point support formulation by specification of the auto-covariate according to

$$x_i = \left(\sum_j w_{ij} z_j \right),$$

having weights w_{ij} which would typically be a function of distance. In the case considered previously, $w_{ij} = 1/n_i$ if unit j is a neighbor of i and $w_{ij} = 0$ otherwise. When the spatial process can be viewed as having point support such as locations of trees, or locations of bird nests, then inverse-distance weights have been used (Rathbun and Cressie, 1994) in which:

$$x_i = \sum_j \frac{z_j}{d_{ij}}$$

or, a local density covariate such as

$$x_i = \sum_j z_i I(d_{ij} < \delta)$$

```
model{

alpha ~ dnorm(0,.01)
beta  ~ dnorm(0,.01)
p ~ dunif(0,1)

for(i in 1:nG){
    x[i,1]<-0
    for(j in 1:numnn[i]){
        x[i,j+1]<-x[i,j]+z[NN[i,j]]
    }
    logit(psi[i])<- alpha + beta*(x[i,numnn[i]+1]/numnn[i])
    z[i]~dbern(psi[i])
    mu[i]<-z[i]*p
    y[i]~dbin(mu[i],J)
    }
}
```

Panel 9.4. WinBUGS model specification for an auto-logistic model with observation of the state variable z subject to imperfect detection.

for some threshold distance δ. A conceptual issue arises if the locations of *all* elements of the population are not known. In this case, variation in x_i can be largely a result of nuisance sampling processes. In general, one would then need a model describing the probability that an element appears in the sample as a function of its location.

If a geographic area is 'discretized,' then the scale of the lattice relative to the observation network becomes an important issue. The scale of observation has to be roughly consistent with the scale of the lattice under consideration in the sense that we can't have too sparse of an observation network or parameters are poorly identified. When the grid is too fine, then a first order neighborhood won't exhibit much spatial structure. But to increase the neighborhood size induces considerable computational expense. As an example, consider the Swiss bird survey, which is based on sampling several hundred of approximately 41000 1 km quadrats. Using a first order rooks neighborhood would almost certainly produce $\beta \approx 0$ since there is basically no information about the value of neighboring states from such a sparse observation network. Thus, the coarseness of the lattice has to be consistent in some way with the spatial structure in the process being modeled.

9.6 SPATIO-TEMPORAL DYNAMICS

The spatial model considered in the previous section is not a dynamic model but, rather, is purely descriptive, similar to the models considered in Chapter 3. Conversely, the model described in Sections 9.1 and 9.3 is dynamic, but there is no spatial influence on the dynamics. We would like to achieve a conceptual unification of these two ideas, so that the occupancy state is evolving over time in response to spatially local information. We thus seek to devise models of $z(i, t)$ such that the state-transitions are functions of previous values of z *and* previous states of each z in some spatial vicinity. As we did in the previous section we focus on describing the state model, as it is instructive to contemplate sensible behavior for the state process in terms of how it is related to the state variable's value and location at previous times. We can then tack on an observation model which is, as it always has been, a simple Bernoulli model.

Obvious areas of relevance for such models include modeling invasive species and studying range dynamics of species in response to such things as variation in weather or the effects of climate change on species distributions. Despite this, there has not been a great deal of work on statistical modeling of spatio-temporal occupancy systems, although some recent efforts can be found in Zhu et al. (2005); and Hooten and Wikle (2007). The development presented here derives from work with our colleague F. Bled (Univ. Toulouse, France) in the context of several dynamic occupancy systems (including invasive species). We don't provide those examples here but describe some of the basic modeling considerations which can be implemented in **WinBUGS**.

9.6.1 Model Formulation

As before, we suppose that patches or sample units are organized in a discrete lattice and let $\mathcal{N}_i$ denote the collection of patches (sites) that are neighbors of site i and let n_i be the cardinality of $\mathcal{N}_i$. It is therefore natural to define the spatio-temporal auto-covariate:

$$x(i, t-1) = \frac{1}{n_i} \left(\sum_{j \in \mathcal{N}_i} z(j, t-1) \right).$$

The basic approach is to develop logistic regression-like models for $z(i, t)$ that allow the parameters of that model to depend on this auto-covariate. As before, the main technical difficulty is that the $x(i, t-1)$ are not generally observed, which we accommodate easily within a Bayesian framework for inference. Note that $x(i, t-1)$ is basically a measure of local density at the previous time and so we can entertain the notion of diffusive spread by considering $x(i, t-1)$ in a model for occurrence probability at time t. The key to developing space-time models of occupancy dynamics lies in the specification of the model for $\Pr(z(i, t) = 1 | x(i, t-1))$. We focus on a describing a model in which these conditional probabilities have sensible interpretations in terms of metapopulation dynamics.

9.6.1.1 Spatial models of survival
Consider the following model for survival

$$\Pr(z(i, t) = 1 | z(i, t-1) = 1) = \phi + (1 - \phi)\, \alpha\, x(i, t-1).$$

This admits an explicit partitioning of the survival process into two components where $\Pr(z(i, t) = 1 | z(i, t-1) = 1)$ is 'net' survival rate, i.e., it is the probability that a site is occupied at time t given that it was occupied at $t-1$. The parameter ϕ is what we will refer to as the *intrinsic survival probability* – the probability that a local population sustains itself in the absence of any support from neighboring local populations, i.e., when there are no occupied neighbors so that $x(i, t-1) = 0$. The parameter α could be interpreted as a *'rescue effect'* (Brown and Kodric-Brown, 1977). It is the probability that a local population that goes extinct after $t-1$ is recolonized prior to the next occasion t. The probability of a local population being 'rescued' increases as the number of occupied neighbors increases (from 0 to n_i).

The implication of this formulation is that there are *two distinct mechanisms* that lead to consecutive, occupied states, i.e., two consecutive, say $(z(i, t-1), z(i, t)) = (1, 1)$, can arise either because a site was occupied and remained occupied, or because a site becomes unoccupied and then recolonized prior to t. That both parameters are identifiable is a result of the additional information afforded by spatial structure in the data viz. the auto-covariate $x(i, t)$.

9.6.1.2 Dynamic colonization model

We describe the net colonization probability according to

$$\Pr(z(i, t) = 1 | z(i, t-1) = 0) = \gamma + (1 - \gamma)\beta x(i, t-1).$$

This formulation of the model makes a distinction between 'random' colonization, as might be expected in a stable metapopulation, and dynamic or diffusive spread, such as might be expected in a growing (or invading) population. Specifically, γ is the probability that a previously unoccupied site with no occupied neighbors is colonized and β describes the increase in colonization probability due to occupancy of neighbors. The parameter β embodies diffusive or dynamic spread due to gradients in local density or occupancy.

9.7 SUMMARY

Ecological systems are fundamentally dynamic, due to basic survival, recruitment, and movement processes operating on individuals. These processes affect occurrence of species and are manifest in metapopulation dynamics that are usually referred to as local extinction and colonization. Models of occupancy that were introduced in Chapter 3 extend directly to include spatial and temporal variation in the state variable $z(i, t)$. Temporal variation in occupancy can be parameterized directly in terms of extinction and colonization probabilities. Models containing explicit occupancy dynamics are widely used in metapopulation ecology, but they are also relevant to modeling disease, invasive species, temporal dynamics of range and distribution, and other spatio-temporal systems.

Dynamic models of occupancy are naturally formulated in terms of a hierarchical model. As in Chapter 3, the model is described by two component models: one for the latent (unobserved, or only partially so) occupancy state variable (the 'process model') and another for the observations conditional on the state variable (the 'observation model'). Both constituent models have remarkably simple forms (being characterized by Bernoulli probability models), yielding a clear segregation of parameters governing ecological processes of interest from those that are responsible for dealing with nuisance sampling artifacts. The hierarchical formulation yields a flexible framework for modeling occupancy dynamics. The conditional probability structure of the hierarchical formulation is ideally suited for Bayesian analysis, and implementation of these models can be achieved in **WinBUGS**.

In the case of the static occupancy model (Chapter 3) there was not a significant benefit to the hierarchical formulation of the model except, in that case, we believe that it is a more natural way to formulate the model. However, in dynamic systems, we believe the case for hierarchical formulations is more compelling. The hierarchical formulation of the model yields several important inferential advantages over the likelihood-based approach (MacKenzie et al., 2003). Importantly, it yields a generic, flexible, and practical framework for modeling individual ('site') effects, or other latent structure in parameters (e.g., random year effects), in dynamic occupancy models. Models with explicit spatial dynamics are necessarily formulated as individual effects models, where the individual effect is a function of occupancy states at nearby spatial units. For example, in the spatial auto-logistic model, we introduce a site-specific covariate that is related to the occupancy status of neighbors. This model does not have a 'non-hierarchical' formulation.

10

MODELING POPULATION DYNAMICS

The use of data from marked individuals in studies of animal populations is widespread in contemporary ecology. Capture–recapture models represent one of the most useful classes of statistical methods in population ecology. Many useful models have been devised for the case where the population being sampled can be viewed as a closed population (see Chapters 5 and 6). That is, when sampling occurs during a period in which mortality and recruitment processes are not occurring. There also exists considerable interest in models of open populations, i.e., those that allow for explicit consideration of survival and recruitment.

One class of models in widespread use are the so-called Cormack–Jolly–Seber (CJS) models (see Chapter 11). In CJS models, attention is focused on the survival process and the model is developed conditional on the initial capture of individuals. As such, information about the recruitment process is discarded. The CJS models represent a restriction of a broader class of models known commonly as Jolly–Seber (JS) models. The Jolly–Seber model allows for recruitment, essentially by 'unconditioning' upon entry into the sample. That is, entry into the sample is regarded as the outcome of a random variable, and the parameters that describe entry into the sample are related to recruitment. There are a number of parameterizations of JS models that vary primarily in the manner in which the addition of new individuals (recruits) to the population is parameterized. Schwarz and Arnason (2005) recognize at least 5 practically or conceptually distinct representations of the JS model beginning with the original formulation devised by Jolly (1965) and Seber (1965), the 'reverse time' formulation of Pradel (1996), the superpopulation model of Schwarz and Arnason (1996) and, most recently, that of Link and Barker (2005). Schwarz and Arnason (2005) provide a good description of these various parameterizations and their relationships to one another.

In this chapter, we offer a new parameterization of capture–recapture models that allow for both survival and recruitment. We draw on the structural similarity between the dynamic occupancy models of the previous chapter, and models that act on animal populations. Survival and recruitment are fundamental processes in both systems, and the models for both systems are fundamentally equivalent. As we have seen in other contexts, the main distinction is that, in occupancy models, the size of the list of sites is fixed whereas, in JS type models, the size of the list is not fixed. We resolve this problem by the use of data augmentation (Royle et al., 2007a). Using data augmentation, we augment a list of individuals obtained in a traditional capture–recapture study with a large number of all-zero encounter histories. These uncaptured individuals represent the pool available for recruitment. We establish the relationship between this parameterization and existing parameterizations. In particular, this 'occupancy model' parameterization is very closely related to the 'superpopulation' formulation of the Jolly–Seber model put forth by Schwarz and Arnason (1996), in which the superpopulation size parameter is replaced in the model by a zero-inflation parameter, or its complement, what we have referred to previously as an inclusion (probability) parameter.

We believe that the formal duality between Jolly–Seber type models and metapopulation occupancy models represents the main conceptual contribution of our approach. However, the primary practical contribution is that these parameterizations readily allow for the inclusion of individual effects, such as individual heterogeneity or individual covariates. Because of the simple conditional structure of the model parameterization that is induced by data augmentation, they yield easily to a Bayesian analysis by Markov chain Monte Carlo techniques and can be easily implemented in **WinBUGS**.

10.1 DATA AUGMENTATION

Before proceeding with the formal specification of the hierarchical formulation of the Jolly–Seber model, we provide the conceptual formulation of our approach to the analysis of the class of Jolly–Seber models by data augmentation.

The key difficulty in developing the Jolly–Seber type model is parameterization of recruitment. In the context of the JS model, this is essentially related to the basic problem – the unknown size of the data set – which we have confronted in all closed population models. We will define N to be the total number of individuals ever alive. If N was known, then we would have a data set having a number of rows corresponding to all-zero encounter histories and (we assert without developing this idea) the model specification would be relatively simplified. For example, we might put a Poisson prior on each N_t and then, conditional on the total (i.e., N), the joint distribution of all N_t parameters is multinomial with equal cell probabilities.

To address that N is not known, we adapt the data augmentation approach which we have previously applied to similar classes of models. Using data augmentation, we introduce a pseudo-population[1] of $M >> N$ individuals, where M is fixed *a priori*. Conceptually, M is a population of pseudo-individuals from which the population of individuals exposed to sampling is drawn. Precisely, we suppose that $N \sim \text{Bin}(M, \psi)$ where ψ is the *inclusion probability* – the probability that an individual on the list of size M is a member of the superpopulation of size N. The model is greatly simplified having fixed M because this introduces observations $y(i, t)$ for $i = n + 1, \ldots, M$ that are known (zeros), and we can develop the model as a zero-inflated version of the known–N model.

Formally, the parameter N is replaced as an estimand by the inclusion probability ψ (or equivalently, the zero-inflation parameter $1 - \psi$). To accomplish this, suppose that $f(\mathbf{y}|N, \theta)$ is the probability of the observations conditional on N and additional parameters θ, and suppose that $N \sim \text{Bin}(M, \psi)$. That is, the prior distribution for N is binomial based on a sample size of M, and parameter ψ. The joint likelihood of the parameters (θ, N, ψ) is:

$$\mathcal{L}(\theta, N, \psi | M, \mathbf{y}) = f(\mathbf{y}|N, \theta)\text{Bin}(N|M, \psi). \qquad (10.1.1)$$

The parameters N and ψ are, in this case, redundant (note that this is not the case in metapopulation models of patch occupancy, see Royle and Kéry (2007) and Section 10.3.1), and it is therefore natural to focus on the resulting integrated likelihood, removing N by summation. This yields:

$$\mathcal{L}(\theta, \psi | M, \mathbf{y}) = \sum_{N=n}^{\infty} f(\mathbf{y}|N, \theta)\text{Bin}(N|M, \psi), \qquad (10.1.2)$$

where n is the total number of individuals captured during the T samples. The resulting likelihood is the zero-inflated version of $f(\mathbf{y}|N, \theta)$.

The binomial prior for N (with a sample size of M and inclusion probability ψ) is non-informative in the sense that if we place a uniform$(0, 1)$ prior on ψ, then the resulting marginal prior for N is discrete uniform on $(0, M)$ (Royle et al., 2007a). This is natural prior for N in Bayesian treatments of capture–recapture models. Thus, analyses of the Jolly–Seber type models described here begin with the addition of $M - n$ all-zero histories to the data set comprising n observed individuals.

10.1.1 Implementation

There are a number of general strategies for implementing data augmentation. The first is to directly zero-inflate the conditional-on-N likelihood, $f(\mathbf{y}|N, \theta)$, by doing

---

[1] In Royle et al. (2007) we used the term 'superpopulation' for M, but in the present context its use conflicts with its use for N by Schwarz and Arnason (1996). Therefore, here we will use the term 'superpopulation' in reference to N.

the calculation in Eq. (10.1.2). This calculation is straightforward. Consider the Schwarz and Arnason (1996) parameterization in which $f(\mathbf{y}|N, \theta)$ is multinomial having unknown index N. When we carry out the summation in Eq. (10.1.2), the result is a multinomial with index M instead of N (see Section 5.1). Specifically, let y_h be the frequency of individuals having encounter history h. Then the observed frequencies have a multinomial with index N and cell-probabilities $\{\pi_h^{(sa)}\}$. See Schwarz and Arnason (2005) for the precise form of $\pi_h^{(sa)}$ in terms of the survival and entrance probability parameters. After marginalizing N out of $f(\mathbf{y}|N, \theta)$, the result is that the frequencies y_h are multinomial with cell probabilities $\pi_h^{(da)} = \psi \pi_h^{(sa)}$ for the observable encounter histories and $\psi \pi_0^{(sa)} + (1 - \psi)$ for the unobserved encounter history. The parameter ψ replaces N under data augmentation. We demonstrated this substitution analytically for a distance sampling example given in Section 7.1.6 as well as in Section 5.6.

Data augmentation in this case leads to an analytic reparameterization of the model, but does not yield any particular advantage, e.g., with respect to developing models of individual effects. To resolve this, it is possible to devise a data augmentation strategy focused on the state-space decomposition of $f(y|N, \theta)$, i.e., its representation as $f(y|z, p)g(z|\theta)$. We develop this approach in the following sections.

10.2 STATE-SPACE PARAMETERIZATION OF THE JOLLY–SEBER MODEL

We suppose that a population is sampled on T occasions yielding encounter histories on n unique individuals. The *encounter history* for individual i, is the vector $\mathbf{y}_i = (y(i, 1), y(i, 2), \ldots, y(i, T))$ where $y(i, t) = 1$ if individual i is captured during sample t and 0 otherwise. As done in Chapter 9, we adopt a state-space formulation of the model for these individual encounter histories, in which the model is composed of a model for the unobserved or partially observed state process, and a model for the observations conditional on the latent process. In the case of the Jolly–Seber type models, the process model proves to be conceptually simple, being equivalent to a reduced form of the dynamic occupancy model considered in Chapter 9 (see Section 10.3.1).

For each encounter history $\mathbf{y}_i$, there is a corresponding *state history* given by $\mathbf{z}_i$, a vector of binary state variables describing whether individual i is alive ($z(i, t) = 1$) or not ($z(i, t) = 0$). Both vectors have length T whether or not individual i entered the population at $t = 1$. In that case, it will be understood (and parameterized as such in the model), that entries prior to recruitment are fixed zeros.

Suppose for a moment that the number of individuals ever alive during the T samples was known to be N. Then, there exists a sequence of all-zero encounter histories, $\mathbf{y}_i = \mathbf{0}$ for $i = n + 1, \ldots, N$, in addition to the n observed encounter histories. The observations then consist of the two-dimensional array $y(i,t)$ for $i = 1, 2, \ldots, n, n + 1, \ldots, N$ and $t = 1, 2, \ldots, T$, and a corresponding array of state variables $\{z(i,t)\}$.

10.2.1 The Observation Model

We begin with a statement of the observation model that is conditional on the partially observed state process, z. One of the advantages of the state-space formulation in general, i.e., independent of the data augmentation framework, is that it yields remarkably simple formulations of both observation and process model. For example, conditional on the state process, the binary observations are independent Bernoulli random variables,

$$y(i,t)|z(i,t) \sim \mathrm{Bern}(p_t\, z(i,t)). \qquad (10.2.1)$$

Thus, if $z(i,t) = 0$ (individual i has either died, or has not yet recruited) then $y(i,t) = 0$ with probability 1, otherwise $y(i,t)$ is a Bernoulli trial with parameter p_t. We might replace p_t with p_{it} in the above expression, and extend the model statement to include certain effects that vary by individual or time (see Section 10.5).

10.2.1.1 Pollock's robust design

To accommodate data collected according to the robust design (Pollock, 1982) requires a minor change to the observation model. Under the robust design, K secondary samples are obtained for each primary period. Then, the observation model is

$$y(i,t)|z(i,t) \sim \mathrm{Bin}(K, p_t\, z(i,t)). \qquad (10.2.2)$$

Bayesian analysis of either observation model is fundamentally the same (i.e., in terms of MCMC implementation). Note that, under the classical design in which $K = 1$, not all p_t; $t = 1, 2, \ldots, T$, parameters are identifiable (see Section 10.3.3).

10.3 PROCESS MODEL FORMULATIONS

Here we develop models for the demographic processes of survival and recruitment operating on the augmented data set of size M. Individuals that are alive at time

t have an opportunity to survive to $t + 1$. This is the event that $z(i, t + 1) = 1|z(i, t) = 1$. Additional individuals (from among the M that are available for recruitment) may also be recruited into the population, and this is the event that $z(i, t + 1) = 1|z(i, t) = 0$.

As has been demonstrated by the diversity of existing likelihood-based formulations in widespread use, there is not a unique (nor even preferred) way to describe recruitment. Similarly, under data augmentation there are a number of formulations of the state model. In what follows, we describe one formulation that is precisely equivalent to the dynamic occupancy model of Chapter 9. As we have noted elsewhere in this book, the duality between models of occupancy and closed population models has been exploited in a number of contexts and so, naturally, we might seek to benefit from this duality in other contexts, including the present. While this leads to a new parameterization of Jolly–Seber type models, it is one that has a simple relationship to the Schwarz and Arnason (1996) parameterization, the equivalence arising under a reparameterization of the birth model. This is discussed in Section 10.3.2.

10.3.1 Jolly–Seber Model as a Restricted Occupancy Model

Suppose that individuals could die and then re-enter the population. This analogous process occurs in metapopulation models of occupancy, in which sites or patches are recolonized. A model describing the state process is:

$$z(i, t+1) \sim \text{Bern} \left\{ z(i, t)\phi_t + (1 - z(i, t))\gamma_{t+1} \right\}. \tag{10.3.1}$$

For $t = 1$, the initial state is determined by:

$$z(i, 1) \sim \text{Bern}(\gamma_1). \tag{10.3.2}$$

In words, if an individual is alive at time t (i.e., $z(i, t) = 1$) then its status at time $t + 1$ is the outcome of a Bernoulli random variable with parameter ϕ_t. If an individual is not a member of the population at time t (i.e., $z(i, t) = 0$), then the outcome is a Bernoulli trial with parameter γ_{t+1}. That is, the available zeros (at time t) can be recruited into the population (or a landscape patch may become colonized). Note that a site may become unoccupied but become recolonized at some subsequent time. Obviously, this model is conceptually, indeed structurally, very similar to the Jolly–Seber type models in the sense that ϕ_t is precisely a survival probability and γ_t are very similar to recruitment in the sense that they control the number of new additions to the population (or newly occupied sites).

There are two important differences between this occupancy model and classical Jolly–Seber model formulations that we must confront. First, in the former, the list of individuals (sites or patches) contains all-zero encounter histories. That is, they are observed as sites that never appear to be occupied. Secondly, re-colonization does not have a sensible interpretation in population demography. The first issue is resolved by data augmentation, in which the observed encounter histories are augmented with $M - n$ additional, all-zero, encounter histories. In effect, data augmentation only yields a redefinition of the resulting recruitment/colonization parameters (as described in Section 10.3.2). To remedy the re-colonization problem (i.e., that it cannot occur in animal populations), note that we can extend the occupancy model to differentiate between 'initial colonization' and 're-colonization.' Define the auto-covariate $A(i,t) = 1$ ('availability') if the individual (site) has never been colonized prior to t, and $A(i,t) = 0$ if the site has previously been colonized. Set $A(i,1) \equiv 1; i = 1, 2, \ldots, M$. Formally, we can express $A(i,t)$ as the indicator function $A(i,t) = \prod_{k=1}^{t-1}(1 - z(i,k))$. Then, sites that are presently unoccupied ($z(i,t) = 0$) have different colonization rates depending on whether $A(i,t) = 1$ or $A(i,t) = 0$. An expression for the more general state model is

$$z(i,t+1) \sim \text{Bern}(\pi(i,t+1)),$$

where

$$\pi(i,t+1) = \phi_t z(i,t) + \gamma_{t+1}(1-z(i,t))A(i,t) + \theta_{t+1}(1-z(i,t))(1-A(i,t)).$$

The Jolly–Seber state process model corresponds to the restriction $\theta_t = 0$ for all t. As such, under data augmentation, the basic Jolly–Seber model can be described concisely by the following 3 Bernoulli model components. The state model is:

$$z(i,t+1) \sim \text{Bern}\left\{\phi_t z(i,t) + \gamma_{t+1}\left\{\prod_{k=1}^{t}(1-z(i,k))\right\}\right\} \qquad (10.3.3)$$

with the initial state given by

$$z(i,1) \sim \text{Bern}(\gamma_1), \qquad (10.3.4)$$

and the observation model:

$$y(i,t)|z(i,t) \sim \text{Bern}(p_t\, z(i,t)). \qquad (10.3.5)$$

In this formulation, there are T initial colonization probabilities, $\gamma_t; t = 1, 2, \ldots, T$, which are unconstrained. Conversely, Schwarz and Arnason (1996) (henceforth 'SA') recognize $T - 1$ free 'entrance probabilities,' and one additional superpopulation size parameter, N (the size of the list of individuals ever alive), for a total of T parameters to control the number and distribution of individuals across sample periods. We establish the precise linkage between γ_t in the present formulation and the SA parameters in Section 10.4. Note that the indexing of γ_t is inconsistent with the development of Chapter 9, which we have done to allow for an extra γ parameter – individuals can enter at any of the T samples. The additional recruitment parameter takes the place of the initial occurrence probability in the occupancy models.

10.3.2 The Implied Recruitment Model

It is instructive to understand precisely the implied model for births under the model for augmented data. The implied recruitment model has 'births' being generated from the pool of available zeroes, which begins at some arbitrarily large number, M, and diminishes over time. The parameters γ_t under this formulation are the probabilities that an *available* individual from the augmented list is recruited at time t. Under the state-space formulation, this recruitment model is manifested *indirectly* by the sequence of Bernoulli state distributions $z(i, t)|z(i, t-1)=0 \sim \text{Bern}(\gamma_t)$.

In fact, the state model Eq. (10.3.3) implies a factorization of a model for recruits that is product-binomial. That is, if B_t are the number of births in period t, then the process model implied by Eqs. (10.3.3) and (10.3.4) has the following hierarchical structure (supposing for clarity that $T = 3$):

$$g_{da}(B_1, B_2, B_3|M, \gamma_1, \gamma_2, \gamma_3) = \text{Bin}(B_1|M, \gamma_1)\text{Bin}(B_2|M - B_1, \gamma_2)$$
$$\times \text{Bin}(B_3|M - B_1 - B_2, \gamma_3). \qquad (10.3.6)$$

That is, at $t = 1$, B_1 is binomial with sample size M, and parameter γ_1. And, in subsequent periods, B_t is binomial with sample size equal to the remaining pool of available pseudo-individuals (e.g., $M - B_1$ for $t = 2$, etc.). Thus, γ_t is the entrance probability of individuals remaining in the pool of available pseudo-individuals (those on the augmented list that have not yet been recruited).

Note that in the construction given by Eq. (10.3.6), there is a remainder term, the individuals that are not recruited, say B_0. An equivalent representation that keeps track of B_0 is the multinomial

$$g_{da}(B_0, B_1, B_2, B_3|M, \cdot) = \frac{M!}{B_1!B_2!B_3!B_0!}\gamma_1^{B_1}\left\{\gamma_2(1-\gamma_1)\right\}^{B_2}$$
$$\times \left\{\gamma_3(1-\gamma_2)(1-\gamma_1)\right\}^{B_3}\left\{1-\psi\right\}^{B_0}. \quad (10.3.7)$$

Here, ψ is the sum of the first 3 terms (recall that ψ, the inclusion probability, is the probability that an element of the list of size M is a member of the superpopulation of size N). The duality between Eq. (10.3.6) and Eq. (10.3.7) is the same as exists in a classical 'removal' type model in which a population is sequentially sampled and individuals are removed from the population. That is, they are simply alternative parameterizations of the sequential removal process.

Equations (10.3.6) and (10.3.7) provide the bridge to establishing the relationship between the occupancy-derived formulation of the Jolly–Seber type model and the SA parameterization which we take up in Section 10.4.

10.3.3 Parameter Identifiability

In the classical formulation of the Jolly–Seber model (i.e., when $K = 1$), with time-varying parameters $p_t; t = 1, 2, \ldots, T$, at least two of the p_t are regarded as being unidentifiable, and these are usually fixed so that $p_1 = p_2$, and $p_T = p_{T-1}$, or $p_1 = 1$ and $p_T = 1$. Strictly speaking, the identifiable parameters represent functions of other parameters with p_1 and p_T, and the particular constraints introduced to estimate the remaining parameters are not necessarily innocuous (Link and Barker, 2005). The best solution to the broader identifiability problem is to collect more data (e.g., according to the robust design), or to standardize protocols so that a particular constraint might be reasonable (e.g., $p_1 = p_2$). See Schwarz (2001), Schwarz and Arnason (2005) and Link and Barker (2005) for extensive discussions of identifiability and parameter constraints.

Setting $p_1 = 1$ has the effect that all individuals which were not detected at $t = 1$ appear as recruits at time $t = 2$. Thus, γ_2 is not a clean estimate of recruitment *per se*. If $p_1^{(\text{true})}$ is the true value of p_1, and then let ψ_1 be the probability that an individual in the list of size M is an individual exposed to sampling at $t = 1$. That is, under data augmentation, suppose that $N_1 \sim \text{Bin}(M, \psi_1)$. Then, the apparent probability of recruitment at time $t = 2$ is $\gamma_2 = (1 - p_1^{(\text{true})})\psi_1\phi_1 + \gamma_2^{(\text{true})}(1 - \psi_1)$. Thus, apparent recruitment has a component of individuals that were alive at $t = 1$ and survived. It seems like this bias would be greatly reduced by setting $p_1 = p_2$. On the other hand, it probably doesn't matter so much if we just acknowledge that γ_2 is practically uninterpretable. A secondary effect of setting $p_1 = 1$ is that the estimated size of the superpopulation, N, is actually short by a few individuals, those that died between $t = 1$ and $t = 2$. However, in this case, we may as well adopt the interpretation of N as being the number of individuals ever alive between $t = 2$ and $t = T$.

Regardless of the ambiguity over the interpretation of γ_2, we typically would interpret γ_2 as apparent recruitment (i.e., including immigration and birth). Thus, with Jolly–Seber type data (i.e., having a single capture occasion in each primary period or year), we basically sacrifice a year of data before we begin learning about real biological parameters in the second and subsequent years (picking up a little information about ϕ_1 along the way).

10.3.4 Bayesian Analysis of the Models

Because of the simple hierarchical structure of the models i.e., as a sequence of 3 Bernoulli random variables (Eqs. (10.3.3), (10.3.4) and (10.3.5)), it is straightforward to devise MCMC algorithms for these models. We require prior distributions for all model parameters. Because they are probabilities under the parameterization induced by data augmentation, we normally choose $U(0, 1)$ priors for all parameters. We discuss prior specification further in Section 10.3.7. The methods for devising and sampling from full-conditional distributions are now considered conventional, and so we will not provide the details. Some details for the occupancy type models, using beta prior distributions, can be found in Royle and Kéry (2007). For more complex models, including those for alternative prior distributions for the model parameters, the MCMC algorithm can become somewhat more complex, requiring generic methods (e.g., Metropolis–Hastings) for sampling non-standard full-conditional distributions. However, analysis of the state-space representation of the model under data augmentation can be implemented directly in **WinBUGS** and that is in the approach we adopt in the examples in the following section.

One of the most remarkable things about the occupancy-derived formulation of the model that arises under data augmentation is the simplicity of the **WinBUGS** model formulation which is shown in Panel 10.1. In the model description, the objects T, M and y are data that must be provided by the user. The specifications shown in Panel 10.1 can serve as a very general template, such as for the inclusion of individual covariates or individual heterogeneity in p or ϕ. These extensions are discussed in Section 10.5. Note that the implementation shown in Panel 10.1 imposes the constraint $p_1 = 1$ and $p_T = 1$. We do not advocate this constraint, but it seems to be common in practice. In Section 10.4, we provide the MLEs under that model.

10.3.5 Abundance and Other Derived Parameters

There are several quantities of interest that are not structural parameters of the model but, rather, arise as functions of the latent state variables $z(i, t)$. For example, the total number of individuals alive at time t is

$$N_t = \sum_{i=1}^{M} z(i, t)$$

whereas, the number of new recruits (births) is

$$B_t = \sum_{i=1}^{M} (1 - z(i, t-1)) z(i, t)$$

and the number of deaths is

$$D_t = \sum_{i=1}^{M} z(i, t-1)(1 - z(i,t)).$$

These can be converted to per capita birth and death rates by appropriate normalization. The superpopulation size under the occupancy derived parameterization is

$$N = \sum_{i=1}^{M} I \left\{ \sum_{t=1}^{T} z(i,t) > 0 \right\}.$$

```
model {
### T, M, and Y are data.
for(j in 1:(T-1)){
    phi[j]~dunif(0,1)
    gamma[j]~dunif(0,1)
}
gamma[T]~dunif(0,1)
p[1]<- 1; p[T]<- 1
for(i in 2:(T-1)){
    p[i]~dunif(0,1)
}
for(i in 1:M){
    z[i,1]~dbern(gamma[1])
    mu[i]<-z[i,1]*p[1]
    y[i,1]~dbern(mu[i])
    recruitable[i,1]<-1
    for(j in 2:T){
        survived[i,j]<- phi[j-1]*z[i,j-1]
        recruitable[i,j]<- recruitable[i,(j-1)]*(1-z[i,j-1])
        mu2[i,j]<-  survived[i,j] + gamma[j]*recruitable[i,j]
        z[i,j]~dbern(mu2[i,j])
        mu3[i,j]<-z[i,j]*p[j]
        y[i,j]~dbern(mu3[i,j])
    }
  }
}
```

Panel 10.1. WinBUGS model specification of the Jolly–Seber type model as a constrained model of patch occupancy. Here, M is the size of the list, including n observed individuals and M–n augmented all-zero encounter histories. The M × T data matrix is y.

Here, $I(arg)$ is an indicator function that evaluates to 1 if individual i was ever alive.

Within an MCMC framework for inference, in which the missing values of $z(i, t)$ are updated, these derived parameters can be computed at each iteration of the algorithm and their posterior distributions estimated by the resulting collection of simulated values. These derived quantities have metapopulation analogs, which we referred to as finite-sample quantities in Chapter 9 (see also Royle and Kéry (2007)).

10.3.6 Analysis of the European Dipper Data

Here we provide an analysis of the classical European dipper data described by Lebreton et al. (1992). These data are from a 7-year study of European dippers (*Cinclus cinclus*), originating from Marzolin (1988), and have been used extensively by others including Brooks et al. (2000) and Royle (2008c). The number of unique individuals first captured in each of the 7 periods was $(22, 60, 78, 80, 88, 98, 93)$. These data appear to be widely available, e.g., from E.G. Cooch's website: `http://www.phidot.org/software/mark/docs/book/`.

Bayesian estimates under the occupancy parameterization are given in Table 10.1. The columns labeled 'prior 1' were obtained under the specification $\gamma_t \sim \mathrm{U}(0, 1)$, as well as uniform priors for the remaining probability parameters. Estimates in Table 10.1 include both the canonical 'colonization' probabilities, γ_t, and also the derived entrance probabilities π_t of the SA parameterization (which we address later).

10.3.7 Prior Distributions for Recruitment

In Section 10.4 we provide estimates of N using alternative parameterizations (and also the MLEs). We will note general inconsistencies between estimates of N using different parameterizations of the model, as well as between the MLEs and the Bayesian estimates. In small samples such inconsistencies are typical in multi-parameter models because posterior means and multi-dimensional modes are different quantities. However, another important consideration has to do with prior specification. Since all of the parameters of the model are probabilities, we used the conventional $\mathrm{U}(0, 1)$ priors for all parameters. This choice of priors is not completely innocuous. Note that

$$\mathrm{E}(N | M, \gamma_1, \gamma_t, \ldots, \gamma_T) = M \left\{ 1 - (1 - \gamma_1)(1 - \gamma_2) \ldots (1 - \gamma_T) \right\}.$$

As such, when we tinker around with priors on γ_t, we are inducing prior structure on N.

Data augmentation was motivated (Section 10.1) as arising under the assumption of a discrete uniform prior for N on the integers $[0, M]$, which was constructed hierarchically as a $\mathrm{Bin}(M, \psi)$ for N and a $\mathrm{U}(0, 1)$ prior for ψ. However, ψ is not specified directly in the occupancy-based parameterization, instead it is a derived parameter. The model is constructed in terms of the recruitment parameters γ_t. Under the occupancy-derived parameterization, $\gamma_t \sim \mathrm{U}(0, 1)$. The result is that the *implied prior* for the inclusion probability, i.e., for $\psi = 1 - (1 - \gamma_1)(1 - \gamma_2)...(1 - \gamma_T)$, is *not* $\mathrm{U}(0, 1)$. As such, the resulting implied prior on N is *not* uniform on the integers $[0, M]$ as desired to justify data augmentation (see Section 5.6). In addition, the implied entrance probabilities are not uniform across the T periods. These facts turn out to have some influence on the posterior of N and, consequently, some of the other parameters as well.

One way to remedy this potential sensitivity to prior distributions is to parameterize the model directly in terms of ψ – place a uniform prior on that parameter, and specify equal entrance probabilities across the T periods (i.e., place a Dirichlet prior having sample sizes $\alpha_t = 1$). This solution arises naturally as the data augmentation version of the SA model (Section 10.4). Alternatively, we can experiment with the priors $\gamma_t \sim \mathrm{Beta}(a_t, b_t)$ such that the fraction of individuals ever recruited (out of M) is 0.5, and an equal number (in expectation) recruit at each time period. For the dipper data, $T = 7$ periods, we would like a prior that allocates $1/14$ of the individuals to each period, so that $7/14$ of the individuals on the augmented list are members of the superpopulation. One way to achieve this is to set $a_t = S/(14 - (t - 1))$ and $b_t = S - a_t$, where $S \approx 1.5$ seems to yield approximately a $\mathrm{U}(0,1)$ prior for ψ. These results are shown under 'prior 2' in Table 10.1. Compared to 'prior 1' in Table 10.1, we see a slight difference between the posterior of N as well as the other parameter estimates. We provide context for these estimates in Section 10.4.2. The **WinBUGS** model specification for this parameterization, and its implementation in **R**, are provided on the Web Supplement.

10.4 SCHWARZ AND ARNASON'S FORMULATION

.

Conditional on N, it seems natural to allocate the N individuals among the T sample periods according to the multinomial:

$$g_{sa}(B_1, B_2, B_3 | N, \boldsymbol{\pi}) = \mathrm{Multin}(N, (\pi_1, \pi_2, \pi_3)).$$

Then, we seek to estimate N and the parameters $\{\pi_t\}$ (constrained to sum to 1). This is precisely the SA parameterization of the birth process. We noted previously that it can be motivated by assuming that $N_t \sim \mathrm{Po}(\lambda)$ in which case, conditional on

Table 10.1. Posterior summaries of model parameters for the dipper data under the occupancy derived parameterization of the Jolly–Seber model. Results labeled 'prior 1' have $\gamma_t \sim U(0,1)$ whereas 'prior 2' has $\gamma_t \sim \mathrm{Beta}(a_t, b_t)$, where a_t and b_t are chosen to yield approximately $U(0,1)$ prior on ψ, the inclusion probability.

Parameter	prior 1			prior 2		
	mean	SD	median	mean	median	SD
N	334.1	19.41	330	326.5	16.64	323
ϕ_1	0.763	0.134	0.770	0.741	0.135	0.744
ϕ_2	0.443	0.072	0.439	0.454	0.071	0.452
ϕ_3	0.480	0.061	0.479	0.481	0.061	0.480
ϕ_4	0.628	0.061	0.628	0.628	0.060	0.628
ϕ_5	0.602	0.058	0.602	0.602	0.057	0.602
ϕ_6	0.525	0.050	0.525	0.526	0.050	0.526
p_2	0.571	0.150	0.566	0.618	0.154	0.625
p_3	0.851	0.091	0.867	0.863	0.085	0.879
p_4	0.871	0.067	0.881	0.876	0.065	0.886
p_5	0.869	0.059	0.877	0.873	0.058	0.880
p_6	0.900	0.053	0.909	0.902	0.052	0.911
γ_1	0.058	0.012	0.057	0.056	0.012	0.055
γ_2	0.257	0.083	0.239	0.233	0.081	0.212
γ_3	0.153	0.050	0.155	0.148	0.055	0.154
γ_4	0.206	0.041	0.203	0.197	0.038	0.194
γ_5	0.236	0.050	0.231	0.222	0.045	0.218
γ_6	0.340	0.074	0.329	0.316	0.063	0.308
γ_7	0.399	0.122	0.371	0.358	0.094	0.340
π_1	0.068	0.014	0.068	0.068	0.014	0.067
π_2	0.281	0.076	0.268	0.263	0.077	0.245
π_3	0.131	0.051	0.133	0.135	0.055	0.142
π_4	0.143	0.029	0.141	0.146	0.028	0.145
π_5	0.129	0.026	0.128	0.132	0.026	0.131
π_6	0.141	0.025	0.140	0.145	0.025	0.144
π_7	0.107	0.019	0.106	0.111	0.020	0.111

$N = \sum_t N_t$, the vector $(N_1, \ldots, N_T)$ has a multinomial distribution with cell probabilities $\lambda_t / \sum_t \lambda_t$. Thus, estimating π_t is equivalent to estimating λ_t parameters.

The dynamic occupancy formulation described in the preceding section is closely related to the SA parameterization of the model. For the list of individuals augmented by $M - n$ 'all-zero' encounter histories, assume that $N \sim \mathrm{Bin}(M, \psi)$. Then, we may marginalize N out of the multinomial likelihood (see Section 5.1) which yields a $T + 1$ cell multinomial (still with T parameters), according to

$$g_{\mathrm{occ}}(B_1, B_2, B_3 | M, \gamma_1, \gamma_2, \gamma_3, \psi) = \sum_{N=n}^{M} g_{sa}(B_1, B_2, B_3 | N, \boldsymbol{\pi})$$
$$\times \mathrm{Bin}(N | M, \psi), \qquad (10.4.1)$$

where the cell probabilities of g_{occ} are those described by Eq. (10.3.7). Specifically,

$$
\begin{aligned}
\pi_1^{(occ)} &= \gamma_1 = \pi_1 \psi \\
\pi_2^{(occ)} &= \gamma_2(1 - \gamma_1) = \pi_2 \psi \\
\pi_3^{(occ)} &= \gamma_3(1 - \gamma_2)(1 - \gamma_1) = \pi_3 \psi.
\end{aligned}
$$

As such, we see that, under data augmentation the multinomial cell probabilities of the occupancy-derived birth model – which are functions of the recruitment (colonization) parameters γ_t – are related directly to the entrance probabilities of the SA parameterization and the parameter ψ. Under data augmentation, the parameter ψ formally replaces the superpopulation size parameter of the original SA parameterization.

10.4.1 Implementation

The SA formulation is somewhat more complicated to implement than the occupancy-derived formulation. This is because entry of individuals into the population is conditional on N. Given the augmented list of individuals of size M, we must choose N of them from M and then distribute individuals among the T samples. An alternative strategy, which we adopt, is to distribute all M pseudo-individuals among the T samples, and then specify the observation model as a mixture of the M individuals and zeros. We describe the model explicitly here, and its implementation in **WinBUGS** is given in the Web Supplement.

10.4.1.1 The process model

Data augmentation produces a population of M individuals that are allocated among the T periods according to a Dirichlet prior on the entrance probabilities

$$
\{\pi_t\} \sim \text{Dirichlet}(\boldsymbol{\alpha}),
$$

where $\boldsymbol{\alpha}$ is a vector of prior sample sizes of length T. We set $\alpha_t = 1$ for all t, which allocates individuals uniformly across periods.

For implementation, it is convenient to construct the prior in terms of independent gamma random variables, $\beta_t \sim \text{Gamma}(1,1)$ and then set $\pi_t = \frac{\beta_t}{\sum_{j=1}^{t} \beta_j}$. In addition, it is also convenient to re-express these entrance probabilities as conditional probabilities, the probability of entry at t given that it is a member of N and has not yet entered the population, i.e.,

$$\pi_1^{(c)} = \pi_1$$
$$\pi_t^{(c)} = \frac{\pi_t}{1 - \sum_{j=1}^{t-1} \pi_j}; \quad t = 2, \ldots, T.$$

This is convenient because it allows us to retain the sequential specification of the state model:

$$z(i, 1) \sim \text{Bern}(\pi_1)$$
$$z(i, t)|z(i, t-1) \sim \text{Bern}\left\{ \phi_{t-1} z(i, t-1) + \pi_t^{(c)} \prod_t (1 - z(i, t)) \right\} \quad \text{for } t = 2, \ldots, T,$$

where $\prod_t (1 - z(i, t))$ is an indicator of recruitability, as before.

10.4.1.2 The observation model

Having created a population of size M individual state histories, or $\mathbf{z}_i$, we suppose that each pseudo-individual has associated with it the latent variable $w_i \sim \text{Bern}(\psi)$. Those with $w_i = 1$ are exposed to sampling if they are presently alive in the population. Thus, the observation model has two components:

$$y(i, t) \sim \text{Bern}(p_t z(i, t)) \quad \text{if } w_i = 1$$
$$y(i, t) \sim \text{Bern}(0) \quad \text{if } w_i = 0.$$

This formally admits the zero-inflation of the augmented data set.

10.4.2 Analysis of the Dipper Data

For the **WinBUGS** implementation of the SA parameterization, see the Web Supplement. Bayesian estimates of the SA parameterization under data augmentation are given in Table 10.2. These are broadly consistent with the estimates obtained under the occupancy-derived formulation, using the 'prior 2' specification (See Table 10.1). The posterior mean of N is slightly higher because the Dirichlet prior on the multinomial cell probabilities $\boldsymbol{\pi}$ is not precisely equivalent to the sequential beta prior for which results are summarized in Table 10.1.

For comparison, the MLEs for the SA parameterization (using the program MARK, (White and Burnham, 1999)), are given in Table 10.2.

Table 10.2. Maximum likelihood estimates and standard errors of parameters under the Schwarz and Arnason parameterization of the Jolly–Seber model fitted to the Dipper data. Right columns are the posterior summaries of model parameters for the dipper data using the SA parameterization of the Jolly–Seber model under data augmentation.

	MLEs			Bayes	
Parameter	MLE	Std. Er.	mean	SD	median
N	310.433	9.816	321.2	11.9	319.0
ψ	–	–	0.590	0.030	0.589
ϕ_1	0.698	0.147	0.720	0.131	0.720
ϕ_2	0.439	0.068	0.454	0.071	0.451
ϕ_3	0.478	0.060	0.482	0.060	0.481
ϕ_4	0.626	0.059	0.628	0.061	0.628
ϕ_5	0.599	0.056	0.601	0.058	0.601
ϕ_6	0.531	0.050	0.526	0.050	0.526
p_1	1.000	–			
p_2	0.734	0.157	0.667	0.129	0.670
p_3	0.927	0.069	0.864	0.083	0.878
p_4	0.915	0.057	0.876	0.065	0.886
p_5	0.902	0.053	0.872	0.058	0.879
p_6	0.934	0.045	0.903	0.052	0.912
p_7	1.000	–			
π_1	0.070	–	0.070	0.014	0.069
π_2	0.214	0.048	0.235	0.051	0.228
π_3	0.156	0.035	0.151	0.040	0.152
π_4	0.152	0.025	0.148	0.028	0.147
π_5	0.138	0.024	0.135	0.025	0.133
π_6	0.150	0.024	0.147	0.025	0.146
π_7	0.120	0.020	0.114	0.019	0.113

10.5 MODELS WITH INDIVIDUAL EFFECTS

One of the main benefits of the hierarchical formulation of the model induced by data augmentation is that parameterization of individual effects is relatively straightforward. For example, a model of some interest in evolutionary biology (Cam et al., 2002a) is that where individual heterogeneity exists in ϕ. A natural representation of the model is the mixed logit model (Coull and Agresti, 1999), in which

$$\text{logit}(\phi_{it}) = \alpha_t + \eta_i,$$

where $\eta_i \sim \text{N}(0, \sigma_\phi^2)$.

For this model, analysis by MCMC is slightly more involved as the model prior for η_i is not conjugate. However, this is done conditional on each individual's history, $\mathbf{z}_i$, and so the problem here is no more complicated than it is in standard mixed binomial regression or Bayesian GLMs in general under non-conjugacy (Gilks and

Wild, 1992; Clayton, 1996). For example, we can sample each η_i using a Metropolis–Hastings step of the MCMC algorithm, the full-conditional being proportional to

$$[\eta_i|\cdot] \propto \left\{ \prod_{t=f_i+1}^{l_i} \phi_{it}^{z(i,t)}(1-\phi_{it})^{1-z(i,t)} \right\} \mathrm{N}(0,\sigma_\phi^2),$$

where f_i and l_i are, respectively, the period of entry and the period of mortality. After having updated the latent state variables $z(i,t)$, we need to keep track of entry into and exit out of the population for each individual. Given η_i, sampling missing values of $z(i,t)$ is simple because the full-conditional distribution of the latent state variables is Bernoulli.

While the development of MCMC algorithms for these models using data augmentation is straightforward, so, too, is their implementation in **WinBUGS**. The **WinBUGS** model specification containing individual heterogeneity in survival, is given in Panel 10.2. The Markov chain for this model exhibits slow mixing and long chains are necessary to produce tolerable levels of Monte Carlo error. This particular implementation makes use of the informative prior on the γ parameters described in Section 10.3.7, so as to induce a (approximately) uniform prior on ψ, the sample inclusion probability. This model was fitted to the dipper data, and the results are presented in Table 10.3. Because there is little indicated heterogeneity in survival, the estimates are little changed from Table 10.2.

We believe that analysis of these models by data augmentation will also facilitate implementation of individual covariate models. An example in closed population models can be found in Royle (2008a). In open populations, one has to be concerned with unobserved individuals (and hence their covariate values) over time, which can require a model for the covariate that describes its temporal variation. One such extension of the CJS model was presented by Bonner and Schwarz (2006), which, because the CJS model is conditional on entry into the sample, there is no need for data augmentation (there is no recruitment). We believe the Jolly–Seber version of their model could be implemented without difficulty using data augmentation.

10.6 SUMMARY

Drawing on the analogy between open population systems and models of occupancy (Chapter 9), we have presented a novel, hierarchical formulation of Jolly–Seber type models. Under this formulation, the basic Jolly–Seber model can be described by the following sequence of 3 Bernoulli models: For the state process,

$$z(i,t+1)|z(i,t) \sim \mathrm{Bern}\left\{ z(i,t)\phi_t + \gamma_{t+1}(1-z(i,t))A(i,t) \right\},$$

```
model {
sigma~dunif(0,10)
tau<-1/(sigma*sigma)
for(j in 1:(T-1)){
     phi0[j]~dunif(0,1)
     lphi[j]<-log(phi0[j]/(1-phi0[j]))
}
p[1]<- 1; p[T]<- 1
for(i in 2:(T-1)){
     p[i]~dunif(0,1)
}
for(j in 1:M){
     eta[j]~dnorm(0,tau)
     for(i in 1:(T-1)){
          logit(phi[j,i])<-lphi[i]  + eta[j]
     }
}
for(i in 1:T){
gamma[i]~dbeta(a[i],b[i])  # a[i], b[i] are data
}
for(i in 1:M){
    z[i,1]~dbin(gamma[1],1)
    mu[i]<-z[i,1]*p[1]
    y[i,1]~dbin(mu[i],1)
    recruitable[i,1]<-1
        for(j in 2:T){
             survived[i,j]<- phi[i,j-1]*z[i,j-1]
             recruitable[i,j]<- recruitable[i,(j-1)]*(1-z[i,j-1])
             mu2[i,j]<-  survived[i,j] + gamma[j]*recruitable[i,j]
             z[i,j]~dbin(mu2[i,j],1)
             mu3[i,j]<-z[i,j]*p[j]
             y[i,j]~dbin(mu3[i,j],1)
        }
    }
}
```

Panel 10.2. WinBUGS model specification for Jolly–Seber type model with individual heterogeneity in survival using data augmentation. In this case, M is the size of the augmented data set. This particular implementation uses an informative prior distribution on the entry probabilities as described in Section 10.3.7. The parameters of the prior, the vectors a and b, are specified by the user.

Table 10.3. Parameter estimates for the dipper data under a model allowing for individual heterogeneity in survival probability. Results are with the data set augmented by 100 all-zero encounter histories. The parameters γ_t are entrance probability with respect to the pool of available pseudo-individuals and π_t are the Schwarz–Arnason entrance probability parameters.

Parameter	mean	SD	median
σ	0.342	0.207	0.302
N	324.6	15.33	321.0
ϕ_1	0.732	0.135	0.735
ϕ_2	0.451	0.074	0.448
ϕ_3	0.474	0.063	0.473
ϕ_4	0.619	0.063	0.619
ϕ_5	0.589	0.061	0.590
ϕ_6	0.511	0.054	0.512
p_2	0.640	0.149	0.649
p_3	0.863	0.085	0.879
p_4	0.877	0.065	0.887
p_5	0.875	0.057	0.882
p_6	0.904	0.051	0.913
γ_1	0.056	0.012	0.055
γ_2	0.222	0.074	0.204
γ_3	0.154	0.051	0.158
γ_4	0.195	0.037	0.192
γ_5	0.219	0.043	0.216
γ_6	0.311	0.059	0.305
γ_7	0.350	0.084	0.336
π_1	0.068	0.014	0.067
π_2	0.252	0.071	0.237
π_3	0.142	0.052	0.147
π_4	0.147	0.028	0.146
π_5	0.133	0.026	0.132
π_6	0.146	0.025	0.145
π_7	0.112	0.020	0.112

where $A(i,t) = \prod_{k=1}^{t-1}(1-z(i,k))$. The initial state is given by

$$z(i,1) \sim \text{Bern}(\gamma_1),$$

and the observation model is:

$$y(i,t)|z(i,t) \sim \text{Bern}(p\, z(i,t)).$$

Not surprisingly, Bayesian analysis of the hierarchical formulation can be achieved directly by Gibbs sampling, and using **WinBUGS**. Within the broader hierarchical formulation of the Jolly–Seber type models, a number of extensions can be obtained, including robust design (Pollock, 1982), inclusion of individual covariates, and individual heterogeneity. For example, under the robust design, the resulting observation model is modified from a Bernoulli distribution with parameter p_t to a Binomial distribution with sample size T and parameter p_t. The underlying state process is unaffected.

There are two essential technical features of the formulation described here: First is the state-space representation, that has proved useful in various contexts (e.g., multi-state models (Dupuis, 1995); and CJS models with individual heterogeneity (Royle, 2008c)). This state-space representation is convenient because it decouples the observation model from the process model, yielding 2 very simple model components that yield to Bayesian analysis by MCMC. The second element of the formulation proposed here is the use of data augmentation (Royle et al., 2007a), which serves to fix the size of the observation matrix. The idea is to augment the observed matrix of size n with $M - n$ additional, all-zero, encounter histories. These pseudo-individuals are members of the population of size N with probability ψ, the inclusion probability. Formally, data augmentation expands the *model* so that $N \sim \text{Bin}(M, \psi)$, where M is fixed. In effect, N is replaced as a formal parameter of the model by the parameter ψ. In a hierarchical, Bayesian framework for inference, our contention is that dealing with ψ is a simpler proposition. Given a state-space formulation of the model *and* data augmentation, we have shown that Jolly–Seber type models arise under a formal restriction of a general form of dynamic occupancy model, the restriction being that re-colonization probability is set to 0 (in context, once an individual dies it cannot be subsequently recruited).

There is a formal linkage between the occupancy-based formulation and that of Schwarz and Arnason (1996), under data augmentation. The latter is parameterized in terms of the 'superpopulation' size, N, and multinomial cell probabilities (entrance probabilities), π_t, which distribute the individuals among the T periods. The SA parameterization yields naturally to data augmentation under which $N \sim \text{Bin}(M, \psi)$, where M is chosen to be sufficiently large (perhaps by trial and error). Upon marginalizing N out of the model, this yields new entrance probabilities $\psi \pi_t$ which are related to the 'colonization probabilities,' γ_t, by a reparameterization of the multinomial:

$$\gamma_1 = \pi_1 \psi$$
$$\gamma_2(1 - \gamma_1) = \pi_2 \psi$$
$$\gamma_3(1 - \gamma_2)(1 - \gamma_1) = \pi_3 \psi$$
$$\cdots = \cdots .$$

This relationship is that of relating the multinomial and product-binomial representations of a sequential removal process.

11

MODELING SURVIVAL

Survival is one of the fundamental processes governing animal population dynamics, and understanding factors that influence survival is an important focus of many studies of animal populations. As with all other classes of problems described in this book, one of the main practical problems in modeling survival is that we typically cannot make a direct measurement of the state variable in question – whether an individual is alive or dead at some point in time. While we may be able to follow individuals over time, there are some that we never see again, and we cannot know if they are alive or have died at any point after their last observation. This classical problem is resolved using methods of capture–recapture, which were considered in the context of closed populations (i.e., those not experiencing mortality, etc.) in Chapters 5, 6 and 7. The use of data from marked individuals in studies of animal populations is widespread in contemporary ecology (Seber, 1982; Williams et al., 2002).

For modeling animal survival in the presence of imperfect detection of individuals, the class of models known as 'Cormack–Jolly–Seber' (CJS) models (Cormack, 1964; Jolly, 1965; Seber, 1965; Lebreton et al., 1992) is widely used, and constitutes one of the most popular classes of statistical models in all of ecology. While such models have been applied to dozens of taxa, they are commonly used in studies of bird populations for which large banding (or ringing) schemes exist in many countries. These programs collect vast quantities of individual encounter data, and CJS-type models are used for the analysis of such data. Many countries actively band harvested populations of birds, which generates large numbers of recoveries of hunter-killed individuals. So-called 'band-recovery models' (Brownie et al., 1985) for the analysis of dead-encounter data are an important, special class of CJS models for studying many species.

Understanding factors that influence nest survival is also very important in avian population biology, and modeling avian nest survival constitutes another application area for a class of models closely related to CJS models (Mayfield, 1975; Johnson, 1979; Hensler and Nichols, 1981; Stanley, 2000; Dinsmore et al., 2002; Rotella et al., 2004). Like CJS models, in nest survival models, entry of nests into the sample is not the outcome of a random variable. That is, models are constructed 'conditional on first encounter'. The main distinction between nest survival and CJS models is that $p = 1$ in subsequent nest samples (a nest, once found, is detectable for the remainder of its lifespan). The unifying characteristic of both classes of models is uncertainty over the 'terminal state' of marked individuals (or nests). In the case of CJS models, an individual may never again appear in the sample after a certain point and there is uncertainty about whether it is alive or dead at any period subsequent to its last encounter. Conversely, in studies of nest survival, the fate of each nest is often known, either successfully fledged or failed, but the observed failure represents a right-censored failure date, not a precise observation. Because of the similarities between the two classes of models, the underlying state process model is equivalent.

In this chapter we describe the traditional formulation of CJS models for survival and also a hierarchical formulation under which the model is described explicitly by two distinct components: (1) a model for the unobserved, or partially observed state process (whether an individual is alive at sample t), and (2) a model for the observations, conditional on the state process (whether the individual is encountered in the sample). Fundamental to this formulation is the definition of a set of individual state variables, say $z(i, t)$, which describe the state process – whether individual i is alive and available for capture at time t. In the classical CJS situation, the process model, that governs the dynamics of $z(i, t)$, is described by a Bernoulli probability mass function, conditional on the previous state value. The observation model consists of a collection of independent Bernoulli trials that are conditional on $z(i, t)$. The CJS state model is formally a constrained version of the Jolly–Seber type model described in Chapter 10. The reduction arises as a result of 'conditioning on first capture'. First capture of individuals in period t is not assumed to be the outcome of a random process. Rather, they enter the sample with probability 1.

The hierarchical formulation of models for survival is advantageous for the same basic reasons we discussed in previous chapters (e.g., Chapter 9) – a hierarchical approach yields easily to many extensions including modeling individual effects in the form of either random effects ('heterogeneity') or individual covariates. For example, we consider an individual effect in the analysis of data from a constant-effort mist netting program, where each individual has associated with it an indicator variable of residency status. We also develop a model to describe spatial variation in survival rate. Under this model, we place a prior distribution on the survival probability parameters that includes a spatially-indexed, spatially-correlated random effect. Finally, we consider an analysis (Royle, 2008c) of a classic data set (the European Dipper data) to evaluate individual heterogeneity in model parameters.

11.1 CLASSICAL FORMULATION OF THE CJS MODEL

Suppose a population is sampled on T occasions (typically years, corresponding to the breeding interval of many vertebrates) yielding capture histories of n unique individuals. The capture or encounter history for individual i, first captured at time f_i, is the vector $\mathbf{y}_i = (y(i, f_i), y(i, f_i + 1), \ldots, y(i, T))$. The observations are individual encounter histories we have worked with previously in connection with closed population sampling. However, here, we assume that individuals are exposed to mortality or permanent emigration between sample periods.

Let ϕ_t be the probability that an individual survives from t to $t+1$. As with closed population models, we have encounter probabilities p_t, the probability of capture in period t. We note that each encounter history represents a discrete outcome, and the frequencies of all $2^T - 1$ possible outcomes have a multinomial distribution with cell probabilities π_k - the probability of encounter history k. Under the assumption that successive events for each individual are independent Bernoulli trials, the π_k can be constructed as the product of the constituent Bernoulli probabilities. As an example, suppose $T = 3$ and consider the encounter history (101) for an individual captured in the first and last period but not in period 2. By assumption, the individual enters at sample 1 with probability 1, must have survived the first interval, was not captured in period 2, survived the second interval and was captured in period 3. The resulting probability is: $\Pr(\mathbf{y} = (1, 0, 1)) = \phi_1 (1 - p_2) \phi_2 p_3$. We emphasize again that the first capture is not regarded as the outcome of a random process. Thus, the encounter history probability is constructed beginning at the first capture.

The difficulty in devising the encounter history probabilities arises when an individual is last captured in some period t prior to the last sampling occasion. In this case, there are multiple events that make up the terminal event 'last seen on sample t'. Consider the encounter history $(1, 1, 0)$ for an individual last seen at sample $t = 2$ for a study comprised of $T = 3$ samples. In this case, we don't know whether the individual died prior to the third sample or whether the individual was alive but not captured. Thus, the terminal zero observation is represented by the sum of two possible events: (1) the individual was alive in period 3 but not encountered; and, (2) the individual did not survive until period 3. Thus, we can write (based on the Law of Total Probability)

$$\Pr(\mathbf{y} = (1, 1, 0)) = \phi_1 p_2 \left\{ (1 - \phi_2) + \phi_2 (1 - p_3) \right\}.$$

Calculation of the terminal probabilities is not difficult, but does become a bookkeeping problem when T is large and $t << T$. Fortunately, there is a simple recursive relationship that can be used to compute each of the terminal probabilities.

Let χ_t denote the probability that an individual is never encountered again given that it was released at time t, i.e.,

$$\chi_t = \Pr(\text{never encountered again}|\text{released at } t).$$

Then, $\chi_T = 1$, and we can obtain $\chi_{T-1}, \ldots, \chi_1$ recursively (Lebreton et al., 1992; Williams et al., 2002) according to

$$\chi_t = (1 - \phi_t) + \phi_t(1 - p_{t+1})\chi_{t+1}.$$

The likelihood for each encounter history can be calculated as a function of the parameters ϕ_t, p_t and the terminal probabilities χ_t. Thus, conceptually, we form the likelihood as the product over the n observed encounter histories:

$$L(\boldsymbol{\phi}, \mathbf{p}|\mathbf{y}_1, \ldots, \mathbf{y}_n) \propto \prod_{i=1}^{n} \Pr(\mathbf{y}_i|\boldsymbol{\phi}, \mathbf{p})$$

which can be maximized easily using standard numerical methods. The encounter history frequencies have a multinomial distribution and so, in practice, it is not necessary to compute the probability of each encounter history individually. Estimation based on maximum likelihood is implemented in popular software packages such as SURVIV (White, 1983; Hines, 1996), MARK (White and Burnham, 1999), and MSURGE (Choquet et al., 2004).

11.1.1 Parameter Identifiability

Naturally, it is desirable to consider year specificity of parameters p and ϕ. As the model has been described, there are T samples and $T - 1$ survival intervals, and so hypothetically there are $2T - 1$ parameters under the fully year-specific model. However, not all of these year effects are identifiable. In particular, in the classical formulation of the fixed effects model, only $p_2, \ldots, p_{T-1}$, $\phi_1, \ldots, \phi_{T-2}$ and the product $\theta = p_T \phi_{T-1}$ are identifiable. In practice, it is often the case that p_T is set to one and the interpretation of the estimated ϕ_{T-1} is qualified, or vice versa. Thus, a fully parameterized model has $T - 2$ detection probability parameters, $T - 2$ survival probabilities, and the remaining product, for a total of $2T - 3$ parameters.

11.2 MODELING NEST SURVIVAL

As we have done in previous chapters when developing specific methods, we begin here by considering a situation in which $p = 1$. This allows us to focus on describing the basic state process model and developing an analysis framework for the model. The class of models commonly employed for modeling avian nest survival is closely related to survival models of animals, but where individual nests can be observed with probability 1 (an unusual circumstance for most animal population studies). Therefore, we provide a brief development of models for avian nest survival. Historical background and recent developments can be found in Mayfield (1975), Johnson (1979), Hensler and Nichols (1981), Stanley (2000), Dinsmore et al. (2002) and Rotella et al. (2004). We note also that survival models with $p = 1$ are often encountered in the analysis of radio telemetry data (White and Garrott, 1990). We develop models of nest survival using data from a study of American redstarts (*Setophaga ruticilla*) conducted in Michigan's Upper Peninsula (Hahn and Silverman, 2006, 2007; Hahn, 2007) to evaluate the effects of social behavior on elements of population dynamics. The specific data set considered here includes 115 nests initiated during the 2006 nesting season.

We suppose that nests are found over time and thus enter the sample at various points (days, samples or times) during the nesting season. For these data, we assume that when a nest is first discovered its age can be determined precisely, and the outcome of the nest (success or failure) is determined perfectly, as in Dinsmore et al. (2002). The models can be generalized to accommodate uncertainty in nest age (Stanley, 2004), but we consider only the simpler case here. After each nest is found and enters the sample it is detected with probability 1 during subsequent visits, since the investigator knows the nest location. Each nest is revisited at least once, perhaps irregularly, and all nests might have different revisitation dates. At the revisit, the status 'active' or 'failed' is recorded for each nest. If a nest is determined to have failed then this represents a sort of censored fate, since the day of failure is not always known precisely. Rather, it is only known that failure occurred between the previous and last visits.

Encounter histories of 10 nests are depicted in Panel 11.1. Columns of this data table index successive days of the study. Uncertain states are indicated by '−,' terminal 0 indicates nests that were checked and found to have failed. The interior dashes preceding that zero correspond to uncertain states. Leading x's indicate states prior to nest entry into the sample. Consider the first encounter history in Panel 11.1. This nest was first located on day 2 and known to have fledged by day 25. Let f_i and l_i be the day of entry and last check, respectively, for nest i, and let m_i be the day at which the nest was checked prior to l_i. In the development of the likelihood (described below), we will only be concerned about m_i for failed nests. Thus, for the nest corresponding to row 1 of Panel 11.1, $f_i = 2$ and $l_i = 25$. For the encounter history in the second row, $f_i = 3$, $l_i = 10$ and $m_i = 7$.

Let ϕ be the probability that at least one egg or individual survives an interval (defined here to be a day). We assume that ϕ varies by individual and day in response to measurable covariates. We consider first describing the probability of observing a particular encounter history $\mathbf{y}_i$ for a nest that entered the sample at f_i, was observed alive at m_i and then observed to have failed by l_i. Thus, it is known to have failed between m_i and l_i. Here, $\mathbf{y}_i$ is the vector of encounter observations of length $l_i - f_i$, such as from Panel 11.1, without the leading and trailing dashes. An example is nest 2 in Panel 11.1, for which $\mathbf{y}_i = (1, 1, 1, 1, 1, -, -, 0)$, $f_i = 3$, $l_i = 10$ and $m_i = 7$. If successive nest states are independent Bernoulli trials, then the probability of observing a string of survival events from f_i to m_i is simply the product of the interval-specific survival probabilities between f_i and m_i. However, the terminal portion of the encounter history (that subsequent to m_i) is not known with certainty. Rather, it is composed of a number of possible events, and we will have to sum the probabilities of these constituent events in order to obtain the probability of the terminal state, which we will also refer to as χ_t for a nest last known to have been alive on day t. As such, the probability of an encounter history is

$$\Pr(\mathbf{y}_i) = \left\{ \prod_{t=f_i}^{m_i - 1} \phi(i, t - 1) \right\} \chi_{m_i}.$$

When a nest failure is observed, the terminal component of the encounter history is the event that nest i is 'failed by l_i given that it was known to be active at m_i.' Thus, define $\chi_{m_i} = \Pr(\text{observed to have failed by } l_i | \text{known alive at } m_i)$. Under the

```
nest  1 2 3 4 5 6 7 ---------- day --------------------------------|
 1       x 1 1 1 1 1 1 1 1 1 1 1 1 1 1 1 1 1 1 1 1 1 1 1 1
 2       x x 1 1 1 1 1 - - 0
 3       x x x 1 1 1 1 1 1 1 1 1 1 1 1 1 1 1 1 1 1 1 1 1 - 0
 4       x x x x x x x 1 1 1 1 1 1 1 1 1 1 1 1 1 1 1 1 1 1 1 1 1
 5       x x x x x x x 1 1 1 1 1 1 1 1 1 1 1 1 1 1 1 1 1 1 1 - - 0
 6       x x x x x x x 1 - - 0
 7       x x x x x x x x 1 1 1 1 1 1 1 1 1 1 1 1 1 1 1 1 1 1 1 - - - 0
 8       x x x x x x x x x x x x x x x 1 1 1 1 1 1 1 1 1 1 1 1
 9       x x x x x x x x x 1 1 1 1 1 1 1 1 1 1 1 1 1 1 1 1 1 1 1 1 1
10       x x x x x x x x x x x x x x x x x x x 1 1 1 1 1 1 - - 0
```

Panel 11.1. Sample of 10 nest encounter histories for the American Redstart (*Setophaga ruticilla*) data (Hahn and Silverman, 2006). Leading x's indicate days prior to nest entry into the sample. Interior dashes indicate uncertain states prior to a check in which the nest was observed to have failed.

assumption that successive nest states are independent Bernoulli trials, this equals (Dinsmore et al., 2002):

$$\chi_{m_i} = 1 - \prod_{t=m_i}^{l_i-1} \phi(i,t).$$

Therefore, the probability of a particular encounter history for a nest that failed between m_i and l_i is:

$$\Pr(\mathbf{y}_i) = \left\{ \prod_{t=f_i}^{m_i-1} \phi(i,t-1) \right\} \left\{ 1 - \prod_{t=m_i}^{l_i-1} \phi(i,t) \right\}. \qquad (11.2.1)$$

Now consider an encounter history for a nest that was observed at l_i and determined to have fledged successfully between m_i and l_i. If we are uncertain about the precise day of fledging, the encounter history probability will have a similar form

$$\Pr(\mathbf{y}_i) = \left\{ \prod_{t=f_i}^{m_i-1} \phi(i,t-1) \right\} \chi_{m_i}^*,$$

where the probability of the terminal state in this case is $\chi_{m_i}^* =$ Pr(known to have fledged by l_i|known alive at m_i). Note that the two events 'known to have fledged by l_i given known alive at m_i' and 'observed to have failed by l_i given known alive at m_i' are mutually exclusive and represent the only two possible outcomes given that the nest was alive at m_i. Thus, their probabilities must sum to 1. That is, $\chi_{m_i} + \chi_{m_i}^* = 1$ or $\chi_{m_i}^* = 1 - \chi_{m_i}$. As such,

$$\Pr(\mathbf{y}_i) = \left\{ \prod_{t=f_i}^{m_i-1} \phi(i,t-1) \right\} (1 - \chi_{m_i})$$

which is precisely equivalent to

$$\Pr(\mathbf{y}_i) = \left\{ \prod_{t=f_i}^{l_i-1} \phi(i,t-1) \right\}. \qquad (11.2.2)$$

The likelihood for a collection of n encounter histories can be constructed from the two expressions given by Eq. (11.2.1) (for failed nests) and Eq. (11.2.2) (for nests known to have survived). As with the classical CJS-type models, obtaining the MLEs for model parameters is straightforward in principle. We provide an illustration in the next section.

11.2.1 Analysis of the Redstart Data

For the redstart data, we consider models for variation in survival probability that potentially include the day of the season ('Day'), the age of the nest ('Age'), and the density of neighbors ('Dense') which we assume to be known as a result of extensive nest searching. We allowed for the possibility for a quadratic response to day-of-season. The models considered are of the form

$$\text{logit}(\phi(i,t)) = \beta_0 + \beta_1 \text{Age}(i,t) + \beta_2 \text{Day}(i,t) + \beta_3 \text{Day}^2(i,t)$$
$$+ \beta_4 \text{Dense}(i,t). \tag{11.2.3}$$

A sample implementation in **R** for a model containing only nest age is shown in Panel 11.2. The vectors `first` and `last` are composed of the elements f_i and l_i, respectively, whereas `last1` contains m_i.

The results of fitting a number of models to the redstart data are summarized in Table 11.1. We see that the preferred model, by AIC, is that containing a quadratic day-of-season effect and nest age, and there is a strong positive effect of nest age on survival probability. This is plausible if the conspicuous nests – in a sense, the worst nest locations – are discovered first by predators. The AIC of the full model, including an effect of neighbor density, is only slightly lower.

```
lik<-function(parms){
beta0<-parms[1]
beta1<-parms[2]
lik2<-rep(1,nind)
lik1<-rep(NA,nind)
for(i in 1:nind){
phi<- expit(   beta0 + beta1* (AGE[i,]-10)/5   )
   lik1[i]<-sum(log(phi[(first[i]+1):last1[i]]))
  if(last1[i]<last[i])
  lik2[i]<- log(1-prod( phi[ (last1[i]+1):last[i]]))
  }
-1*sum(lik1   + lik2)
}
```

Panel 11.2. Likelihood construction for a nest survival with a single covariate, nest age, given by `AGE`. The vectors `first`, `last1` and `last` correspond to f_i, m_i and l_i in the text. Implementation is provided in the Web Supplement.

Table 11.1. Parameter estimates (MLEs) and AIC of nest survival models fitted to the American Redstart (*Setophaga ruticilla*) data. For explanation of parameters see text and Eq. (11.2.3).

β_0	age	day	day^2	dense	AIC
2.257	0.346	−0.147	0.366	–	286.54
2.092	0.322	−0.144	0.344	0.147	286.62
2.541	–	–	–	0.209	292.27
2.554	–	0.023	0.232	–	293.01
2.542	0.093	–	–	0.200	293.35
2.808	0.106	–	–	–	294.96
2.819	–	–	–	–	294.20
2.777	–	−0.078	–	–	295.78

11.2.2 Hierarchical Formulation

Let $z(i,t)$ for $i = 1, \ldots, n$ and $t = 1, \ldots, T$ be Bernoulli random variables describing whether nest i is active ($z(i,t) = 1$) or not ($z(i,t) = 0$) at time t. We specify the probability of observing a particular encounter history conditional on the time of first encounter of each nest, f_i. That is, $z(i, f_i) = 1$ with probability 1. The nest survival process is described by the Markovian state model:

$$z(i,t)|z(i,t-1) \sim \text{Bern}(z(i,t-1)\phi(i,t-1)) \qquad (11.2.4)$$

for $t = f_i + 1, \ldots, T$. In other words, if a nest is active at time $t - 1$, its survival outcome is a Bernoulli random variable with parameter $\phi(i, t - 1)$. If a nest has failed by time $t-1$, then $z(i,t)$ is Bernoulli with success probability 0, i.e., $z(i,t) = 0$ with probability 1.

Since $p = 1$ in nest survival studies, we take observations such as from Panel 11.1 as being equivalent to z. The main consideration is that observations for the uncertain interior state variable values – those preceding the last 0 – be regarded as missing values. In a Bayesian analysis of the model, these uncertain states are regarded the same as parameters of the model and drawn from the posterior distribution by MCMC.

11.2.2.1 Redstart nest survival
We return to the nest survival data and provide an analysis of the hierarchical formulation under an extension of the model to accommodate a random effect. As is common in many ecological field studies, observations in the study of American redstart nest survival were obtained on spatial units, each containing a number of

nests. In this study, the spatial units were 9 ha plots, and nests were monitored on twelve such units. We expect nest fates within a plot to be correlated and thus, failure to account for variation among plots might adversely influence inference about things we care about (in the present case, age or density). Naturally, we view plot as a random effect since the plots were approximately randomly selected from a large number of ostensibly similar plots and the scope of inference is beyond the specific units that were sampled.

The model described in the previous section was extended to include a random plot effect, say α_{plot}, assumed to be a mean-zero normal random effect with variance σ^2_{plot}. The **WinBUGS** model description is shown in Panel 11.3. Note that, due to the similarity of the state model, this specification is derived directly from the dynamic occupancy model from Chapter 9 with slight modifications (e.g., there is no recruitment/colonization component of the model). In particular, the kernel of the state model specification is the following Markovian structure:

```
model {
beta0~dunif(-5,5)
beta1~dunif(-5,5)
deneff~dunif(-5,5)
dayeff~dunif(-5,5)
dayeff2~dunif(-5,5)
sigmaplot~dunif(0,7)
tauplot<-sqrt(1/(sigmaplot*sigmaplot))
for(i in 1:12){
alpha[i]~dnorm(0,tauplot)
}
for(i in 1:nrec){
  for(j in (first[i]+1):last[i]){
     logit(phi[i,j])<-beta0 + beta1*((AGE[i,j]-10)/5)
                   + dayeff*DAY[i,j] + dayeff2*DAY2[i,j]
                   + deneff*dense[i] +alpha[plot[i]]
       mu[i,j]<-phi[i,j]*y[i,j-1]
       y[i,j]~dbern(mu[i,j])
     }
  }
}
```

Panel 11.3. WinBUGS model specification for fitting the nest survival model with a random plot effect.

```
for(j in (first[i]+1):last[i]){
    mu[i,j]<-phi[i,j]*y[i,j-1]
    y[i,j]~dbern(mu[i,j])
}
```

If a nest is alive at time $j-1$, then it can survive with probability phi[i,j] whereas, once it fails, its subsequent survival probability evaluates to 0. As we have done previously, we prefer prior distributions having compact support and therefore the regression parameters are given uniform priors. This is usually innocuous when covariates are suitably scaled. Specification of the random plot effects (the vector alpha) should be clear. Elements of alpha are referenced by the data vector plot which contains an integer index to plot membership of each nest (an integer between 1 and 12). Posterior summaries from fitting the model to the American redstart data are given in Table 11.2. These summaries are based on approximately 60000 posterior draws. We see that, contrary to our expectations, the variance among plots is relatively small, indeed the posterior mass for σ_{plot} is concentrated near 0, and the results are broadly similar to those of the model fitted by maximum likelihood (Table 11.1).

11.3 HIERARCHICAL FORMULATION OF THE CJS MODEL

We adopt a hierarchical formulation of the CJS model that may be more versatile for certain situations such as the inclusion of individual effects or modeling individual covariates (Bonner and Schwarz, 2006; Royle, 2008c). For the hierarchical formulation we describe the state model – the individual 'alive state' – and then the observation process is conditional on an individual being alive and available to be encountered. As in the previous Section, we denote the state process by the binary state variable $z(i,t)$ indicating whether individual i is alive at time t. In the present case, the $z(i,t)$ state variables are governed by a simple survival process (described subsequently). The second component model

Table 11.2. Posterior summaries of nest survival model with random plot effects (Panel 11.3) fitted to the Redstart data.

effect	mean	sd	2.5%	50%	97.5%
density	0.163	0.118	−0.069	0.162	0.397
age	0.320	0.134	0.056	0.321	0.579
day	−0.098	0.188	−0.449	−0.104	0.290
day^2	0.407	0.146	0.140	0.400	0.712
intercept	2.096	0.294	1.533	2.088	2.699
σ	0.397	0.424	0.021	0.276	1.518

is a model for the binary observations $y(i,t)$ conditional on the latent state variables $z(i,t)$.

11.3.0.2 The state model

Let $z(i,t)$ for $i = 1, \ldots, n$ and $t = 1, \ldots, T$ be Bernoulli random variables describing whether individual i is alive ($z(i,t) = 1$) or dead ($z(i,t) = 0$) at time t. The CJS model is developed conditional on the time of first capture of each individual, f_i. That is, $z(i, f_i) = 1$ with probability 1. The survival process is given by the conditional model

$$z(i,t)|z(i,t-1) \sim \text{Bern}(z(i,t-1)\phi_{t-1}) \qquad (11.3.1)$$

for $t = f_i+1, \ldots, T$. In other words, if an individual is alive at time $t-1$, its survival outcome is a Bernoulli random variable with parameter ϕ_{t-1}. If an individual is not alive at $t-1$, then $z(i,t)$ is Bernoulli with success probability 0, i.e., $z(i,t) = 0$ with probability 1.

11.3.0.3 The observation model

Conditional on the state process, $y(i,t)$ are independent Bernoulli random variables,

$$y(i,t)|z(i,t) \sim \text{Bern}(p_t \, z(i,t)). \qquad (11.3.2)$$

Thus, if $z(i,t) = 0$ then $y(i,t) = 0$ with probability 1, otherwise $y(i,t)$ is a Bernoulli trial with parameter p_t.

The models have been described here with time-varying parameters. While this is a common model, so, too, are age-structured models, or models in which either p_t or ϕ_t vary in response to known covariates. These and other modifications are both straightforward to construct and practical to implement in **WinBUGS** as we describe subsequently.

11.3.1 Bayesian Analysis

The hierarchical formulation of the CJS model is easily described in **WinBUGS** pseudo-code as shown in Panel 11.4 for a simple case in which both p and ϕ are constant. Here, the data are the encounter history matrix y, the vector `first` and the fixed constants `nind` and `nyear`. The line `z[i,first[i]]` $\sim$ `dbern(1)` admits that the first latent state variable of each individual is fixed, whereas the meaning of the remaining model specification should be self-evident. In particular, note that

$U(0, 1)$ priors are assumed for both ϕ and p, and the distributions of $z(i, t)|z(i, t-1)$ and $y(i, t)|z(i, t)$ are those described by Eqs. (11.3.1) and (11.3.2), respectively.

We illustrate the fitting of this model using **WinBUGS** with a small data set shown in Table 11.3. These data are encounter histories of 10 yellow warblers (*Dendroica petechia*) obtained from a constant-effort mist netting station that is part of the Monitoring Avian Productivity and Survival (MAPS) program (see Section 11.4 for additional details). These individuals were first captured in 1992 and the data include recaptures through 2000. The model described in Panel 11.4 was fitted to these data (an **R** script is provided in the Web Supplement). Posterior summary statistics for the warbler data are given in Table 11.4. These summaries are based on 60000 Monte Carlo draws after discarding 2000 burn-in samples.

11.3.2 Representation as a Constrained Jolly–Seber Model

In Chapter 9 we described the related class of models known as Jolly–Seber models which allow for both recruitment and survival. In the CJS model, by conditioning

```
model {

## Prior distributions
phi~dunif(0,1)
p~dunif(0,1)

  for(i in 1:nind){
     ### Individuals enter sample with probability 1
      z[i,first[i]]~dbern(1)
     ### Definition of state and observation models
      for(j in (first[i]+1):nyear){
         mu1[i,j]<-p*z[i,j]
         y[i,j]~dbern(mu1[i,j])
         mu2[i,j]<-phi*z[i,j-1]
         z[i,j]~dbern(mu2[i,j])
      }
  }
}
```

Panel 11.4. WinBUGS model specification for fitting the CJS model with constant p and ϕ.

Table 11.3. Yellow warbler (*Dendroica petechia*) encounter history data for a sample of 10 individuals from a constant-effort mist netting station. Individuals were all first captured in 1992.

$t = 1$	$t = 2$	$t = 3$	$t = 4$	$t = 5$	$t = 6$	$t = 7$	$t = 8$	$t = 9$
1	0	0	0	0	0	0	0	0
1	0	0	0	0	0	0	0	0
1	0	0	0	0	0	0	0	0
1	1	1	1	0	0	0	0	0
1	1	0	0	1	1	1	1	1
1	0	0	0	0	0	0	0	0
1	0	0	0	0	0	0	0	0
1	0	0	0	0	0	0	0	0
1	0	0	0	0	0	0	0	0
1	0	0	0	0	0	0	0	0

Table 11.4. Posterior summary statistics from analysis of the warbler data (*Dendroica petechia*) set under a constant-p, constant-ϕ model. These data are all individuals encountered at a single MAPS station over a nine year period.

parameter	mean	SD	$q_{0.025}$	$q_{0.50}$	$q_{0.975}$
ϕ	0.571	0.105	0.361	0.573	0.767
p	0.648	0.149	0.342	0.657	0.905

on capture, or entry into the sample, we lose information about recruitment parameters. The relationship between the two models can be seen by considering the Jolly–Seber state model, as described in Chapter 9, which is:

$$z(i, t + 1) \sim \text{Bern}\left(\phi_t z(i, t) + \gamma_{t+1}\left\{\prod_{k=1}^{t}(1 - z(i, k))\right\}\right),$$

where $\prod_{k=1}^{t}(1 - z(i, k))$ is an indicator of previous recruitment. That is, an individual can be recruited with probability γ_{t+1} if it has never previously been recruited. Evidently the CJS model is a formal reduced version of this model, which we obtain by setting $\gamma_t = 0$. In addition, entry into the sample is regarded as a random variable under the Jolly–Seber model. Thus, the observation model will contain additional probability structure for the initial encounter, such as $y(i, 1) \sim \text{Bern}(p)$, for each $i = 1, 2, \ldots, n$.

11.4 MODELING AVIAN SURVIVAL FROM MIST-NET DATA

We develop a simple analysis of data from the Monitoring Avian Productivity and Survival (MAPS) program (DeSante et al., 1995, 2004). MAPS operates a continental-scale network of constant-effort mist netting stations that yields capture–recapture data on many species of passerines annually at approximately 500 sites. There have been over 1000 stations operated since 1989 and more than 700 of these have been operated for at least 4 years (J. Saracco, pers. comm.). The program is ongoing but we consider data from the period 1992–2003 in our analyses here. We use data for the yellow warbler (*Dendroica petechia*) from 139 distinct MAPS stations shown in Figure 11.1.

One of the features of the data is the presence of 'transients'. These are individuals who are not residents of the local population. They result in all-zero encounter histories (subsequent to initial capture). If they are regarded as being equivalent to resident individuals, these all-zero encounter histories cause apparent survival to be lower than nominal survival of resident birds. So, when estimation of the survival probability that is relevant to the local population of resident birds is of interest, transients need formal consideration in a model. The formalization of this notion originates from Pradel et al. (1997); also see Hines et al. (2003) for additional development. We thus begin with a development of the CJS model with transients before moving on to hierarchical elaborations of the model in which we have spatial variation in survival rates.

For MAPS data, the encounter history data are spatially-indexed with most MAPS stations contributing data for only a few individuals of most species, i.e., the data are sparse. This provides some practical motivation for pursuing a hierarchical modeling strategy and, in Section 11.5, we consider an extension of the basic model described here to further a model-based aggregation of these data, while retaining some degree of spatial resolution in the estimated survival probabilities.

11.4.1 CJS Model with Pre-determined Residents

To obtain information on the fraction of residents in the population exposed to sampling, some individuals are 'pre-determined' as residents. Mist nets are set approximately every 10 days and if an individual is captured on more than one sampling day during the season in which it was initially captured, then it is pre-determined to be resident. The remaining individuals represent an unknown mixture of transients and residents. To clarify the meaning of this pre-determined resident label, let R_i denote an individual's *true* residency status. We suppose that

$$R_i \sim \text{Bern}(\pi_r),$$

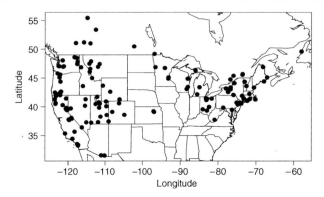

Figure 11.1. 139 Monitoring Avian Productivity and Survivorship (MAPS) stations used in our analysis of the yellow warbler data.

where π_r is the probability that an individual susceptible to sampling is a resident. Note that we are not able to observe R_i except for certain resident individuals. The parameter π_r is a biological parameter of some relevance as it could pertain to dispersal, local population dynamics, source–sink dynamics, and other ecological considerations. Thus, estimation of π_r might generally be of some interest and we need to acknowledge that the sample quantity (the number of predetermined residents in the sample divided by n) is biased by imperfect detection. Now, let $r_i = 1$ if individual i is a pre-determined resident and $r_i = 0$ otherwise. We suppose that

$$r_i|(R_i = 1) \sim \text{Bern}(\tau),$$

where τ is the probability that an individual is pre-determined to be a resident given that it is a resident. For $R_i = 0$, then $r_i = 0$ with probability 1. Thus, we may express the model as we have previously done in similar situations:

$$r_i|R_i \sim \text{Bern}(R_i\tau).$$

The parameter τ is largely a nuisance parameter, in the sense that it is solely a function of the detection probability of the sampling program. As p goes to 1, then τ must also tend to 1 (note that we could specify the model explicitly in terms of p using within-year capture–recapture data, which we have not done in these analyses). Our main interest in introducing these parameters τ and π_r is that this formulation of the model clarifies the relationship between observations r_i and the true residency state R_i and, furthermore, we require R_i in the specification of the state model (as described subsequently). As specified here, neither the parameter

π_r nor τ has the same interpretation as the parameter introduced by Pradel et al. (1997) and Hines et al. (2003), who define a parameter (also labeled by them as τ) which is the probability that an unmarked bird captured and released is a transient. As the residence probability (or its complement – probability of transience) is not likely to be constant, both Pradel et al. (1997) and Hines et al. (2003) allowed year-specificity on the additional parameter. We neglect that extension in our analysis below.

We require definition of R_i because the state model is different for residents and non-residents. In particular, for non-residents, the encounter history is comprised of all-zero observations subsequent to first capture. However, some of the observed all-zero encounter histories may correspond to residents that simply were not recaptured. Thus, R_i partitions these all-zeros into two groups, those corresponding to residents and those corresponding to non-residents. In fact, the parameter π_r is precisely a zero-inflation parameter, allowing for excess all-zero encounter histories subsequent to initial capture. As we have done in similar problems, we can accommodate this zero inflation with the following **WinBUGS** specification of the model, for the case where ϕ and p are constant:

```
R[i]~dbern(pi)
for(j in (first[i]+1):nyear){
    muy[i,j]<-p*z[i,j]
    y[i,j]~dbern(muy[i,j])
    muz[i,j]<-z[i,j-1]*phi*R[i]
    z[i,j]~dbern(muz[i,j])
}
```

An alternative way to admit the pre-determined resident information into the model is as a mixture on the parameter π_r. We note that $\pi_r = 1$ whenever $r_i = 1$ and otherwise θ, where θ in this case is the probability that an individual is a resident *given that it is not pre-determined to be a resident*. We can express π_r as

$$\pi_r = 1^{r_i}\theta^{1-r_i}. \tag{11.4.1}$$

The **WinBUGS** specification is:

```
mu[i]<- pow(1,r[i])*pow(tau,1-r[i])
R[i] ~ dbern(mu[i])
```

This parameterization is implemented in the model described by Panel 11.6. This is somewhat inconvenient because we may wish to estimate π_r, or even to model it (e.g., as a function of covariates), and this formulation renders it a derived parameter. Nevertheless, we can recover the residency probability, π_r, according to

$$\pi_r = \mathrm{E}[r] + \theta(1 - \mathrm{E}[r]),$$

where $\mathrm{E}[r]$ can be estimated as the sample proportion of individuals that were pre-determined to be residents. This is an application of the Law of Total Probability as follows:

$$\Pr(R = 1) = \Pr(R = 1 | r = 1) \Pr(r = 1) + \Pr(R = 1 | r = 0) \Pr(r = 0)$$
$$= \mathrm{E}[r] + \theta \, \mathrm{E}[r].$$

To the best of our knowledge, there has not been any attention given to modeling and inference about the residency probability π_r. However, with the model parameterized explicitly in terms of π_r, it is straightforward to develop models to test-specific hypotheses about this parameter, and its relationship to other model parameters such as the survival probability.

11.4.2 Analysis of the Warbler Data

We analyze the MAPS warbler data from Bird Conservation Region (BCR) 15 (Sierra Nevada), for which we have data from 634 individuals. BCRs originate from the North American Bird Conservation Initiative[1] (NABCI) (Sauer et al., 2003) and are now widely used to summarize bird population status and trends. For these data, there were 145 pre-determined residents over the 12-year period. The complete **WinBUGS** specification for a model having year-specific parameters is shown in Panel 11.5. This implementation has uniform priors on all parameters. Recall that ϕ_{T-1} is confounded with p_T which is admitted in the model by explicitly setting the last p to 1.

    The results from fitting this model are summarized in Table 11.5. These results are based on approximately 40000 Monte Carlo draws after discarding 2000 burn-in samples. Recall that τ is the probability that a resident is pre-determined to be a resident upon entry into the sample. Thus, the posterior mean of τ, 0.5038, suggests that we only properly identify about half of all resident individuals. The population parameter that is potentially of biological interest is π_r, the probability that an individual in the population exposed to sampling is a resident. The posterior mean of π_r is 0.4509. Note that the product $\tau \, \pi_r$ is 0.227, which is roughly in line with the sample proportion of pre-determined residents (in the sample, we had 145 out of 634 individuals pre-determined to be a resident).

---

[1]Sierra Nevada, see `http://www.nabci-us.org/bcrs.html`.

11.5 MODELING SPATIAL VARIATION IN SURVIVAL

One practical consideration in developing statistical models of survival from MAPS data is that there are many stations with relatively sparse data. As such, it is not usually feasible to obtain estimates at the level of individual MAPS stations, or even at finer than multi-state or regional scales. Historically, this has been dealt with by pooling data within coarse geographic strata for purposes of estimation (Saracco

```
model {
for(t in 1:(nyear-2)){
   phi0[t]~dunif(0,1)
   lphi0[t]<-log(phi0[t]/(1-phi0[t]))
   p[t]~dunif(0,1)
   p2[t+1]<-p[t]
}
p2[1]<-1; p2[nyear]<-1
phi0[nyear-1]~dunif(0,1)
lphi0[nyear-1]<-log(phi0[nyear-1]/(1-phi0[nyear-1]))

pi~dunif(0,1)
tau~dunif(0,1)
for(i in 1:nind){
 R[i]~dbern(pi)
 mu[i]<-R[i]*tau
 r[i]~dbern(mu[i])
    for(j in 1:first[i]){
       z[i,j]~dbern(1)
     }
    for(j in (first[i]+1):nyear){
       logit(phi[i,j-1])<- lphi0[j-1]
       muy[i,j]<-p2[j]*z[i,j]
       y[i,j]~dbern(muy[i,j])
       muz[i,j]<-z[i,j-1]*phi[i,j-1]*R[i]
       z[i,j]~dbern(muz[i,j])
    }
  }
}
```

Panel 11.5. WinBUGS model specification of a CJS model allowing for transients, using data on pre-determined residents, for the MAPS yellow warbler data.

Table 11.5. Posterior summary statistics from analysis of the yellow warbler data under a model with year-specific p and ϕ. The data include individuals captured from 1992–2003 from a number of MAPS (DeSante et al., 1995) stations in BCR 15 (Sierra Nevada). Adopting the conventional parameterization of CJS-type models, the last survival probability, ϕ_{11} is confounded with p_{12}.

parameter	mean	SD	0.025	0.50	0.975
p_2	0.645	0.118	0.410	0.648	0.865
p_3	0.629	0.151	0.324	0.636	0.894
p_4	0.568	0.145	0.289	0.569	0.841
p_5	0.682	0.135	0.405	0.691	0.914
p_6	0.563	0.130	0.314	0.563	0.812
p_7	0.756	0.112	0.511	0.767	0.939
p_8	0.608	0.114	0.383	0.610	0.821
p_9	0.456	0.106	0.262	0.453	0.673
p_{10}	0.742	0.093	0.542	0.750	0.902
p_{11}	0.444	0.084	0.290	0.439	0.621
ϕ_1	0.842	0.109	0.594	0.860	0.993
ϕ_2	0.473	0.134	0.253	0.459	0.782
ϕ_3	0.551	0.151	0.299	0.539	0.882
ϕ_4	0.588	0.141	0.335	0.579	0.888
ϕ_5	0.597	0.140	0.349	0.588	0.900
ϕ_6	0.536	0.108	0.342	0.530	0.763
ϕ_7	0.568	0.112	0.375	0.559	0.819
ϕ_8	0.595	0.131	0.367	0.583	0.881
ϕ_9	0.498	0.090	0.335	0.494	0.686
ϕ_{10}	0.831	0.107	0.605	0.843	0.991
ϕ_{11}	0.347	0.064	0.233	0.344	0.481
π_r	0.451	0.034	0.389	0.450	0.519
τ	0.504	0.042	0.422	0.504	0.587

et al., 2008a). An alternative approach is to consider model-based approaches to aggregation, wherein data from separate MAPS stations are combined via explicit models on survival and other parameters (Saracco et al., unpublished). While a model-based approach to aggregation could benefit estimation of model parameters, as a scientific matter, interest in these data is focused on assessing demographic causes of population trends (Saracco et al., 2008a). Thus, spatial model-based aggregation of these data is consistent with scientific objectives that involve understanding factors that influence variation in survival of species, which we might expect to operate at relative fine spatial scales.

There are a number of viable strategies for formulating spatial models based on data such as those arising from the MAPS program. The basic conceptual formulation is to express the model for survival probability according to

$$\text{logit}(\phi_i) = \mu_\phi + \alpha_i, \tag{11.5.1}$$

where μ_ϕ is the overall mean, and α_i is a zero-mean, site-specific, random effect assumed to be spatially correlated. Then, we seek parameterizations of spatial dependence in the random effects α. Note that this model could also have temporal effects and covariates, or other embellishments, but we avoid that level of complexity here. There are a number of ways to parameterize spatial dependence in this random effect. In general, the strategy is to describe the multivariate distribution of the vector of random effects corresponding to the observation locations $\alpha = (\alpha_1, \ldots, \alpha_M)$. Most applications of spatial models rely on an assumption of multivariate normality:

$$\alpha \sim \mathrm{MN}(\mathbf{0}, \Sigma),$$

where Σ is an $M \times M$ variance-covariance matrix. Typically Σ is described by some covariance function having parameter(s) θ. For example, the exponential covariance function is

$$\Sigma_{ij} = \exp(-||s_i - s_j||/\theta),$$

where $||s_i - s_j||$ denotes the distance between locations s_i and s_j. This basic formulation – a multivariate normal model with a parametric covariance matrix – underlies the method usually referred to as 'kriging'. Diggle et al. (1998) adapted this formulation in the development of spatial models for counts within a generalized linear modeling framework. The main deficiency with this approach is that it can become computationally prohibitive as the number of spatial samples increases owing to the dimensionality of Σ.

An efficient class of models that has seen widespread use are the conditional autoregressive (CAR) models. Like the auto-logistic models introduced in Section 9.4, the CAR models are ideally suited for situations in which the spatial process is inherently discrete, such as in analyses involving states, counties or other geopolitical units. The CAR model relates α_i to α at other locations in the conditional mean, according to:

$$\mathrm{E}[\alpha_i | \alpha_{-i}] = \gamma \sum_j C_{ij} \alpha_j,$$

where γ is a correlation parameter, C_{ij} are weights reflecting spatial association between units and α_{-i} denotes the vector of random effects except the ith. As described here, the permissible range of γ is constrained by the eigenvalues of the matrix of weights $\mathbf{C} = \{C_{ij}\}$. For highly irregular lattices, this can limit the strength of the correlation significantly. A more flexible version of this model is the *intrinsic CAR model* (Besag et al., 1991), which has the form

$$\alpha_i | \boldsymbol{\alpha}_{-i} = \mathrm{N}\left(\frac{1}{n_i}\sum_{j\in\mathcal{N}_i}\alpha_j, \sigma^2/n_i\right),$$

where $\mathcal{N}_i$ is the neighborhood of cell i, which is the collection of grid-cell identities of neighboring cells, and n_i is the number of neighbors of grid cell i, i.e., $n_i = \dim(\mathcal{N}_i)$. In this analysis we used a first order 'queen's neighborhood', so the 8 cells surrounding a cell are its neighbors. Banerjee *et al.* (2004) provide an accessible treatment of CAR models.

11.5.1 Analysis of the MAPS Data

To apply the intrinsic CAR model to the yellow warbler data from the MAPS program, we created a 2-degree grid (2 deg. longitude by 2 deg. latitude) over North America, and assigned each MAPS station to a grid cell. The CJS model for the warbler data was extended to include a 'degree-block effect,' which was assumed to be a spatially-correlated random effect. The degree-block effects were assigned a CAR prior. As such, the basic model is similar in structure to the American redstart nest survival example (e.g., Panel 11.3), having random spatial effects with (potentially) multiple observations being associated with each random spatial effect. Essentially, this construction produces a relatively fine-scale stratification scheme (relative to BCRs), but not so fine so as to produce many empty cells (i.e., without data).

We consider a model with year-specific detection probabilities and a spatially varying survival probability, governed by a CAR prior distribution on the logit-transformation of ϕ. In particular, let ϕ_k be the survival probability for a bird located within degree block k. Then, the model for survival probability for a MAPS station i within degree-block k is the additive model

$$\mathrm{logit}(\phi_{ik}) = \mu_\phi + \alpha_k,$$

where the vector of random effects $\boldsymbol{\alpha}$ is given the intrinsic CAR prior described in the previous section.

The **WinBUGS** description of this model for the warbler data is shown in Panel 11.6. Specification of the CAR model requires 3 data objects. First `weights` which is defined (for the intrinsic CAR model) to be a vector of 1s. The length of `weights` is `sumNumNeigh` (also data) which is equal to the number of adjacent pairs of grid cells. The data object `adj`, describes the adjacency structure of the grid, and `num`, is a vector having elements equal to the number of neighbors of each grid cell. These can be produced in **WinBUGS** using the adjacency tool under the `maps` menu. We provide an example of their calculation using **R** functions on the Web Supplement. Finally, the data object `gridid` is a vector having length equal to the number of individual birds in the data set. Each element of `gridid` indicates the grid-cell the individual is associated with which.

The model was fitted in **WinBUGS** and the resulting map of survival probability is shown in Figure 11.2. Each point on the map represents a prediction made at the center of a 2-degree grid cell. The shading of each point is related to the magnitude of the estimated survival probability, whereas the *size* of each point is inversely proportional to the posterior standard deviation. Thus, more precise predictions are indicated with larger circles and *vice versa*. This map of predicted survival probability shows strong gradients in survival probability over North America. Survival is very low in the east and southwest coast, and relatively higher in the interior of North America. We note that areas having higher MAPS station densities yield more precise predictions of local survival probability, as a result of the relatively greater contribution of data (i.e., relative to the spatial model).

11.6 SURVIVAL MODELS WITH INDIVIDUAL EFFECTS

Models for animal demography that contain individual effects are important in many situations. Two broad classes of such models exist: those in which heterogeneity is modeled as unstructured variation, due to an individual 'random effect' (Norris and Pollock, 1996; Pledger, 2000; Dorazio and Royle, 2003) and models which seek to describe heterogeneity explicitly, by the use of individual covariates (Pollock, 2002; Royle, 2008a). Methods of inference for both classes of problems have, individually, received considerable attention, primarily focused on the case of closed populations (Chapter 6). In this section, we adopt the hierarchical formulation of the CJS model in order to allow for individual effects on parameters in open population models. Considerable attention has recently been devoted to extending open population models to allow for individual effects. Cam et al. (2002a) provide an analysis of the special case in which individuals are detected with probability 1. Pledger et al. (2003) developed a CJS model containing heterogeneity in both detection probability and survival probability using finite-mixture models. Bayesian analysis of individual effects models has recently been considered by Bonner and Schwarz (2006), Royle (2008c), Gimenez et al. (2007) and King et al. (2008), and an early treatment of the problem was given by Dupuis (1995). The analysis presented here follows closely that of Royle (2008c).

11.6.1 Model Formulation

There is considerable theoretical interest in the existence and magnitude of
heterogeneity in survival and other vital rates among individuals (Cam et al., 2002a,
2004). In addition, heterogeneity masks other effects that might be the focus of

```
model {

for(t in 1:nyear){
  p[t]~dunif(0,1)
}
phi0~dunif(0,1)
lphi0<-log(phi0/(1-phi0))
theta~dunif(0,1)

for(j in 1:sumNumNeigh) { weights[j] <- 1}
spacetau~dgamma(.1,.1)
alpha[1:ngrid] ~ car.normal(adj[], weights[], num[], spacetau)

for(i in 1:nind){
    logit(phi[i])<- lphi0 + alpha[gridid[i]]
    tmp[i]<- pow(1,resident[i])*pow(theta,1-resident[i])
    R[i] ~ dbern(tmp[i])

    for(j in 1:first[i]){
      z[i,j]~dbern(1)
     }
    for(j in (first[i]+1):nyear){
        mu1[i,j]<-p[j]*z[i,j]
        x[i,j]~dbern(mu1[i,j])
        mu2[i,j]<-z[i,j-1]*phi[i]*R[i]
        z[i,j]~dbern(mu2[i,j])
    }
  }
}
```

Panel 11.6. **WinBUGS** model specification of a CJS model allowing for transients, and
with spatially-correlated survival probability parameters. For illustration, this model uses the
alternative parameterization for transients having parameter θ that is related to residency
probability by Eq. (11.4.1).

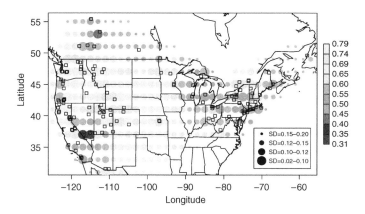

Figure 11.2. Map depicting spatial variation in yellow warbler survival. Values are point-wise posterior means. The shading reflects the magnitude of predicted survival probability whereas the size of the plot symbol is proportional to the inverse of the posterior standard deviation. Larger circles indicate smaller posterior standard deviations. The 139 Monitoring Avian Productivity and Survivorship (MAPS) stations used in the analysis are indicated with square symbols.

investigation is widely known (e.g., senescent decline in survival rates (Service, 2000; Cam et al., 2002a)). To elucidate these effects, it is necessary to account for heterogeneity explicitly in models of survival. However, as discussed in Royle (2008c), because the CJS model conditions on entry of individuals into the sample, it is not clear whether, or under what conditions, the estimated heterogeneity distribution is an estimate of the population quantity that is most relevant to ecological and evolutionary theory. Intuitively, individuals with higher intrinsic survival rates will appear in the sample in higher proportions than individuals with lower intrinsic survival rates, because they have more net exposure to sampling through time. Because the accumulation of individuals over time favors good surviving individuals appearing in the sample, the interpretation of estimates of heterogeneity should be considered carefully.

We consider a model in which both survival and capture probability vary by year and by individual. Using customary notation for describing capture–recapture models, this model would be indicated as $\{p(t+h), \phi(t+h)\}$. Specifically, we assume that

$$u_{it} \equiv \text{logit}(p_{it}) = a_t + \alpha_i$$

and

$$v_{it} \equiv \text{logit}(\phi_{it}) = b_t + \beta_i,$$

where p_{it} is the probability of capture of individual i in year t and ϕ_{it} is the probability of survival of individual i over the interval $(t, t+1)$, a_t and b_t are the annual fixed effects and α_i and β_i are latent individual effects. The individual effects α_i and β_i are assumed to be mean zero random effects with variances σ_p^2 and σ_ϕ^2, respectively. Not all fixed effects are identifiable under this model. In particular, in the classical formulation of the fixed effects model with $T = 7$, only $p_2, \ldots, p_6$, $\phi_1, \ldots, \phi_5$ and the product $\theta = p_7 \phi_6$ are identifiable (typically, p_7 is set to one and the interpretation of ϕ_6 is so qualified, or vice versa). Thus, only the corresponding logit-scale fixed effects are identifiable in the present construction.

Specification of prior distributions for the model parameters merits some discussion. Priors with bounded support typically yield better-mixing Markov chains. Thus, for all parameters, we assume proper uniform priors: $U(0, 1)$ priors on the inverse-logit of the annual fixed effects, $\text{logit}^{-1}(a_t) \sim U(0, 1)$, $\text{logit}^{-1}(b_t) \sim U(0, 1)$. Gelman (2006) recommends a proper uniform prior for the standard deviation parameters σ_p and σ_ϕ in similar classes of models due to potential sensitivity to the variance component prior in models with latent variables. This was considered in the present context by Royle (2008c), and we address the salient points below.

11.6.2 Analysis of the European Dipper Data

We fit this model to data on European dippers (*Cinclus cinclus*) originally from Marzolin (1988) and used also by Lebreton et al. (1992) and many others. The particular analysis is from Royle (2008c). The data and **R** and **WinBUGS** functions for analyzing these data can be found in the Web Supplement. The **WinBUGS** model specification for the heterogeneity model is given in Panel 11.7, and posterior summaries of model parameters are given in Table 11.6. Due to relatively slow mixing of the Markov chains, these summaries are based on approximately 1.8 million posterior draws. The results deviate slightly from Royle (2008c) due to Monte Carlo error. Based solely on the magnitude of the variance components (the posterior mean of σ_p is 1.395), the results might appear to indicate considerable heterogeneity in p, but a lesser degree of heterogeneity in ϕ. We consider this in the context of a formal model-selection procedure in the following subsection.

It would be straightforward to adapt this model specification to a model in which there was an individual covariate influencing survival or detection probability. An interesting case is where the individual covariate is time varying (e.g., body mass). This is a difficult statistical problem because the covariate can only be measured when the individual is captured. Such models have received some attention recently by Bonner and Schwarz (2006) and King et al. (2008). To specify such a model

Table 11.6. Posterior summaries of the year-effects model with individual heterogeneity fitted to the European dipper data. q_m is the mth percentile of the posterior distribution.

parameter	mean	SD	$q_{0.025}$	$q_{0.50}$	$q_{0.975}$
ϕ_1	0.745	0.134	0.474	0.750	0.976
ϕ_2	0.462	0.079	0.319	0.458	0.628
ϕ_3	0.492	0.070	0.363	0.489	0.637
ϕ_4	0.623	0.065	0.495	0.623	0.751
ϕ_5	0.599	0.064	0.473	0.599	0.726
p_1	0.671	0.153	0.349	0.682	0.926
p_2	0.877	0.087	0.662	0.897	0.987
p_3	0.878	0.074	0.697	0.893	0.977
p_4	0.898	0.058	0.760	0.908	0.979
p_5	0.919	0.053	0.789	0.930	0.988
σ_p	1.359	0.955	0.013	1.204	3.655
σ_ϕ	0.349	0.233	0.027	0.315	0.878

in **WinBUGS**, we can take the basic template given in Panel 11.7 and modify the construction of the individual and year-specific survival probability according to:

```
logit(phi[i,t])<- logitphi[t] + beta*covariate[i,j]
```

We would then also require a prior distribution on the individual covariate (see Bonner and Schwarz (2006)).

11.6.3 Model Selection

As seen in Table 11.6, the posterior distributions of the variance components are very diffuse, with, in particular, much of the mass of the posterior for σ_ϕ being close to zero. This raises the obvious question as to whether the individual heterogeneity components should be included in the model at all. Royle (2008c) computed the posterior model probabilities for each of four models: Model 1 (M1) has year effects on p and ϕ (i.e., no heterogeneity); Model 2, is the year effects model plus heterogeneity in p; Model 3 has year effects plus heterogeneity in ϕ, and Model 4 has year effects plus heterogeneity in both parameters.

As we have done previously in Bayesian model selection problems (e.g., Section 3.4.3), we can compute posterior model probabilities in **WinBUGS** by specifying a set of latent indicator variables, one for each model effect, say w_k for effect k, and imposing a Bernoulli prior on each w_k, say having parameter π_k. This notion was suggested by Kuo and Mallick (1998). Specification of the π_k dictates the prior probability for each model. In the present problem, we introduce two indicator

```
model {
for(j in 1:(nyear-2)){
   yeffp[j]~dunif(0,1)
   logitp[j]<-log(yeffp[j]/(1-yeffp[j]))
   yeffphi[j]~dunif(0,1)
   logitphi[j]<-log(yeffphi[j]/(1-yeffphi[j]))
}
yeffp[nyear-1]~dunif(0,1)
logitp[nyear-1]<-log(yeffp[nyear-1]/(1-yeffp[nyear-1]))
logitphi[nyear-1]<-0

sigma.phi~dunif(0,10)
sigma.p~dunif(0,10)
tauphi<-1/(sigma.phi*sigma.phi)
taup<-  1/(sigma.p*sigma.p)

for(i in 1:nind){
  eta[i]~dnorm(0,taup)
  delta[i]~dnorm(0,tauphi)
  for(t in 1:(nyear-1)){
     logit(p[i,t])<- logitp[t] + eta[i]
     logit(phi[i,t])<- logitphi[t] + delta[i]
  }
}

for(i in 1:nind){
  z[i,first[i]]~dbern(1)
  for(j in (first[i]+1):nyear){
    mu2[i,j]<-phi[i,j-1]*z[i,j-1]
    z[i,j]~dbern(mu2[i,j])
    mu1[i,j]<-p[i,j-1]*z[i,j]
    y[i,j]~dbern(mu1[i,j])
   }
 }
}
```

Panel 11.7. WinBUGS model specification for CJS model with year-specific parameters and individual heterogeneity in both detection and survival probability.

Table 11.7. Posterior model probabilities for the four models considered for the dipper data.

Model	Posterior probability
Model 1: $p(t) + \phi(t)$	0.711
Model 2: $p(t + h) + \phi(t)$	0.229
Model 3: $p(t) + \phi(t + h)$	0.044
Model 4: $p(t + h) + \phi(t + h)$	0.016

variables, say w_1 and w_2, having Bern(0.5) prior distributions and we multiply the additive random effects by these indicator variables (see online Appendix D of Royle (2008c), located at `http://www.biometrics.tibs.org/datasets/060233.pdf`). The relevant modification to the **WinBUGS** model specification is given by

```
for(t in 1:(nyear-1)){
    logit(p[i,t])<- logitp[t] +  w1*eta[i]
    logit(phi[i,t])<- logitphi[t] + w2*delta[i]
}
```

Under this formulation of the model selection problem, Model 1 is indicated if both $w_1 = 0$ and $w_2 = 0$, Model 2 is indicated if $w_1 = 1$ but $w_2 = 0$, etc. The posterior probabilities of the four possible events were computed from the MCMC histories (Table 11.7). The results suggest that heterogeneity in ϕ is probably unimportant, whereas heterogeneity in p is not so negligible.

11.6.4 Prior Sensitivity

The use of ostensibly vague inverse-gamma prior distributions for σ^2 (i.e, IG(ϵ, ϵ) for small ϵ) is fairly conventional in Bayesian analysis of models containing random effects or other latent structure. Gelman (2006) considered the appropriateness of this prior distribution, noting that "...for datasets in which low values of σ are possible, inferences become very sensitive to ϵ in this model, and the prior distribution hardly looks non-informative...." (p. 522). Royle (2008c) provided a similar analysis of prior sensitivity under the CJS model with heterogeneity, using various IG(ϵ,ϵ) priors for σ_p^2. The posterior of σ_ϕ appeared relatively insensitive to the prior specification.

In Figure 11.3, the three panels correspond to the estimated posterior under a uniform(0,5) prior (left panel) for σ_p, the estimated posterior under an IG(1,1) prior for σ_p^2 (middle panel) and under an IG(.01,.01) prior for σ_p^2 (right panel). In all cases, the posterior is restricted to $[0,5]$ where most of the posterior mass occurs (the histograms were restandardized to sum to 1 over that range). These results are

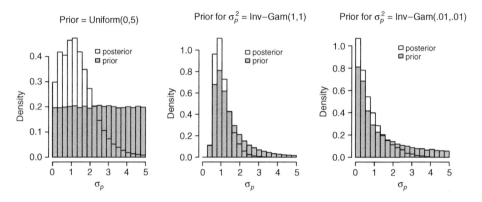

Figure 11.3. Estimated posterior distributions of σ_p and corresponding histograms obtained under several prior distributions. Reproduced from Royle (2008c).

strikingly similar to those reported by Gelman (2006). In particular, note that the uniform prior does not affect the posterior so much, whereas for the conventional vague conditionally-conjugate prior, the posterior does appear to be influenced by choice of ϵ . As summarized by Gelman (2006): "... the inverse-gamma(ϵ,ϵ) prior is not at all 'non-informative' for this problem since the resulting posterior distribution remains highly sensitive to the choice of ϵ". Similarly, Figure 11.3 suggests that a proper uniform prior for the variance component is justified.

11.7 SUMMARY

Survival is a fundamental process in populations and understanding survival is therefore one of the main objectives of many animal population studies. The class of models known as the Cormack–Jolly–Seber (CJS) model is the standard framework for modeling and inference about animal survival. In this chapter we provided a brief introduction to the classical likelihood inference under CJS-type models. We also provided a hierarchical formulation of the model based on individual-specific observation and state models. We mainly advocate the hierarchical formulation of the CJS-type models because it yields easily to several extensions including the modeling of individual covariates, or heterogeneity, which is important in many studies of evolutionary biology and conservation and management (Nichols et al., 1982; Cam et al., 2002a, 2004; Cooch et al., 2002). Moreover, the binary Markovian state model is closely related to the occupancy state models considered in Chapter 9. Therefore, as a practical matter, analysis of the different classes of models is largely equivalent. Inference under the hierarchical formulation can be achieved easily in **WinBUGS** and a myriad of useful extensions can therefore be considered directly.

We provided a brief characterization of models for avian nest survival. These models are structurally similar to CJS-type models in the sense that the state process is a simple survival process. Nest survival models differ primarily in that detection of nests is perfect after initial entry into the sample. As with CJS models, there is uncertainty regarding the terminal states (i.e., censoring), since nests are not typically visited each day. Formal accounting for this terminal state uncertainty is perhaps the unifying conceptual theme between the two classes of models. A number of other situations give rise to a survival state process model similar to those considered in this chapter. One situation in which $p = 1$ arises is in the analysis of radio telemetry data which produces known-fate individuals (White and Garrott, 1990). Survival models are also useful in modeling stopover duration in migratory birds (Schaub et al., 2001).

We considered several useful extensions of the basic survival models including models with random spatial effects. In the context of modeling nest survival, we considered a field experiment where nests were monitored on 12 plots, which were regarded as random effects in the model for survival. For the MAPS warbler data, we developed a spatial model for survival probability by including a spatially-correlated random effect in the model, assumed to have a conditional autoregressive prior distribution. Modeling and inference of data from the MAPS program presents a good case study of the need and utility for hierarchical models, as the program generates sparse encounter history data for a large number of sites. Historically, data have been analyzed by stratification and pooling, but hierarchical spatial models provide a flexible framework for modeling and inference from MAPS data. Finally, we considered several variations of models with individual effects. In developing models of survival for the warbler data, we allowed for the possibility that individuals were transients, which is a latent binary variable. We also considered individual heterogeneity in the form of random effects on both detection and survival probability – essentially a variation of 'Model M_h' that was described in Chapter 6.

12

MODELS OF COMMUNITY COMPOSITION AND DYNAMICS

In this chapter we illustrate how the single-species, site-occupancy models described in Chapter 3 may be extended to estimate characteristics of *communities* of species. The size of a community, usually called *species richness*, is perhaps the most important of these characteristics. Species richness is used in quantitative assessments of biological diversity and in the development of ecological theory (MacArthur and Wilson, 1967; Hubbell, 2001). Our definition of an ecological community is consistent with Hubbell's definition of a *metacommunity*. That is to say, we view a community as a collection of spatially distinct, local communities composed of 'trophically similar, sympatric species that actually or potentially compete in a local area for the same or similar resources'. We note that the spatial separation of local communities implies that not all members of the metacommunity are in direct competition with one another.

Generally speaking, the number of species in a community cannot be observed directly because a complete enumeration of every species is often infeasible. This is particularly true in communities composed of rare or elusive species. In practice, a sample of the community is collected or observed, and species richness is estimated from the data obtained in the sample. Therefore, the accuracy of an estimate of species richness depends on both the method of data collection (which includes the sampling design) and the statistical model used to analyze the data.

Our view of this ecological inference problem is that species richness, species accumulation, and other attributes of community structure are most naturally formulated using a model of individual species *occurrence* that explicitly accounts for the imperfect detection of individuals during sample collection. We believe that community-level and species-level attributes should be combined in the same modeling framework – that is, the framework should allow community-level *or* species-level characteristics to be estimated as needed in specific problems. In published work on this subject (Dorazio and Royle, 2005a; Dorazio et al., 2006; Kéry and Royle, 2008a,b), we have noted that this versatility is not shared by alternative methods for estimating species richness, such as the extrapolation of empirical species-accumulation curves (Gotelli and Colwell, 2001; Ugland et al., 2003; Colwell et al., 2004) or the application of capture–recapture models to species detections (Burnham and Overton, 1979, Dorazio and Royle, 2003, Mao et al., 2005, also see Chapter 6).

Of course, our solution does come with a price in the sense that additional (within-site) sampling is required to eliminate the confounding between probabilities of species occurrence and species detection. Recall from Chapter 3 that both spatial and temporal replicates are required to resolve the ambiguity of an observed zero in single-species models of site occupancy. That is, a zero may indicate that a species is absent or that it is present but undetected. However, if each sample location is visited repeatedly and if the duration of the survey is short enough that the occupancy status of a species remains unchanged at that location, then the occurrence of a species can be estimated while accounting for its detectability.

To develop the concepts and assumptions that underlie our occupancy-based approach to modeling communities of species, we begin this chapter with an analysis wherein species richness is assumed to be known. We then extend this analysis to the more typical situation where species richness must be estimated. As in previous chapters we show how data augmentation may be used to simplify the analysis of this more complicated modeling problem. We also illustrate the use of spatial covariates of occurrence and detection and the influence of these covariates on community-level and species-level characteristics. We conclude the chapter with a description of dynamic models wherein the composition and richness of species is assumed to change with time.

12.1 MODELS WITH KNOWN SPECIES RICHNESS

Suppose a community of species is sampled using a protocol which is identical to that used in the estimation of mean occurrence of a single species. That is to say, suppose each sample location (or site) is visited on $J > 1$ different occasions and the list of species detected on each occasion is recorded. Recall from Chapter 3 that this sampling protocol is identical to the repeated-measures protocol used in occupancy surveys. The total duration of the survey is kept sufficiently short so that the occurrence of each species may safely be assumed to be constant at each site. At the end of the survey we can determine the number of distinct species (n) that have been detected. Here, we assume that n is equivalent to the size N of the community; however, it is more typical to observe samples of communities for which $n < N$. We consider this more typical situation in the next section.

Based on the protocol used to sample the community, we observe the number of occasions in which each species is detected in J visits to each site. Thus, for each species, let y_i denote the number of detections in J visits to site i. Though J need not be identical for all sites, we assume here that J is fixed to simplify notation. Recall from Chapter 3 that one of the simplest models of site occupancy for a single species may be summarized as follows:

$$y_i | J, p, z_i \sim \text{Bin}(J, pz_i)$$
$$z_i | \psi \sim \text{Bern}(\psi),$$

wherein z_i denotes a latent state variable for the occurrence of the species at the ith site (i.e., $z = 1$ denotes species presence and $z = 0$ denotes species absence). In Chapter 3 we described how the probabilities of occurrence ψ and detection p may be estimated from the set of detection frequencies $\{y_1, y_2, \ldots, y_R\}$ observed at R sample locations.

Our *multi-species* model of site occupancy is based on the same set of detection frequencies, but now we observe n of these sets – one for each of n species in the community. For reasons that will become apparent shortly, in this chapter we index site by k and species by i and use a $n \times R$ matrix $\boldsymbol{Y}$ to denote the observed detection frequencies of every species in the community, i.e.,

$$\boldsymbol{Y} = \begin{pmatrix} y_{11} & y_{12} & \cdots & y_{1R} \\ y_{21} & y_{22} & \cdots & y_{2R} \\ \vdots & \vdots & & \vdots \\ y_{n1} & y_{n2} & \cdots & y_{nR} \end{pmatrix}$$

where each row of $\boldsymbol{Y}$ corresponds to the set of detection frequencies of a single species.

Our model of the species detection frequencies in $\boldsymbol{Y}$ is identical to that described earlier, i.e.,

$$y_{ik} | J, p_i, z_{ik} \sim \text{Bin}(J, p_i z_{ik}) \tag{12.1.1}$$
$$z_{ik} | \psi_i \sim \text{Bern}(\psi_i) \tag{12.1.2}$$

except that each observation y_{ik} is now indexed by site (k) *and* by species (i). Note also that our model includes a $n \times R$ matrix $\boldsymbol{Z}$ of latent state variables for species occurrence

$$\boldsymbol{Z} = \begin{pmatrix} z_{11} & z_{12} & \cdots & z_{1R} \\ z_{21} & z_{22} & \cdots & z_{2R} \\ \vdots & \vdots & & \vdots \\ z_{n1} & z_{n2} & \cdots & z_{nR} \end{pmatrix}.$$

Later in this chapter we describe how estimates of $\boldsymbol{Z}$ may be used to calculate community-level characteristics of scientific interest, such as site-specific species richness.

In the model summarized in Eqs. (12.1.1) and (12.1.2), the probabilities of occurrence ψ_i and detection p_i are assumed to be species-specific. Natural communities are composed of species whose abundances vary widely, so it stands to reason that the average occurrence will vary among species. Similarly, differences in abundance, appearance (e.g., size, coloration), or behavior (e.g., time of activity, habitat preferences) of species will almost certainly induce heterogeneity in their detectabilities.

We could treat the species-specific probabilities of occurrence and detection as fixed parameters and compute estimates of ψ_i and p_i by analyzing each row of Y separately. However, there are several reasons for not adopting this approach. First, the estimates may be unstable or lack precision, particularly for rare species that may be present in low abundance at occupied sites and difficult to detect. Second, this approach does not yield a parsimonious solution because the number of parameters to be estimated increases with the size of the community. That is, in a community of n species there are $2n$ parameters to estimate. Another reason to avoid this kind of analysis – actually the most compelling reason – is because this approach does not provide a mechanism for estimating species richness of communities in which N is unknown, as we demonstrate in Section 12.2.

These reasons provide ample motivation for extending the species-specific models of y_{ik} and z_{ik} given in Eqs. (12.1.1) and (12.1.2). In particular, we require modeling assumptions that characterize the heterogeneity in probabilities of occurrence and detection among species. Thus, let $u_i = \text{logit}(\psi_i)$ and $v_i = \text{logit}(p_i)$ denote a logit-scale parameterization of these probabilities. We assume that heterogeneity (among species) in these parameters may be specified using a bivariate normal distribution

$$\begin{pmatrix} u_i \\ v_i \end{pmatrix} \sim \text{N}\left(\begin{pmatrix} \beta \\ \alpha \end{pmatrix}, \Sigma \right), \tag{12.1.3}$$

where Σ denotes a 2×2 symmetric matrix whose diagonal elements, σ_u^2 and σ_v^2, specify levels of the variation in u_i and v_i, respectively, and whose off-diagonal elements, σ_{uv}, equal the covariance between u_i and v_i. The parameters β and α specify the mean logit-scale probabilities of occurrence and detection, respectively, among all species in the community. We anticipate that estimates of σ_{uv} will almost surely be positive because probabilities of occurrence and detection are both expected to increase as the abundance of a species increases. For example, if we let N_{ij} denote the abundance of individuals of species i at site j and assume $N_{ij} \sim \text{Po}(\lambda_{ij})$, then the probability of occurrence is expected to increase with mean abundance λ_{ij} as follows: $\psi_{ij} = \text{Pr}(N_{ij} > 0) = 1 - \exp(-\lambda_{ij})$. Similarly, if we let q_{ij} denote the probability of detecting each of the N_{ij} individuals present at site j, and assume that such detections are independent, then the probability of detecting these individuals increases with N_{ij} as follows: $p_{ij} = 1 - (1 - q_{ij})^{N_{ij}}$.

This relationship is identical to the underlying assumption of the abundance-based occupancy models described in Chapter 4. Our expectation of positive estimates of σ_{uv} is therefore based on simple, yet entirely reasonable, arguments.

This model-based characterization of variation in species occurrence and detection probabilities (in Eq. (12.1.3)) is certainly not unique. Other distributional assumptions are possible and, in fact, may exert a strong influence on estimates of species occurrence, species detection, and species richness (Dorazio and Royle, 2005a; Dorazio et al., 2006). Our view is that this modeling assumption, as with any other, should be based on information relevant to the inference problem and can always be revised as needed in the context of the problem.

12.1.1 Example: North American Breeding Bird Survey

We illustrate the multi-species occupancy model by fitting data collected from an avian community as part of the North American Breeding Bird Survey (BBS). We described the BBS earlier (see Chapters 4 and 8). Here we consider an analysis of detections of birds observed at 50 sampling locations along BBS route number 017 in New Hampshire. This route was visited by the same observer on 11 different days in June and early July of 1991.

During the 11 days of sampling, 99 distinct species of birds were detected. However, there is considerable variation among species in both apparent occurrence (proportion of 50 sites where a species was detected) and apparent detectability (relative frequency of detections within sites that appear to be occupied) (Figure 12.1).

We conducted a Bayesian analysis of these data using non-informative priors for the model's parameters ($\text{expit}(\alpha) \sim \text{U}(0,1)$, $\text{expit}(\beta) \sim \text{U}(0,1)$, $\sigma_u \sim \text{U}(0,10)$, $\sigma_v \sim \text{U}(0,10)$, and $\rho \sim \text{U}(-1,1)$ where $\rho = \sigma_{uv}/(\sigma_u \sigma_v)$). Panel 12.1 contains **WinBUGS** code for implementing this model. The posterior summaries of these parameters, which are given in the first two columns of Table 12.1, indicate that logit-scale parameters for occurrence and detection vary quite a bit among species. In addition species occurrence and detection appear to be positively correlated ($\rho = 0.42$), which is to be expected because, for any species, both ψ and p often increase as abundance of individuals increases (Royle and Nichols, 2003; Dorazio and Royle, 2005a; Dorazio ct al., 2006).

The posterior summary in Table 12.1 also indicates that the average probability of occurrence is only 0.25 ($=\text{expit}(-1.11)$), so we might suspect that the avian community actually comprises additional species that were not detected. This situation is, by far, more common in natural systems. In the next section we consider elaborations of the model that can be used to estimate the number of species in the community that were missed during sampling.

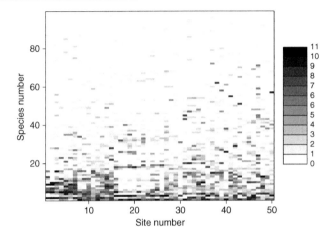

Figure 12.1. Number of times that each bird species was detected in 11 visits to each site of BBS route 017 in New Hampshire, USA. Species are numbered from the most detectable to the least detectable. Gray scale indicates number of detections.

Table 12.1. Posterior summary statistics for two multi-species occupancy models fitted to the BBS data from route 017 in New Hampshire. In one model species richness N is assumed to equal the number of distinct species observed in the survey (i.e., $N = n$). In the other model species richness is estimated.

Parameter	Known N		Estimated N	
	Mean	SD	Mean	SD
β	−1.11	0.18	−2.19	0.53
α	−1.70	0.12	−2.05	0.22
σ_u	1.68	0.15	2.36	0.34
σ_v	1.00	0.11	1.15	0.15
ρ	0.42	0.11	0.58	0.12
Ω			0.55	0.08
N			138	19

12.2 MODELS WITH UNKNOWN SPECIES RICHNESS

Here we develop the modeling assumptions needed to account for situations where the richness of species in a community is believed to exceed the number of species observed in the sample. As noted earlier, such situations are quite common, both taxonomically and geographically. We begin by noting that it is entirely possible to estimate species richness N without additional modeling assumptions. The set of latent state variables $\{z_{ik}\}$ can be eliminated (by summation) to produce a zero-inflated, binomial model for each detection frequency y_{ik} that conditions on the occurrence and detection parameters, u_i and v_i. Then, by assuming conditional independence among observations from different sites, a multinomial likelihood that depends only on the n observed species and the species-specific model parameters

```
model {

psi.mean ~ dunif(0,1)
beta <- log(psi.mean) - log(1-psi.mean)

p.mean ~ dunif(0,1)
alpha <- log(p.mean) - log(1-p.mean)

sigma.u ~ dunif(0,10)
sigma.v ~ dunif(0,10)
tau.u <- pow(sigma.u,-2)
tau.v <- pow(sigma.v,-2)
rho ~ dunif(-1,1)
var.eta <- tau.v/(1.-pow(rho,2))


for (i in 1:n) {
    phi[i] ~ dnorm(beta, tau.u)
    mu.eta[i] <- alpha + (rho*sigma.v/sigma.u)*(phi[i] - beta)
    eta[i] ~ dnorm(mu.eta[i], var.eta)
    logit(psi[i]) <- phi[i]
    logit(p[i]) <- eta[i]

    for (k in 1:R) {
      Z[i,k] ~ dbern(psi[i])
      mu.p[i,k] <- p[i]*Z[i,k]
      Y[i,k] ~ dbin(mu.p[i,k], J)
    }
}

}
```

Panel 12.1. WinBUGS code for implementing the multi-species occupancy model without estimating species richness.

can be formulated and analyzed using either frequentist or Bayesian methods. This approach is described by Dorazio and Royle (2005a).

One disadvantage of this approach is that the occurrence and detection parameters of the *unobserved* species must be eliminated by numerical integration, which is computationally demanding when used in conjunction with numerical optimization or Gibbs sampling. Another difficulty with this approach is that N and other derived parameters also can be cumbersome to calculate. Motivated by these difficulties, we developed an alternative approach based on data augmentation (Dorazio et al., 2006; Royle et al., 2007a), which we describe here.

Note that if species richness N was known, the hierarchical model for estimating species occurrence and detection parameters would be complete without requiring special considerations in the likelihood function for an 'undetected' portion of the community. Furthermore, there would be no difficulty in fitting the model using MCMC. The difficulty with N being unknown is that the length of the parameter vectors associated with species occurrence and detection (say, $\boldsymbol{u} = (u_1, \ldots, u_N)$ and $\boldsymbol{v} = (v_1, \ldots, v_N)$) changes each time another MCMC draw of the parameter N is computed. Therefore, to obtain a model in which the number of parameters is constant, we create a supercommunity of species, one that comprises the n observed species and an arbitrarily large, but known, number of unobserved species for which $\boldsymbol{y}_i = \boldsymbol{0}$ ($i = n+1, n+2, \ldots, N, N+1, \ldots, M$). The supercommunity size M is fixed, and thus the length of the parameter vectors is constant (i.e., not a function of N). In taking this approach we do not directly estimate N as a parameter. Instead, we introduce an additional latent indicator variable, say w_i, which takes the value 1 if a species in the supercommunity is a member of the N species that are vulnerable to sampling; otherwise $w_i = 0$. We assume that $\{w_i\}$ are independent, Bernoulli-distributed random variables indexed by parameter Ω. Obviously, w_i is observed for $i = 1, \ldots, n$, but not otherwise. By introducing the latent indicator variables $\{w_i\}$ into the model, we effectively transform the problem of estimating N into the equivalent problem of estimating $\sum_{i=1}^{M} w_i$, which, of course, depends on the estimated value of Ω.

Our reparameterized model does require that M be assigned a sufficiently high value; however, in practice it is a simple matter to assess the adequacy of any particular choice of M. Recall that $\mathrm{E}(N) = M\Omega$ and that Ω is bounded between 0 and 1; therefore, estimates of Ω will necessarily decline as higher values of M are chosen. If the assigned value of M is too low, the posterior distribution of Ω will be concentrated near the upper limit of its support, and we risk underestimating the true value of N. The obvious solution is to increase M until the posterior of Ω is centered well below its upper limit. However, higher values of M also imply higher computational costs, so some care is advised in assigning too high a value to M.

12.2.1 Modeling Augmented Data

The inclusion of 'data' from unobserved species in the supercommunity requires some minor revision of the modeling assumptions summarized in Eqs. (12.1.1) and (12.1.2). In particular, we now assume a hierarchical model with 3 levels:

$$y_{ik}|J, p_i, z_{ik} \sim \text{Bin}(J, p_i z_{ik}) \tag{12.2.1}$$

$$z_{ik}|\psi_i, w_i \sim \text{Bern}(\psi_i w_i) \tag{12.2.2}$$

$$w_i|\Omega \sim \text{Bern}(\Omega). \tag{12.2.3}$$

Implementing this model in **WinBUGS** requires only minor changes to the code in Panel 12.1. A complete set of **R** and **WinBUGS** code for fitting this model to the BBS data is provided in the Web Supplement.

Table 12.2 illustrates how the observable detection frequencies and the latent state variables for species occurrence and community membership are organized in our conceptual supercommunity. A practical benefit of this organization is that quantities of scientific interest are relatively easy to compute from a MCMC sample of the joint posterior. For example, in addition to species richness N, we might want to estimate the number of species present at a particular sample location or at a particular collection of locations, as in small-area estimation. Such calculations are trivially easy to make, e.g., $N_k = \sum_{i=1}^{M} z_{ik}$ yields the number of species in the community that are present at the kth sample location. We also might want to estimate the similarity of species present at two different sample locations, say k and l. For this problem, we can estimate the number of species that are present at *both* locations quite simply, $N_{kl} = \sum_{i=1}^{M} z_{ik} z_{il}$, and then combine this estimate with estimates of richness at each site individually to compute similarity as follows: $S_{kl} = 2N_{kl}/(N_k + N_l)$, which is based on Dice's (1945) index of similarity in species composition. Calculations of more abstract community-level quantities, such as predictions of species in unsampled locations or species-accumulation curves, are similarly easy to make, as illustrated by Dorazio et al. (2006).

12.2.2 Example: North American Breeding Bird Survey

To fit the model described in this section to the BBS data analyzed in Section 12.1.1, we augmented the detection matrix of the $n = 99$ observed species with 151 all-zero rows to produce a supercommunity of $M = 250$ species. The posterior distribution of species richness N, which is illustrated in Figure 12.2, indicates there are many more species in the community than those observed in the sample (mean $= 138$, SD $= 19$). In fact, the posterior probability that $N = n$ is essentially zero.

Table 12.2. Conceptualization of the supercommunity of M species used in data augmentation. $\boldsymbol{Y}$ denotes the matrix of observable detection frequencies. $\boldsymbol{Z}$ denotes the partially observed matrix of state variables for species- and site-specific occurrence. $\boldsymbol{w}$ denotes the partially observed vector of state variables that indicates membership in the community of N species vulnerable to sampling.

					Site k				
	Observed				Partially observed				
species i	1	2	$\cdots$	R	1	2	$\cdots$	R	w_i
1	y_{11}	y_{12}	$\cdots$	y_{1R}	z_{11}	z_{12}	$\cdots$	z_{1R}	w_1
2	y_{21}	y_{22}	$\cdots$	y_{2R}	z_{21}	z_{22}	$\cdots$	z_{2R}	w_2
$\vdots$	$\vdots$	$\vdots$		$\vdots$	$\vdots$	$\vdots$		$\vdots$	$\vdots$
n	y_{n1}	y_{n2}	$\cdots$	y_{nR}	z_{n1}	z_{n2}	$\cdots$	z_{nR}	w_n
$n+1$	0	0	$\cdots$	0	$z_{n+1,1}$	$z_{n+1,2}$	$\cdots$	$z_{n+1,R}$	w_{n+1}
$\vdots$	$\vdots$	$\vdots$		$\vdots$	$\vdots$	$\vdots$		$\vdots$	$\vdots$
N	0	0	$\cdots$	0	z_{N1}	z_{N2}	$\cdots$	z_{NR}	w_N
$N+1$	0	0	$\cdots$	0	$z_{N+1,1}$	$z_{N+1,2}$	$\cdots$	$z_{N+1,R}$	w_{N+1}
$\vdots$	$\vdots$	$\vdots$		$\vdots$	$\vdots$	$\vdots$		$\vdots$	$\vdots$
M	0	0	$\cdots$	0	z_{M1}	z_{M2}	$\cdots$	z_{MR}	w_M

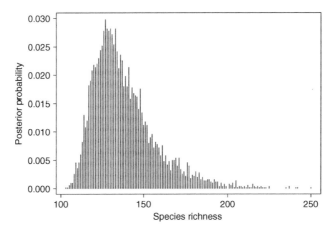

Figure 12.2. Posterior distribution of species richness N for the avian community along BBS route 017 in New Hampshire, USA.

The posterior summaries of this model's parameters are given in Table 12.1. Notice that the average (among species) probabilities of occurrence and detection

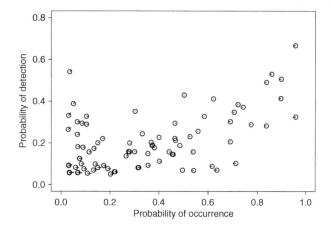

Figure 12.3. Posterior means of occurrence and detection probabilities for the 99 species detected in the BBS survey along route 017 in New Hampshire, USA. 'Tails' on each point indicate how each estimate changes if species richness N is not estimated (i.e., if $N = n$ is assumed).

(on the logit scale) are smaller than those estimated by assuming that N equaled the number of species observed in the sample. This seems like a sensible result because species that were missing from the sample are likely to be those that were present in low numbers or were difficult to detect. Including these lower values of ψ and p would reduce the average values of the community. These lower values of ψ and p are also likely to be responsible for the increase in estimates of heterogeneity and correlation parameters $(\sigma_u, \sigma_v, \rho)$. In other words, the species that were missed had lower values of both ψ and p.

The probabilities of occurrence and detection estimated for the $n = 99$ observed species probabilities were not dramatically different from estimates computed under the simpler model, wherein $N = n$ was assumed (Figure 12.3). Estimates of species occurrence appear to be positively correlated with those of species detection, although, interestingly, there also appear to be a few uncommon species (with low ψ values) that have relatively high probabilities of detection.

12.3 COVARIATES OF OCCURRENCE AND DETECTION

In many surveys of natural communities, covariates that are thought to be informative of species occurrence or detection may be available. Such covariates may include measurements taken during the survey, such as sampling effort, or they may have come from preexisting databases, possibly maintained in geographic information systems. Gelfand et al. (2006) provide a nice example, but their model conditions on a fixed number of *observed* species; consequently, their parameters for species occurrence and detection have a slightly different interpretation than the parameters which we use. Here, we extend the model developed in the previous section, wherein species richness is not known, to include site- and replicate-level covariates in the analysis of community structure.

12.3.1 Modeling Avian Species in Switzerland

We describe here a model extension developed by Kéry and Royle (2008b) for the analysis of data collected in the national Swiss breeding bird survey during 2001. In this survey each of 254 quadrats (=sites) was visited on 2 or 3 separate occasions during the breeding season by a collection of volunteer observers. Each volunteer attempted to cover as much of a quadrat as possible, while traversing an irregular route. Both route length and sampling duration were recorded for each visit, as these were thought to be useful in constructing covariates of detection. Another important covariate of detection was the sampling date because the entire survey required about 3 months (15 April–15 July) to complete. Because the survey included quadrats located throughout Switzerland, covariates which were thought to most influence occurrence of species were forest cover and elevation. An additional covariate, route length, was included to adjust for the increased chance of finding a species along a longer route.

We must extend the notation developed in the previous section to include the possibility that detection differs not only among species, but also among sites and sampling occasions. Therefore, let y_{ijk} denote a binary observation that indicates whether the ith species was detected ($y_{ijk} = 1$) or not ($y_{ijk} = 0$) during the jth visit to site k. Using similar notation for detection probability p_{ijk}, we assume the following model of the observations:

$$y_{ijk}|p_{ijk}, z_{ik} \sim \text{Bern}(p_{ijk}z_{ik}),$$

where

$$\text{logit}(p_{ijk}) = a_{i0} + a_{i1}\text{date}_{jk} + a_{i2}\text{date}_{jk}^2 + a_{i3}\text{effort}_{jk}.$$

The covariate `effort` is the ratio of time required to sample a route divided by the length of the route. Both linear and quadratic terms are included for the covariate `date` in an attempt to model differences in activity patterns of species during the breeding season. That is to say, some species may have a peak in activity sometime during the breeding season while others may begin low and steadily increase (or vice versa).

We specify the effects of covariates on species occurrence probabilities similarly, except that the occurrence of each species is assumed to be fixed during all visits to a site. Therefore, our model of species occurrence is

$$z_{ik}|\psi_{ik}, w_i \sim \text{Bern}(\psi_{ik} w_i),$$

where

$$\text{logit}(\psi_{ik}) = b_{i0} + b_{i1}\texttt{elev}_k + b_{i2}\texttt{elev}_k^2 + b_{i3}\texttt{forest}_k + b_{i4}\texttt{length}_k$$

and w_i denotes a latent indicator of whether the ith species is a member of the N species exposed to sampling. This formulation allows occurrence probabilities of each species to differ with elevation (`elev`), percent forest cover (`forest`) and route length (`length`). For numerical reasons, all covariates except route length are standardized to have zero mean and unit variance.

Thus far, we have specified only the contributions of observable sources of heterogeneity in species occurrence and detectability. A component for heterogeneity *among* species must be added to the model to estimate patterns of occurrence of species that are members of the community but are not observed in the sample. For this modeling component we assume a normal distribution (on the logit scale) as was previously done

$$\begin{pmatrix} \boldsymbol{b}_i \\ \boldsymbol{a}_i \end{pmatrix} \sim N\left(\begin{pmatrix} \beta \\ \alpha \end{pmatrix}, \boldsymbol{\Sigma} \right), \tag{12.3.1}$$

where $\boldsymbol{b}_i = (b_{i0}, b_{i1}, b_{i2}, b_{i3}, b_{i4})$ denotes a vector of logit-scale occurrence parameters for the ith species and $\boldsymbol{a}_i$ denotes a vector of logit-scale detection parameters. Heterogeneity among species in occurrence and detection is specified by the symmetric, 9×9 matrix $\boldsymbol{\Sigma}$. The diagonal elements of $\boldsymbol{\Sigma}$ correspond to the variances of $\boldsymbol{a}_i$ and $\boldsymbol{b}_i$, and the off-diagonal elements of $\boldsymbol{\Sigma}$ are all zero except for $\sigma_{16} = \sigma_{61}$, which allows the 'intercept' parameters a_{i0} and b_{i0} to be correlated. This assumption is identical to that described in the previous section where u_i ($\equiv b_{i0}$) and v_i ($\equiv a_{i0}$) were modeled as correlated parameters.

The remaining modeling assumptions are basically identical to those made previously. For the latent variable w_i, which indicates membership in the community of N species, we assume $w_i \sim \text{Bern}(\Omega)$. A set of independent prior distributions is assumed, each of which is intended to be non-informative (see Kéry and Royle (2008b) for details).

12.3.2 Estimates of Species Richness and Geographic Distribution

In the 254 quadrats sampled in 2001, volunteer observers detected a total of 134 different avian species. Estimates of mean probabilities of occurrence and detection of these species indicate that the community is dominated by species that are rare, not simply difficult to detect (Figure 12.4). For example, average probabilities of occurrence were less than 0.10 for more than half of the 134 species observed in the sample. Given this result, it is not surprising that total richness N of all avian species in Switzerland is estimated to have been 170 (95 per cent credible interval = 151–195), well in excess of the number of species actually observed (Figure 12.5).

An important benefit of our occurrence-based model of the community is that estimates of species richness for individual sample locations are relatively easy to compute. Moreover, these estimates automatically account for the habitat characteristics observed at each location. Figure 12.6 illustrates the posterior distribution of the number of species present at each of 3 different sample locations. At one location, the estimated number of species nearly equals the number of species actually detected; however, at the other two locations, many more species are estimated to be present than were observed in the sample. We can summarize these results by plotting the posterior-predictive mean species richness for each of the 41 365 quadrats in Switzerland (Figure 12.7). This map reveals the importance of elevation and forest cover to estimates of species occurrence and richness. Evidently, higher numbers of species are present in forested locations than in unforested ones, and the richness of species also appears to be lowest at highest elevations.

However, these results reflect only general distributional patterns. The estimated relationships between species occurrence and habitat covariates vary greatly among species. As an illustration, Figure 12.8 shows how the estimated average probability of occurrence changes with elevation for each of the 134 species observed in the survey and for *Anthus trivialis* (tree pipit), *Buteo buteo* (common buzzard), and *Oenanthe oenanthe* (northern wheatear) individually. Evidently, common buzzards are more common at low elevations, northern wheatears are more common at high elevations, and tree pipits prefer intermediate elevations.

These results are indicative of the great variety of scientifically relevant quantities that can be estimated quite easily given our conceptualization of the community (Table 12.2). Quantities that are relevant to the entire community of species, to subsets of the community (e.g., species richness at particular locations), or simply to individual species may all be estimated using the matrix of site-specific species occurrences Z and the vector of community membership w.

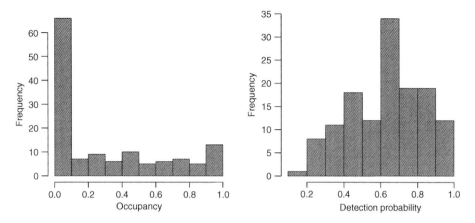

Figure 12.4. Histograms of average probabilities of occurrence and detection for the 134 breeding bird species observed in the 2001 survey of Switzerland.

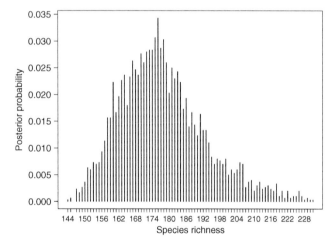

Figure 12.5. Posterior distribution of species richness N for breeding birds sampled in the 2001 survey of Switzerland.

12.4 DYNAMIC MODELS

In the previous sections of this chapter we developed models for situations where the occurrence of each species is assumed to be fixed during repeated visits to

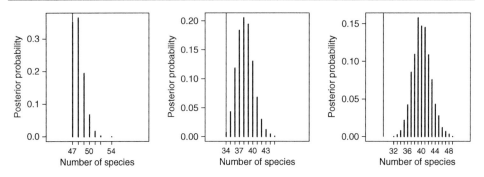

Figure 12.6. Posterior distribution of the number of avian species present at each of 3 locations sampled in the 2001 survey of Switzerland. The thin vertical lines indicate the numbers of species actually detected at these locations.

Figure 12.7. Geographic distribution of the posterior-predictive mean number of avian species present at each of 41,365 locations in Switzerland.

each sample location. In that setting it was natural to consider static features of a community, such as species richness, similarity in species composition among locations, etc. In this section we consider models appropriate for surveys that are undertaken to infer *changes* in a community which may occur as the duration of sampling increases. For example, sites which are occupied by a species at time

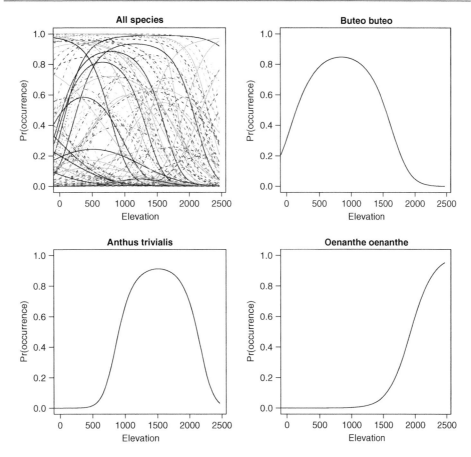

Figure 12.8. Estimated relationship between elevation and average probability of occurrence for each of the 134 species observed in the 2001 survey of Switzerland and for each of 3 individual species.

t may become unoccupied (i.e., experience local extinction) at some future time. Similarly, sites which are unoccupied by a species at time t may become occupied (i.e., locally colonized) by that species at some future time. In other words, we consider models for communities that are 'open' to changes in species composition that occur through demographic processes, such as local extinction and colonization.

Of course, the biological mechanisms that underlie these processes are many and varied. Examples that apply over annual time scales include differences in migratory or activity patterns among species. Over longer time scales, differences in responses of species to the introduction of exotic competitors or to major shifts in climate may induce changes in community composition. Whatever the cause(s), a statistical framework for the analysis of changes in community composition is necessary.

Almost all historical work on this subject has ignored the effects of observation error – that is, an observer's inability to distinguish absence of a species from its non-detection. We anticipate that the development of statistical models which account for observer error will become an expanding area of research. In addition, we envision at least 3 kinds of dynamic models – those based on temporal covariates of species occurrence, those which assume some form of conditional dependence in the site-specific occurrence of a species over time, and hybrids of these two. These approaches are briefly described in the following sections.

12.4.1 Temporal Covariate Models

Suppose a community is surveyed during T primary sampling periods using the sampling protocol described earlier (Section 12.1). Recall that this protocol requires each sample location to be visited J times so that probabilities of occurrence and detection can be estimated for each species. When this protocol is repeated T times, the time between primary sampling periods is assumed to be long relative to the duration of a single survey (or at least long enough that species composition may have changed between primary sampling periods). The addition of T repeated surveys essentially yields a 'robust design' (Pollock, 1982), wherein the J replicate visits correspond to secondary sampling periods nested within each primary sampling period (see Table 9.2).

Now imagine that one or more covariates are observed whose values change between surveys (i.e., temporally) and are thought to be informative of species occurrence or detection. For example, in surveys undertaken to study differences in phenology among species (Kéry et al., 2008), the covariate might be sampling date (Julian day of year). Let x_{kt} denote the value of a covariate observed during the tth primary sampling period $(t = 1, \ldots, T)$ and kth $(k = 1, \ldots, R)$ sample location. We might model the occurrence z_{ikt} of the ith species as a function of this covariate as follows:

$$z_{ikt} | \psi_{ikt}, w_i \sim \mathrm{Bern}(\psi_{ikt}\, w_i),$$

where

$$\mathrm{logit}(\psi_{ikt}) = b_{0i} + b_{1i}\, x_{kt}$$

and the latent variable w_i indicates membership in the community of the N species exposed to sampling during the T primary sampling periods.

Adding other measurable covariates to this model is simple. The main points to note are (1) that temporal differences in species occurrence are modeled conditional on changes in value of the covariate x and (2) that the pattern of these differences is allowed to vary among species through the parameters b_{0i} and b_{1i}. This model of species occurrence is nearly equivalent to that used in Section 12.3, the difference being that occurrence is assumed to vary among primary sampling periods as well as among sites. The definitions of the parameters Ω and N also are slightly different. In this setting N corresponds to the number of distinct species which are present in the community and exposed to sampling sometime during the T primary sampling periods. The number of species that are present during any individual sampling period may, of course, be less than N.

The model of the observed number of detections y_{ikt} at the kth site and tth primary sampling period is also nearly identical to that used in Section 12.3, the difference being that detection is assumed to vary among primary sampling periods as well as among sites:

$$y_{ikt} | J, p_{ikt}, z_{ikt} \sim \text{Bin}(J, p_{ikt} z_{ikt}),$$

where

$$\text{logit}(p_{ikt}) = a_{0i} + a_{1i}\, x_{kt}.$$

Kéry et al. (2008) provide an example of this model wherein the probabilities of occurrence and detection of butterfly species were modeled as functions of sampling date (using linear and quadratic components). The model allowed interspecific differences in phenology of butterfly flight activity (=occurrence) to be estimated and compared. Most butterfly species appeared to have a mid-season peak in their flight period; only a few species exhibited an early - or late-season peak in flight activity.

12.4.2 Temporal Dependence Models

The dynamics of species occurrence may also be modeled as a Markov process, wherein the pattern of temporal dependence in occurrence depends on a set of local colonization and survival/extinction probabilities. This modeling approach is basically identical to the single-species approach described in Chapter 9. To summarize, the initial ($t = 1$) occurrence of species i at site k is modeled as indicated in Eq. (12.2.2):

$$z_{ik1} | \psi_{i1}, w_i \sim \text{Bern}(\psi_{i1} w_i),$$

where ψ_{i1} denotes the probability of occurrence of species i in primary sampling period $t = 1$ and the latent variable w_i indicates membership in the community of the N species exposed to sampling during the T primary sampling periods. Changes from this initial occupancy state are assumed to be Markovian, that is,

$$z_{ik,t+1}|z_{ikt} \sim \text{Bern}(\pi_{it} w_i) \quad \text{for } t = 1, 2, \ldots, T-1 \,, \qquad (12.4.1)$$

where

$$\pi_{it} = \phi_{it} z_{ikt} + \gamma_{it}(1 - z_{ikt}). \qquad (12.4.2)$$

Here, $\phi_{it} = \Pr(z_{ik,t+1} = 1|z_{ikt} = 1)$ denotes the probability that species i 'survives' (continues to be present) at site k between sampling periods t and $t + 1$, and $\gamma_{it} = \Pr(z_{ik,t+1} = 1|z_{ikt} = 0)$ denotes the probability that species i colonizes site k between sampling periods t and $t+1$ given that it was absent at that site in period t. Given these definitions, the occurrence probabilities of species i for sampling periods $2, 3, \ldots, T$ become derived parameters. Specifically, we can show that

$$\psi_{i,t+1} = \phi_{it}\psi_{it} + \gamma_{it}(1 - \psi_{it}) \quad \text{for } t = 1, 2, \ldots, T-1 \,,$$

where $\psi_{i,t+1} = \Pr(z_{ik,t+1} = 1)$ denotes the (unconditional) probability that species i is present during sampling period $t + 1$. Note that this result is identical to that derived for single-species occupancy models in Chapter 9.

Given our formulation of the dynamics of the latent species occurrences, it is straightforward to specify a model of the observed number of detections y_{ikt}. We essentially follow the approach used earlier wherein the observations are modeled conditional on whether each species is present or absent, that is

$$y_{ikt}|J, p_{it}, z_{ikt} \sim \text{Bin}(J, p_{it} z_{ikt}).$$

Many possibilities exist for modeling the species- and time-specific detection probabilities. The effect of species on p_{it} could be modeled as as a collection of random effects conditional on a set of $2T$ fixed parameters for the mean and variance of logit-scale detection probabilities. Alternatively, we might want to specify a trend in detection probabilities, which would require fewer parameters to be estimated. The choice of modeling assumption generally depends on the context of the problem and on the amount of relevant data which might be available.

12.4.3 Hybrid Models

In Section 12.4.1 we described a class of dynamic, occupancy models in which the time-specific state variables for species occurrence (i.e., $\{z_{ikt}\}$) are assumed to be conditionally independent given a set of covariates whose values change with time and possibly location. In Section 12.4.2 we described a class of occupancy models in which the time-specific state variables for species occurrence are assumed to be conditionally *dependent* (in particular, Markovian). We envision a third class of occupancy models that is a hybrid of the first two classes in the sense that its state variables for species occurrence contain both systematic and stochastic sources of variation.

In this hybrid class of models we use an auto-logistic representation of the Markov process, which was described in Chapter 9, for modeling the dynamics of occurrence in one species. In this auto-logistic representation the probabilities of species survival and colonization are parameterized on the logit scale as follows:

$$\text{logit}(\pi_{it}) = a_{it} + b_{it} z_{ikt},$$

where $\text{expit}(a_{it}) = \gamma_{it}$, $\text{expit}(a_{it} + b_{it}) = \phi_{it}$, and $\pi_{it} = \text{Pr}(z_{ik,t+1} = 1 | z_{ikt})$, as in Eqs. (12.4.1) and (12.4.2).

One advantage of this auto-logistic parameterization is that it is easily extended to include systematic sources of variation as measured by temporal or spatial covariates of species occurrence. For example, suppose x_{kt} denotes a covariate whose value differs by site k and sampling period t. We can specify the effects of this covariate on colonization and survival probabilities by adding terms to the previous model as follows:

$$\text{logit}(\pi_{it}) = a_{it} + b_{it} z_{ikt} + \beta_{1i} x_{kt} + \beta_{2i} x_{kt} z_{ikt}.$$

In this expression β_{1i} specifies the effect of x_{kt} on the logit of colonization probability of species i, and $\beta_{1i} + \beta_{2i}$ specifies the effect of x_{kt} on the logit of that species' survival probability. A reduced version of this model, wherein $\beta_{2i} = 0$ is assumed, corresponds to the assumption that the effects of x_{kt} on the the logits of colonization and survival probabilities are identical and equal to β_{1i}.

Additional assumptions are needed, of course, to specify how the parameters a_{it}, b_{it}, β_{1i}, and β_{2i} vary among species. Nonetheless, we think the auto-logistic representation of occupancy models provides a versatile framework for modeling community dynamics and for specifying the effects of spatially and temporally varying covariates.

12.5 SUMMARY

In this chapter we developed models for the analysis of data collected on communities of species. We developed a multi-species, site-occupancy model that accounts for imperfect and variable detection probabilities of species and allows both species-level and community-level characteristics to be estimated in communities of known species richness. We showed that this model can be extended for the more common situation in which species richness is not known and, therefore, must be estimated. This extended model is fitted using data augmentation, a device we have exploited throughout the book. Our occupancy-based framework for modeling species occurrence and detectability accommodates the inclusion of site-level and replicate-level covariates, enabling us to illustrate such covariates can be used to predict the spatial distribution of community-level characteristics, such as species richness.

We also described models for the analysis of data collected on communities that are 'open' to changes in species composition over time. Three classes of models were considered: (1) those in which species occurrence depends on an observable set of temporally varying covariates, (2) those in which the sequence of species occurrences is modeled as a Markov process that depends on local colonization and extinction probabilities, and (3) those in which the sequence of species occurrences is Markovian but with local colonization and extinction probabilities that depend on temporally or spatially varying covariates. The development of statistical models for the analysis of community dynamics is an important area of research, and we anticipate considerable growth in this area.

13

LOOKING BACK AND AHEAD

Biometricians
Vaguely scratching
Numbers wild
Never matching
Counting crows
Or eels electric
Distributions
Parametric?
Badly biased
Or vaguely valid
Significant
Cucumber Salad [1]

Modeling and inference are activities that are fundamental to science. We describe natural systems by models and then attempt to make an assessment of system state, or perhaps a prediction, by fitting the model to data. In this book we outlined a hierarchical modeling framework for inference based on probability models and parametric inference. We have described this approach as 'principled'. By that, we mean an approach which is generic and rigorous in the sense that probability can be used as the basis for modeling and inference in every problem. One size fits all. As such, the framework can be applied in a large and diverse set of problems in ecology – estimating occupancy for a single species, estimating population size in a single closed population, estimating community structure, etc. In this chapter we summarize some of the main themes of the book – the philosophy behind hierarchical modeling and its relevance to the classes of models described in previous chapters. In addition we list specific areas of application of hierarchical models and areas in which we think hierarchical models will be used in the near future.

---

[1] From Droege, S. 2002. *A Heuristic Approach to Validating Monitoring Programs Based on Count Indices (Damn the Statisticians, Full Speed Ahead)* `http://www.pwrc.usgs.gov/naamp3/naamp3.html` .

13.1 THE HIERARCHICAL MODELING PHILOSOPHY

We define a hierarchical model as one that possesses distinct model components for ecological processes and observation processes. In many ecological applications, the process component of the model describes variation in abundance or occurrence over space or time. Sometimes the process component is an 'individual state' – location, or whether an individual is alive or not – which is influenced by population processes of survival and recruitment. In fact, ecological processes that operate at the level of individuals have a conceptual and technical equivalence to occurrence processes that operate on spatially organized populations.

The existence of the process model is central to the hierarchical modeling view. In Chapter 1 we made a distinction between hierarchical models that contain an actual biologically meaningful process model (e.g., abundance, occurrence), which we called an explicit process model, and those for which the process model is basically only descriptive or phenomenological. Such models commonly take the form of GLMs with a collection of random effects describing group structure, typically variance across years or space. We referred to these models as implicit hierarchical models. Usually the implicit process model serves as a surrogate for a real ecological process, but which is difficult to characterize or poorly informed by the data. In this book we focused on hierarchical models of the first kind – those having an explicit process model – and we believe that this is an important distinction between our view of hierarchical models and more conventional views of advocates of hierarchical models. As our objective is typically to gain insight into the form or function of an ecological system or process, we prefer hierarchical models having explicit process models when it is possible to characterize them in a given situation.

The observation component of the hierarchical model often describes the detectability of individuals or species. Detectability has been the focus of considerable attention in statistical ecology for the last several decades. However, many ecologists simply are not concerned with this issue and ignore it in fitting models to observational data. In our view, detectability of individuals (or species) is an important component of the observation model.

13.1.1 Unifying Themes

In this book we tried to achieve broad topical coverage while focusing on classes of problems that could be unified conceptually or technically. Obviously, the hierarchical formulation of models for various ecological systems is the main theme unifying the topics covered in this book. However, the formulation of models and systems under a hierarchical framework is facilitated by two broader themes.

First is the conceptual equivalence of models for these systems, an idea propagated by our colleague Jim Nichols and others over the last two decades. That is, in many problems the observation models and the process models translate across systems. There is a formal equivalence between occupancy models and closed population models, dynamic occupancy models, and models of population dynamics, etc., and we have tried to emphasize this equivalence in several chapters of the book. The result is that the same basic modeling and inference framework is relevant to the 'big three' state variables in ecology: abundance, occurrence, and species richness.

The second theme that ties topics in this book together is the technical device of data augmentation. Indeed, this is motivated by the conceptual equivalence of different modeling and inference problems mentioned in the previous paragraph. Data augmentation formalizes the equivalence between occupancy models and models of repeated sampling of populations of individuals, both in closed population contexts (see Section 3.7.2) and also in open populations (see Chapter 10). We also exploited data augmentation in individual effects models (Chapters 6 and 7) and elsewhere. We believe that data augmentation will prove to be enormously useful in many other classes of models.

13.1.2 Bayesian Hierarchical Models

We generally advocate the Bayesian philosophy of inference because it provides a flexible and coherent framework for statistical inference. However, in the ecological literature Bayesian inference and hierarchical modeling are often viewed together as though one cannot be practiced without the other. For example, it is common to see the term 'Bayesian hierarchical model' used to describe only a model – one for which no inference has been carried out. And, when confronted with a complex problem, some proclaim the need for a 'Bayesian hierarchical model' with the emphasis on the Bayesian part of that. In most cases, what they really mean is that the system is sufficiently complex, so its description requires a modular, hierarchical approach. They are indifferent to whether analysis of the model is given a Bayesian treatment or not.

We agree that a Bayesian approach to inference provides some benefits, but the utility of hierarchical modeling exists independent of the choice of inference framework. We believe that *hierarchical modeling* by itself yields conceptual clarity and practical utility. Unfortunately, in many applications of *Bayesian* hierarchical models, Bayesian is relevant primarily because it comes with a versatile infrastructure of MCMC machinery for estimating parameters under the model. A related misconception in ecology has to do with the importance or emphasis of MCMC in the analysis of models. In many hierarchical modeling applications (i.e., Bayesian ones) there is a strong focus on 'MCMC' as being somehow fundamental to the conceptual development of the problem and how it relates to the science. This focus is not usually stated explicitly, but the manner in which MCMC is emphasized (especially in applied journals) makes it seem more fundamental to the analysis than it is. We don't deny that the choice of parameterization and other issues of statistical modeling can affect the efficiency of MCMC algorithms, and such issues should be the concern of modelers. However, in 10 years we probably will care as little about MCMC algorithms as we care about the algorithms for numerical optimization that are available in commercial software. That is to say, not much. We believe this is the view statisticians should propagate in trying to bring ecologists into the modern era of statistical modeling. MCMC is just an algorithm; it is not fundamental to model construction, or even to inference. We see little point in trying to teach ecologists to write complicated MCMC algorithms in lieu of teaching them the basic concepts of modeling. In focusing on Bayesian computation, ecologists learn little about model construction.

Thus, in our view, Bayesian is mostly just an adjective that is used to describe the method by which a model can be analyzed (although in practice this usually means by MCMC). We believe there is much more to hierarchical modeling than the fact that such models can easily be fitted by adopting a Bayesian approach and using MCMC. We have tried to elucidate this belief in the book by minimizing the relevance of the Bayesian view as it relates to hierarchical models (but not so much as it relates to inference). As a result, we have neglected a thorough literature review focused more directly on Bayesian analysis of hierarchical models. By this omission, we are not trying to short-change or dismiss Bayesianism, we simply choose not to emphasize it because our aim is not to compete with the Bayesian proponents with our view of hierarchical models. By the same token, we think there is much more to the Bayesian approach than its computational convenience in the analysis of hierarchical models. However, most ecologists are relatively unaware of the technical benefits and the subtleties of Bayesian analysis – not because they're ignorant, but because they simply don't care. That the Bayesian confidence interval has a more desirable interpretation is not something that concerns most ecologists.

13.2 APPLICATION OF HIERARCHICAL MODELS

We highlight certain topics covered in the book and try to reiterate some of the concepts that motivated our attention to them in this book.

13.2.1 Occupancy Models

We spent a significant part of this book developing models for occurrence of species. In some cases these models are analytically tractable using a 'non-hierarchical' formulation. For example, the zero-inflated binomial or multinomial models are fairly standard and can be analyzed using classical (frequentist) methods (MacKenzie et al., 2006). However, occupancy models represent a good case study in the development and use of hierarchical models because their hierarchical formulation is exceptionally simple. For the basic case, presented in Chapter 3, the state process is a Bernoulli random variable z_i that indicates the occurrence of the species at site i ($\Pr(z_i = 1) = \psi$). When conditioned on z_i, the corresponding observations also have a simple Bernoulli observation model, having probability $z_i p$.

We have touted the 'conceptual clarity' that hierarchical models bring to a problem. The hierarchical specification of the simple occupancy model can be viewed as a coupled set of logistic regression models: the model for 'z' that would be applied if occurrence could be observed perfectly, and the conditional model for '$y|z = 1$' that is applied to observations at occupied sites.

To obtain information about both model components, we require additional information. We described a design modification (a *repeated measures design*) in which replicate binary observations are made at each of several sites yielding an $M \times J$ matrix of observations. We noted (Section 3.3.3) that this design induces a positive correlation among replicate observations from the same site. In particular, under the basic assumptions used throughout Chapter 3, the correlation between any two observations from site i, say y_{i1} and y_{i2}, is:

$$\text{Corr}(y_{i1}, y_{i2}) = \frac{p(1 - \psi)}{1 - p\psi},$$

where p and ψ are the probabilities of detection and occurrence, respectively.

In this context, we take a moment to elaborate on the distinction we made between implicit and explicit process models. What we called the implicit process model would be consistent with classical models for repeated measurements in which we attempt to describe the joint distribution of the J observations for each site. This model would include a correlation parameter, say ρ, and we might obtain a fairly good description of the variability in the data using this approach. But the explicit process model is much more appealing because the mean, variance, and correlation of the observations can all be derived as functions of parameters that are meaningful in the context of the scientific problem. That is to say, the mean and covariance structure of the observations can be derived as functions of the parameters p and ψ, which have clear biological interpretations.

We single out the occupancy model to illustrate how models may be formulated by 'thinking hierarchically'. However, we believe that the hierarchical approach to modeling is generic and amenable to many modeling extensions that were highlighted in later chapters. Some of those chapters involved the development of models containing individual effects. In other chapters we developed models that are difficult or impossible to formulate non-hierarchically. For example, the spatial, auto-logistic models (Chapter 9) represents a class of models that, to the best of our knowledge, cannot be formulated non-hierarchically.

13.2.2 Metacommunity Models

One important reason we spent so much effort elaborating on the simple occupancy models is that these models form the building blocks of much more complicated models. We considered several extensions of occupancy models to situations where the occurrence state process is indexed by both space and time, say $z(i,t)$. An equally important extension is to the situation of modeling spatial variation in community structure. Models for this situation are naturally formulated as multi-species models of occurrence (Chapter 12).

Inference about community structure has historically been approached using capture–recapture methods for estimating abundance, N, of closed populations of individuals (see Chapter 6). When confronted with a set of spatially-indexed community data from a particular site (i.e., species lists), these models are applied 'site-by-site' to obtain an estimate of species richness N_i for each site ($i = 1, 2, \ldots$). Then, specific hypotheses about species richness are tested by fitting standard regression-type models to the estimated N_i's, treating them as if they were data. This is very much in the style of what we referred to as an observation-driven analysis in Chapter 1 where the focus is on 'adjusting' the data for the observation process in order to regard the result as data in a second-level procedure. In effect, this approach yields a hierarchical procedure, not a hierarchical model. The ecological literature is rife with applications of this type. In a way, this approach resembles a kind of primitive hierarchical model and seems natural to ecologists who think in terms of hierarchical models, but perhaps lack the technical background or tools to analyze such models formally. A slightly more holistic approach would be to acknowledge the structure on the latent variables N_i, as we did for population size in Chapter 8. That is, if N_i is the number of species in some local community i, then assume $N_i \sim \mathrm{Po}(\lambda_i)$ and $\log(\lambda_i) = \beta_0 + \beta_1 x_i$ for some covariate x_i. Given the multinomial observation model implied by Model M_h, the resulting hierarchical model is amenable to formal analysis using integrated likelihood or Bayesian methods. Moreover, these models of N_i fail to preserve species indentity. As such, the resulting *estimates* of N_i have a dependence structure that is not easily incorporated into secondary analyses.

By contrast, in Chapter 12 we described a framework for analysis of metacommunity data based on multi-species occupancy models. These models contain a set of species-specific occurrence and detection probabilities that may depend on measurable covariates. Because the model is developed in terms of species-specific probabilities of occurrence, estimation and prediction of species richness and other summaries of community structure can be obtained. We believe that this versatile, hierarchical formulation of community structure has the potential to revolutionize modeling and inference in metacommunity systems.

13.2.3 Individual Effects and Spatial Capture–Recapture

In Chapter 6 we provided a hierarchical formulation of traditional individual effects models for closed populations. These models include those containing unstructured variation in the form of individual random effects (so-called 'Model M_h') and also models containing individual covariates that influence detection probability. These two classes of models have historically been treated very differently. On one hand, Model M_h has been dominated by 'Burnham's Jackknife estimator' (Burnham and Overton, 1978). On the other hand, inference in individual covariate models is typically practiced using a procedure, colloquially referred to as the 'Huggins–Alho model' (Huggins, 1989; Alho, 1990). These two procedures are conceptually and technically unrelated and are inconsistent with any general theory for the conduct of inference in capture–recapture type problems. We showed that the formulation of these two classes of models is unified within a hierarchical modeling framework and that their analysis is unified by data augmentation.

In Chapter 7 we considered a special class of individual covariate models, models in which the individual covariate is related to spatial location of the individual, which we referred to as spatial capture–recapture models. This class of models includes the traditional distance sampling model, whose hierarchical formulation reveals its essential equivalence to an individual covariate model. In fact, the distance sampling model is an individual covariate model with a single replicate sample of the population. This lack of replication is compensated for by specification of a 0-parameter individual covariate distribution (in addition to a special functional form relating p to the individual covariate). The classical treatment of distance sampling is usually based on 'conditional likelihood' which, while being consistent with a model-based view, is usually justified by design-based arguments.

We also considered formal capture–recapture models wherein some auxiliary information on the spatial location of individuals is available. In the first case we considered a situation in which a fixed area is subjected to uniform search intensity (an 'area search') and locations of captured individuals are measured in J samples. Location in this case is a time-varying individual covariate. We extended the hierarchical model by imposing a prior distribution on individual location. Under this prior distribution, an individual's locations observed during the J samples are related to the individual's home range or territory. The highest level of the hierarchical model consists of a spatial model describing the distribution of individual home ranges (activity centers) over the landscape. Finally, we considered the classical problem of estimating density from trapping arrays. As with the previous model, the hierarchical model consists of a component describing the distribution of individual home range centers on the landscape, a model of exposure to traps, which is essentially an implicit movement model, and a model of capture by traps (capture being, in the cases considered, a Bernoulli trial).

Hierarchical models provide a conceptual and technical unification of these spatial capture–recapture models within the general framework of models containing individual effects.

13.2.4 Dynamic Models

We considered a number of extensions of population and metapopulation models to dynamic systems. In Chapter 9 we described a hierarchical formulation of occupancy models allowing for metapopulation dynamics in the form of local extinction and colonization. We described a new version of this model that distinguishes between colonization and re-colonization and might prove useful in modeling the spread of an invasive species or disease. We also described spatial models and offer some thoughts for parameterization of spatio-temporal dynamics. While these models offer an important extension of the simple models of occupancy considered in Chapter 3, we believe they also yield a profound development in what are usually referred to as Jolly–Seber models. We described these models in Chapter 10 and developed a new parameterization for them based on data augmentation. Under data augmentation, the Jolly–Seber model is precisely equivalent to a dynamic occupancy model. This parameterization illustrates the conceptual and technical unification that is a major theme of our book; however, it also has practical utility. For example, the data augmentation parameterization can be used to specify heterogeneity in survival among individuals, which has not previously been considered in the literature.

Finally, we considered a simple class of dynamic models relevant to modeling survival of individuals – commonly referred to as Cormack–Jolly–Seber models, which are constrained versions of the Jolly–Seber models. Our motive in considering these models is that they represent an especially simple type of hierarchical model, one in which the process model is just a sequence of Bernoulli outcomes. The main benefit that the hierarchical formulation yields is a conceptually simple specification of models with individual effects. We also developed some interesting extensions of these models, including a hierarchical parameterization for populations that contain transient individuals. Under this model, 'residency' status is a latent variable, easily described by a hierarchical model.

13.3 POTENTIAL DEVELOPMENTS

We have shown that hierarchical modeling may be used to analyze data that arise in a variety of commonly used sampling protocols. For example, we used the hierarchical approach to formulate models of occupancy based on repeated measurements of detection/non-detection data at individual sites. Similarly, we applied hierarchical modeling in the analysis of repeated point counts for estimating abundance and for inferring abundance–habitat relationships. We also found the hierarchical approach to be useful for modeling data generated by capture–recapture sampling, double-observer sampling, and other sampling protocols.

Of course, the formal separation of observation and process components of hierarchical models is the reason why they are so versatile. Whatever form the observations may take (detection/non-detection, counts, etc.), they are always modeled *conditional* on the latent, ecological process (state variable) that is the primary object of inference. Thus, we may use removal sampling or double-observer sampling to survey a local (site-specific) population, but the counts are modeled conditionally local abundance state variable.

We anticipate several applications of hierarchical modeling that were not described in earlier chapters, and we briefly describe some of those here.

13.3.1 Occupancy/Abundance

In Chapter 4 we described hierarchical models that link presence/absence data to abundance. Information about abundance is obtained from presence/absence data by apparent heterogeneity in detection probability; sites with high abundance yield more net detections, and vice versa. This is a perfect example of a situation in which hierarchical thinking reveals an interesting model and solves a problem that might otherwise have appeared to be unsolvable. We addressed this topic in the context of static systems, but we believe the linkage between abundance and occurrence (and the induced heterogeneity in detection) can be exploited or developed in a number of other applications.

13.3.1.1 Dynamics

Dynamic models of invasions readily admit a dependence of the rate of spread on abundance (as in diffusion models). Therefore, it follows that metapopulation dynamics of survival and colonization also depend on local population size. A large local population should have a lower probability of extinction and, in a spatially dynamic system, should yield higher local colonization probabilities as a result of dispersal and movement of individuals to neighboring local populations. To illustrate, let z_t denote a binary state variable for occurrence at time t wherein presence and absence are indicated by $z = 1$ and $z = 0$, respectively. The probability of extinction is the complement of $\Pr(z_t = 1|z_{t-1} = 1)$. If the survival of individuals are independent Bernoulli outcomes (with probability ϕ), then

$$\Pr(z_t = 1|z_{t-1} = 1) = 1 - (1 - \phi)^{N_t},$$

where N_t is the local population size at time t. Thus, abundance-induced variation in metapopulation dynamics presents a natural application of hierarchical models.

13.3.1.2 Spatial occupancy models

At sufficiently small scales, spatial occupancy patterns are governed to some extent by the movement of individuals among local populations. For example, when large carnivores, such as bears, are sampled over time using sampling grids on the order of a few kilometers, the same individual can be detected in multiple grid cells. Thus, a single individual can render multiple grid cells occupied. In this case the movements of individuals produce an occupancy state variable that is autocorrelated. We believe it is possible to explicitly parameterize the autocorrelation structure of the binary state process in terms of a model for the distribution and movement of individuals in space (as in Chapter 7). This type of model would allow absolute density of animals to be estimated from spatially organized occupancy data.

13.3.1.3 Design

One practical motivation for pursuing occupancy models is that apparent presence/absence data are simple to collect and can often be obtained at relatively low cost. However, we compromise information content for this efficiency and economy. In surveys of rare or highly clustered species, it can be prohibitively expensive to carry out conventional survey methods (wherein individuals are enumerated) because many sample units may be required to estimate abundance accurately. In Chapter 4 we showed that abundance information can be obtained from simple presence/absence data. Therefore, it stands to reason that hierarchical models may be used to integrate these two types of data.

This leads to consideration of a survey framework wherein we might sample a very large number of units at some low intensity level in order to obtain information about presence. We sample others more intensively to get decent abundance estimates. For example, in the context of a traditional two-stage sampling idea, we might survey a number of plots and record presence/absence. Then we visit a second set and record abundance. An important technical issue is that the probability of recording presence depends on abundance and we have a classical unequal probability sampling problem where the inclusion probabilities are, necessarily, model-based (because we cannot observe the abundance state variable). The second stage sample units will represent a biased sample, tending to have higher abundance than average. Thus, we might consider repeat sampling for occurrence, continuing to accumulate second-stage sample units. Accumulation of the second stage sample in this fashion is analogous to a non-response bias problem (an observation of 'non-detection' being equivalent to non-response). Conceptually, subsequent samples are analogous to the 'call-back' in survey sampling, and they generate information on the non-respondents. A recent paper by Conroy et al. (2008) addresses some sampling issues in the context of the linkage between occupancy and abundance.

13.3.2 Dynamic Models and Related Extensions

Many of the ecological systems considered in the book are for static populations not subject to the dynamics induced by survival, recruitment, and dispersal processes. Dynamic models obviously have many useful applications in ecology (modeling invasive species, disease, and competition among species), but there have been very few statistical treatments of these problems that are consistent with our view of hierarchical models. By that, we mean models that incorporate a realistic observation model and an explicit process model. We considered dynamic systems in some cases (Chapters 9, 10, and 11), but only in the context of a single species. On the other hand, in Chapter 12 we considered multiple-species models, but focused mainly on models for a static community. We feel that incremental extensions of these models to accommodate the additional ecological dimension (species or time) would be enormously useful.

For metapopulation models of abundance, we describe in Chapter 8 extensions of the basic models to accommodate a spatially and temporally indexed state variable, $N(i, t)$, which should be straightforward as there are many variations of such models in widespread use. However, many of the applications regard the model as deterministic, or make unrealistic assumptions of normality (or log-normality) on $N(i, t)$. While such assumptions may be adequate for abundance time-series over large areas, they would typically not be that useful for spatially local-scale models of abundance. Naturally, the extension of metapopulation abundance models to multi-species systems should also be of general interest. We think that such extensions are conceptually straightforward, but one fundamental technical issue that must be confronted is how to devise models for parameterizing high-dimensional interactions (i.e., among species). The multi-species occupancy model described in Chapter 12 is based on a simple null model of independence among species – hardly a realistic assumption in most communities.

13.3.3 Spatial Capture–Recapture

In Chapter 7 we described classes of spatial capture–recapture models. A component of the process model is a spatial model that describes the distribution of individuals in space or, according to our parameterization of such models, the spatial distribution of their activity centers s_i. However, in all of the situations that we considered in Chapter 7, we assumed a simplistic model in which the activity centers are distributed uniformly in space – that is the locations s_i constitute a realization of a homogeneous spatial point process.

This point process is parameterized explicitly in the hierarchical model. We believe that this component of the model can be generalized considerably and in substantive, meaningful directions by making use of standard point-process models. For example, in carnivore populations it is likely that the activity centers are not independent of one another owing both to interactions with neighbors and to habitat variation. To model these interactions, it seems natural to formulate the point process in terms of Markovian dependence, wherein successive points are conditioned on the nearness and even the 'type' (male, female) of neighbors (such models are often referred to as a 'marked point process'). In addition, the possibility exists of formulating point-process models that admit explicit spatio-temporal structure, representing the survival and recruitment of individuals.

13.3.4 Compound Distributions

In many wildlife survey applications, it is not possible to obtain either individual encounter histories or even counts of unique individuals. However, it may be possible to obtain counts of signs of individuals or of detection rate data that are spatially indexed. For example, a wintertime survey of road segments might yield the number of animal track crossings per segment of a given length. Similarly, we might observe scats encountered along sample paths or record the number of encounters detected by motion-sensitive camera traps.

The estimation of abundance or density from sign counts or encounter rate data is a problem of some concern (Carbone et al., 2001; Jennelle et al., 2002; Stanley and

Royle, 2005; Rowcliffe et al., 2008), but one that has not received much attention from a statistical point of view. For such problems, we see great potential for so-called compound distributions (Johnson et al., 2005). As an illustration, consider a model in which there are N_i individuals associated with a local population or sample unit i, and let x_{ij} denote the number of detections (or signs left) for the jth individual in the population ($j = 1, 2, \ldots, N_i$). In a typical situation, we only observe the total number of detections, say y_i, which is

$$y_i = \sum_{j=1}^{N_i} x_{ij}.$$

In this example both N_i and the x_{ij} are unknown; therefore, we require a characterization of the marginal distribution of y_i. Compound distributions provide one such characterization by specifying the component distributions of both N_i and x_{ij}.

An interesting example of a compound distribution is that which arises under the assumption that $N_i \sim \mathrm{Po}(\lambda)$ and $x_{ij} \sim \mathrm{Po}(\theta)$, where θ is the detection rate of individuals (or their signs) in the local population, and λ is the mean size of the local population. This compound distribution is also called a Neyman Type A distribution. The model can be fitted by standard methods of analysis. As Stanley and Royle (2005) noted, simple moment estimators of θ and λ are available: $\hat{\theta} = (s^2 - \bar{y})/\bar{y}$ and $\hat{\lambda} = \bar{y}/\hat{\theta}$, where $\bar{y}$ and s^2 are the sample mean and variance of the observations. Thus, from a single set of spatially referenced sign counts, with no individual identity information, one can obtain information about local population structure and detectability. Investigation of such models would seem to be a fruitful area of research.

13.3.5 Hierarchical Models for Complex Sampling Designs

In some surveys populations are sampled using multiple sampling protocols. This may occur if one protocol is more expensive to implement than another or if multiple protocols are needed to sample different segments of the population. An example of the latter occurs in aerial surveys where groups of individuals are sampled using one protocol and individuals within groups are sampled using another protocol. We believe that hierarchical modeling provides a generic solution to this problem and is sufficiently flexible to accommodate data from multiple sampling protocols.

Other potential applications of hierarchical modeling are surveys with complex sampling designs. In many cases samples are selected from a collection of spatially referenced units, and the selection process may involve clusters or stratifications of those units. The North American breeding bird survey, which we have described in previous chapters, is a good example. In this survey, the sample frame is divided into primary sample units (routes) and observations are made at secondary sample units (stops) that are nested within each primary unit. Thus, a two-stage cluster sample is selected in any single year of the survey. To practice model-based inference, we can account for the effects of the sampling design (e.g., clustering or stratification) by including parameters that are relevant to the data-collection process. In fact, failure to do so can produce misleading inferences (e.g., see Chapter 7 of Gelman et al. (2004)) because sources of variation which are induced by the design will not be accounted for. Fortunately, the hierarchical modeling approach is sufficiently flexible to accommodate the increases in model complexity needed for complex sampling designs.

13.4 NO SUCH THING AS A FREE LUNCH

Hierarchical models offer a generic and rigorous framework for modeling and inference in ecology. But this flexibility and generality does not come for free. We rely heavily on parametric inference in our approach to ecological analysis. This means that we must behave as though the parametric model *is* the data-generating model (i.e., 'truth') when computing inferences or predictions from data. While parametric inference is something of a unifying concept for both Bayesians and frequentists, many practitioners and statisticians are uncomfortable making explicit model assumptions because assumptions of a parametric model are not always testable. Besides, we know they are false a priori ('all models are wrong', right?). In response, this can lead to a focus on procedure-driven analysis, procedures that are poorly defined or contain vaguely stated assumptions, or models that are overly complex and cannot be understood. We think a word of caution is in order against these free-lunch approaches.

One reaction against parametric inference is to adopt procedures that are described as 'robust' or 'non-parametric'. Design-based procedures are also enormously popular. But these procedures are often just red herrings that effectively produce analyses which are irrefutable, unfalsifiable, or immune to criticism. The analysis of Little (2004, p. 550) in the context of the Horwitz–Thompson estimator (HTE) refutes the non-parametric 'free lunch'. Little notes that the HTE has a precise, model-based justification. While the assumptions of the implied model need not be stated explicitly in order to justify the estimator, he notes that "…. the HTE is likely to be a good estimator when [the model] is a good description of the population, and may be inefficient when it is not". The point being that models are not necessarily irrelevant (or their effects innocuous) for procedures that are claimed to be 'model-free' or 'robust'. Because a procedure can be developed absent an explicit statement of a model does not render it robust against model pathologies.

Some statisticians routinely engage in the development of complex models under the guise of 'hierarchical modeling' to introduce complexity – hierarchical modeling for the sake of hierarchical modeling. While the motives are pragmatic, the end result is a model which is unassailable, unrepeatable, unfalsifiable, and beyond comprehension – metaphorically, if we may, a hierarchical Rube Goldberg device. In Chapter 1 we quoted Lindley (2006) from his book 'Understanding Uncertainty:'

> There are people who rejoice in the complicated, saying, quite correctly, that the real world is complicated and that it is unreasonable to treat it as if it was simple. They enjoy the involved because it is so hard for anyone to demonstrate that what they are saying can be wrong, whereas in a simple argument, fallacies are more easily exposed.

Words to live by.

We are enthusiastic proponents of model-based, parametric inference as a general framework for ecological analysis because, in the words of our colleague W. Link (Link, 2003), "Easily assailable but clearly articulated assumptions ought always to be preferable". In matters of scientific inquiry, simplicity is a virtue. Not necessarily procedural simplicity, but *conceptual* simplicity – and clearly assailable assumptions.

BIBLIOGRAPHY

Agresti, A. (2002), *Categorical Data Analysis*, second edition, Hoboken, New Jersey: Wiley.

Airoldi, E. M., Anderson, A. G., Fienberg, S. E., and Skinner, K. K. (2006), "Who wrote Ronald Reagans radio addresses", *Journal of Bayesian Analysis*, 1, 289–320.

Alho, J. M. (1990), "Logistic regression in capture-recapture models", *Biometrics*, 46, 623–635.

Alpízar-Jara, R. and Pollock, K. H. (1996), "A combination line transect and capture-recapture sampling model for multiple observers in aerial surveys", *Environmental and Ecological Statistics*, 3, 311–327.

Anderson, C. J., Wikle, C. K., Zhou, Q., and Royle, J. A. (2007), "Population influences on tornado reports in the United States", *Weather and Forecasting*, 22, 571–579.

Arnold, S. F. (1993), "Gibbs sampling", in *Handbook of Statistics 9: Computational Statistics*, ed. Rao, C. R., Amsterdam: Elsevier Science, pp. 599–625.

Augustin, N. H., Mugglestone, M. A., and Buckland, S. T. (1996), "An autologistic model for the spatial distribution of wildlife", *Journal of Applied Ecology*, 33, 339–347.

Bailey, L. L., Simons, T. R., and Pollock, K. H. (2004), "Estimating site occupancy and species detection probability parameters for terrestrial salamanders", *Ecological Applications*, 14, 692–702.

Barbraud, C., Nichols, J. D., Hines, J. E., and Hafner, H. (2003), "Estimating rates of local extinction and colonization in colonial species and an extension to the metapopulation and community levels", *Oikos*, 101, 113–126.

Basu, D. (1977), "On the elimination of nuisance parameters", *Journal of the American Statistical Association*, 72, 355–366.

Bayes, T. (1763), "An essay towards solving a problem in the doctrine of chances", *Philosophical Transactions of the Royal Society of London*, 53, 370–418.

Baylcy, P. B. and Peterson, J. T. (2001), "An approach to estimate probability of presence and richness of fish species", *Transactions of the American Fisheries Society*, 130, 620–633.

Becker, R. A., Chambers, J. M., and Wilks, A. R. (1988), *The New S Language: A Programming Environment for Data Analysis and Graphics*, Pacific Grove, California: Wadsworth and Brooks/Cole.

Berger, J. O., Liseo, B., and Wolpert, R. L. (1999), "Integrated likelihood methods for eliminating nuisance parameters (with discussion)", *Statistical Science*, 14, 1–28.

Berliner, L. M. (1996), "Hierarchical Bayesian time series models", *Maximum Entropy and Bayesian Methods*, 15–22.

Besag, J. E. (1972), "Nearest-neighbour systems and the auto-logistic model for binary data", *Journal of the Royal Statistical Society. Series B (Methodological)*, 34, 75–83.

Besag, J. E., York, J., and Mollie, A. (1991), "Bayesian image restoration with two applications in spatial statistics", *Annals of the Institute of Statistical Mathematics*, 43, 1–59.

Bibby, C. J., Burgess, N. D., and Hill, D. A. (1992), *Bird Census Techniques*, London, UK: Academic Press.

Bissell, A. F. (1972), "A negative binomial model with varying element sizes", *Biometrika*, 59, 435–441.

Bonner, S. J. and Schwarz, C. J. (2004), "Continuous time–dependent Individual covariates and the Cormack–Jolly–Seber model", *Animal Biodiversity and Conservation*, 27, 149–155.

Bonner, S. J. and Schwarz, C. J. (2006), "An Extension of the Cormack–Jolly Seber model for continuous covariates with application to *Microtus pennsylvanicus*", *Biometrics*, 62, 142–149.

Borchers, D. L., Zucchini, W., and Fewster, R. M. (1998), "Mark-recapture models for line transect surveys", *Biometrics*, 54, 1207–1220.

Borchers, D. L., Buckland, S. T., and Zucchini, W., eds. (2002), *Estimating Animal Abundance: Closed Populations*, London: Springer-Verlag.

Boulanger, J. and McLellan, B. (2001), "Closure violation in DNA-based mark-recapture estimation of grizzly bear populations", *Canadian Journal of Zoology*, 79, 642–651.

Boulinier, T., Nichols, J. D., Hines, J. E., Sauer, J. R., Flather, C. H., and Pollock, K. H. (2001), "Forest fragmentation and bird community dynamics: inference at regional scales", *Ecology*, 82, 1159–1169.

Boulinier, T., Nichols, J. D., Sauer, J. R., Hines, J. E., and Pollock, K. H. (1998), "Estimating species richness: the importance of heterogeneity in species detectability", *Ecology*, 79, 1018–1028.

Boveng, P. L., Bengtson, J. L., Withrow, D. E., Cesarone, J. C., Simpkins, M. A., Frost, K. J., and Burns, J. J. (2003), "The abundance of harbor seals in the Gulf of Alaska", *Marine Mammal Science*, 19, 111–127.

Box, G. E. P. (1980), "Sampling and Bayes inference in scientific modelling and robustness (with discussion)", *Journal of the Royal Statistical Society, Series A*, 143, 383–430.

Boyce, M. S. and McDonald, L. L. (1999), "Relating populations to habitats using resource selection functions", *Trends in Ecology and Evolution*, 14, 268–272.

Brander, S. M., Royle, J. A., and Eames, M. (2007), "Evaluation of the status of anurans on a refuge in suburban Maryland", *Journal of Herpetology*, 41, 52–60.

Brooks, S. P., Catchpole, E. A., and Morgan, B. J. T. (2000), "Bayesian animal survival estimation", *Statistical Science*, 15, 357–376.

Brown, J. H. (1984), "On the relationship between abundance and distribution of species", *American Naturalist*, 124, 255–279.

Brown, J. H. and Kodric-Brown, A. (1977), "Turnover rates in insular biogeography: Effect of immigration on extinction", *Ecology*, 58, 445–449.

Brown, J. H. and Maurer, B. A. (1989), "Macroecology: The division of food and space among species on continents", *Science*, 243, 1145.

Brownie, C., Anderson, D. R., Burnham, K. P., and Robson, D. R. (1985), *Statistical inference from band recovery data: A handbook*, Washington, D.C.: U.S. Fish and Wildlife Service, Resource Publication 156.

Buckland, S. T., Anderson, D. R., Burnham, K. P., Laake, J. L., Borchers, D. L., and Thomas, L. (2001), *Introduction to Distance Sampling: Estimating Abundance of Biological Populations*, New York: Oxford University Press.

Buckland, S. T., Newman, K. B., Thomas, L., and Koesters, N. B. (2004a), "State-space models for the dynamics of wild animal populations", *Ecological Modelling*, 171, 157–175.

Buckland, S. T., Anderson, D. R., Burnham, K. P., Laake, J. L., Borchers, D. L., and Thomas, L. (2004b), *Advanced Distance Sampling*, Oxford: Oxford University Press.

Bunge, J. and Fitzpatrick, M. (1993), "Estimating the number of species: a review", *Journal of the American Statistical Association*, 88, 364–373.

Burnham, K. P. (1972), Estimation of population size in multiple capture-recapture studies when capture probabilities vary among animals, Ph.D. dissertation, Oregon State University, Corvallis.

Burnham, K. P. and Anderson, D. R. (1976), "Mathematical models for nonparametric inferences from line transect data", *Biometrics*, 32, 325–336.

Burnham, K. P. and Anderson, D. R. (2002), *Model Selection and Multi-Model Inference: A Practical Information-Theoretic Approach*, 2nd edition, New York: Springer.

Burnham, K. P. and Anderson, D. R. (2004), "Multimodel inference: Understanding AIC and BIC in model selection", *Sociological Methods & Research*, 33, 261–304.

Burnham, K. P. and Overton, W. S. (1978), "Estimation of the size of a closed population when capture probabilities vary among animals", *Biometrika*, 65, 625–633.

Burnham, K. P. and Overton, W. S. (1979), "Robust estimation of population size when capture probabilities vary among animals", *Ecology*, 60, 927–936.

Cam, E., Link, W. A., Cooch, E. G., Monnat, J. Y., and Danchin, E. (2002a), "Individual covariation in life-history traits: Seeing the trees despite the forest", *American Naturalist*, 159, 96–105.

Cam, E., Monnat, J. Y., and Royle, J. A. (2004), "Dispersal and individual quality in a long lived species", *Oikos*, 106, 386–398.

Cam, E., Nichols, J. D., Hines, J. E., Sauer, J. R., Alpizar-Jara, R., and Flather, C. H. (2002b), "Disentangling sampling and ecological explanations underlying species-area relationships", *Ecology*, 83, 1118–1130.

Cam, E., Nichols, J. D., Sauer, J. R., and Hines, J. E. (2002c), "On the estimation of species richness based on the accumulation of previously unrecorded species", *Ecography*, 25, 102–108.

Carbone, C., Christie, S., Conforti, K., Coulson, T., Franklin, N., Ginsberg, J. R., Griffiths, M., Holden, J., Kawanishi, K., and Kinnaird, M., et al. (2001), "The use of photographic rates to estimate densities of tigers and other cryptic mammals", *Animal Conservation*, 4, 75–79.

Carlin, B. P. and Louis, T. A. (2000), *Bayes and Empirical Bayes Methods for Data Analysis*, second edition, Boca Raton, Florida: Chapman and Hall.

Casella, G. and Berger, R. L. (2002), *Statistical Inference*, second edition, Pacific Grove, California: Duxbury.

Chambers, J. M. (1998), *Programming with Data: A Guide to the S Language*, New York: Springer-Verlag.

Chao, A. (1987), "Estimating the population size for capture-recapture data with unequal catchability", *Biometrics*, 43, 783–791.

Chen, M. H., Dey, D. K., and Ibrahim, J. G. (2004), "Bayesian criterion based model assessment for categorical data", *Biometrika*, 91, 45–63.

Choquet, R., Reboulet, A. M., Pradel, R., Gimenez, O., and Lebreton, J. D. (2004), "M-SURGE: new software specifically designed for multistate capture-recapture models", *Animal Biodiversity and Conservation*, 27, 207–215.

Claeskens, G. and Hjort, N. L. (2003), "The focused information criterion (with discussion)", *Journal of the American Statistical Association*, 98, 900–945.

Clark, J. S. (2003), "Uncertainty and variability in demography and population growth: A hierarchical approach", *Ecology*, 84, 1370–1381.

Clark, J. S. (2007), *Models for Ecological Data: An Introduction*, Princeton, NJ: Princeton University Press.

Clark, J. S. and Gelfand, A. E. (2006), *Hierarchical Modelling for the Environmental Sciences: Statistical Methods and Applications*, New York: Oxford University Press.

Clark, J. S., Carpenter, S. R., Barber, M., Collins, S., Dobson, A., Foley, J. A., Lodge, D. M., Pascual, M., Pielke, R., and Pizer, W., et al. (2001), "Ecological forecasts: an emerging imperative", *Science*, 293, 657–660.

Clark, C. W. and Rosenzweig, M. L. (1994), "Extinction and colonization processes: Parameter estimates from sporadic surveys", *American Naturalist*, 143, 583–596.

Clayton, D. G. (1996), "Generalized linear mixed models", *Markov Chain Monte Carlo in Practice*,.

Collett, D. (1991), *Modelling Binary Data*, London: Chapman and Hall.

Colwell, R. K., Mao, C. X., and Chang, J. (2004), "Interpolating, extrapolating, and comparing incidence-based species accumulation curves", *Ecology*, 85, 2717–2727.

Congdon, P. (2005), *Bayesian Models for Categorical Data*, New York: Wiley.

Conroy, M. J., Runge, J. P., Barker, R., Schofield, M. R. and Fonnesbeck, C. (2008), Efficient estimation of abundance for patchily distributed populations via 2-stage, adaptive sampling, Unpublished manuscript.

Cooch, E. G., Cam, E., and Link, W. A. (2002), "Occam's shadow: levels of analysis in evolutionary ecology–where to next?", *Journal of Applied Statistics*, 29, 19–48.

Cook, R. D. and Jacobson, J. O. (1979), "A design for estimating visibility bias in aerial surveys", *Biometrics*, 35, 735–742.

Cormack, R. M. (1964), "Estimates of survival from the sighting of marked animals", *Biometrika*, 51, 429–438.

Coull, B. A. and Agresti, A. (1999), "The use of mixed logit models to reflect heterogeneity in capture-recapture studies", *Biometrics*, 55, 294–301.

Creel, S., Spong, G., Sands, J. L., Rotella, J., Zeigle, J., Joe, L., Murphy, K. M., and Smith, D. (2003), "Population size estimation in Yellowstone wolves with error-prone noninvasive microsatellite genotypes", *Molecular Ecology*, 12, 2003–2009.

Dahl, K. (1919), "Studies of trout and trout-waters in Norway", *Salmon and Trout Magazine*, 18, 16–33.

Day, J. R. and Possingham, H. P. (1995), "A stochastic metapopulation model with variability in patch size and position", *Theoretical Population Biology*, 48, 333–360.

De Valpine, P. and Hastings, A. (2002), "Fitting population models incorporating process noise and observation error", *Ecological Monographs*, 72, 57–76.

Demidenko, E. (2004), *Mixed Models: Theory and Applications*, Hoboken, New Jersey: John Wiley & Sons.

Dennis, B., Ponciano, J. M., Lele, S. R., Taper, M. L., and Staples, D. F. (2006), "Estimating density dependence, process noise, and observation error", *Ecological Monographs*, 76, 323–341.

DeSante, D. F., Burton, K. M., Saracco, J. F., and Walker, B. L. (1995), "Productivity indices and survival rate estimates from MAPS, a continent-wide programme of constant-effort mist-netting in North America", *Journal of Applied Statistics*, 22, 935–948.

DeSante, D. F., Saracco, J. F., O'Grady, D. R., Burton, K. M., and Walker, B. L. (2004), "Methodological considerations of the monitoring avian productivity and survivorship (MAPS) program", *Studies in Avian Biology*, 29, 28–45.

Dice, L. R. (1938), "Some census methods for mammals", *Journal of Wildlife Management*, 2, 119–130.

Dice, L. R. (1945), "Measures of the amount of ecologic association between species", *Ecology*, 26, 297–302.

Diggle, P. J., Tawn, J. A., and Moyeed, R. A. (1998), "Model-based geostatistics", *Applied Statistics*, 47, 299–350.

Dinsmore, S. J., White, G. C., and Knopf, F. L. (2002), "Advanced techniques for modeling avian nest survival", *Ecology*, 83, 3476–3488.

Dodd Jr., C. K. and Dorazio, R. M. (2004), "Using counts to simultaneously estimate abundance and detection probabilities in a salamander community", *Herpetologica*, 60, 468–478.

Doherty Jr., P. F., Sorci, G., Royle, J. A., Hines, J. E., Nichols, J. D., and Boulinier, T. (2003), "Sexual selection affects local extinction and turnover in bird communities", *Proceedings of the National Academy of Sciences*, 100, 5858–5862.

Dorazio, R. M. (2007), "On the choice of statistical models for estimating occurrence and extinction from animal surveys", *Ecology*, 88, 2773–2782.

Dorazio, R. M., Jelks, H. L., and Jordan, F. (2005), "Improving removal-based estimates of abundance by sampling a population of spatially distinct subpopulations", *Biometrics*, 61, 1093–1101.

Dorazio, R. M., Mukherjee, B., Zhang, L., Ghosh, M., Jelks, H. L. and Jordan, F. (2008), Modeling unobserved sources of heterogeneity in animal abundance using a Dirichlet process prior, *Biometrics*, 64, in press.

Dorazio, R. M. and Royle, J. A. (2003), "Mixture models for estimating the size of a closed population when capture rates vary among individuals", *Biometrics*, 59, 351–364.

Dorazio, R. M. and Royle, J. A. (2005a), "Estimating size and composition of biological communities by modeling the occurrence of species", *Journal of the American Statistical Association*, 100, 389–398.

Dorazio, R. M. and Royle, J. A. (2005b), "Rejoinder to the performance of mixture models in heterogeneous closed population capture-recapture", *Biometrics*, 61, 874–876.

Dorazio, R. M., Royle, J. A., Söderström, B., and Glimskär, A. (2006), "Estimating species richness and accumulation by modeling species occurrence and detectability", *Ecology*, 87, 842–854.

Draper, D. (1996), "Utility, sensitivity analysis, and cross-validation in Bayesian model-checking", *Statistica Sinica*, 6, 760–767.

Dreitz, V. J., Lukacs, P. M., and Knopf, F. L. (2006), "Monitoring low density avian populations: An example using mountain plovers", *The Condor*, 108, 700–706.

Dupuis, J. A. (1995), "Bayesian estimation of movement and survival probabilities from capture-recapture data", *Biometrika*, 82, 761–772.

Edwards, A. W. F. (1974), "The history of likelihood", *International Statistical Review*, 42, 9–15.

Edwards, A. W. F. (1992), *Likelihood, Expanded Edition*, Baltimore: Johns Hopkins University Press.

Efron, B. (1986), "Why isn't everyone a Bayesian?", *The American Statistician*, 40, 1–5.

Engler, R., Guisan, A., and Rechsteiner, L. (2004), "An improved approach for predicting the distribution of rare and endangered species from occurrence and pseudo-absence data", *Journal of Applied Ecology*, 41, 263–274.

Erwin, R. M., Nichols, J. D., Eyler, T. B., Stotts, D. B., and Truitt, B. R. (1998), "Modeling colony-site dynamics: A case study of gull-billed terns (*Sterna nilotica*) in Coastal Virginia", *The Auk*, 115, 970–978.

Farnsworth, G. L., Pollock, K. H., Nichols, J. D., Simons, T. R., Hines, J. E., and Sauer, J. R. (2002), "A removal model for estimating detection probabilities from point-count surveys", *The Auk*, 119, 414–425.

Fienberg, S. E. (1972), "The multiple-recapture census for closed populations and incomplete 2^k contingency tables", *Biometrika*, 59, 591–603.

Frost, K. J., Lowry, L. F., and Ver Hoef, J. M. (1999), "Monitoring the trend of harbor seals in Prince William Sound, Alaska, after the Exxon Valdez oil spill", *Marine Mammal Science*, 15, 494–506.

Gardner, B., Royle, J. A. and Wegan, M. T. (2008), "Hierarchical models for estimating density from DNA mark-recapture studies", *Ecology (in review)*.

Gaston, K. J., Blackburn, T. M., and Lawton, J. H. (1997), "Interspecific abundance-range size relationships: An appraisal of mechanisms", *Journal of Animal Ecology*, 66, 579–601.

Gelfand, A. E. and Ghosh, S. K. (1998), "Model choice: a minimum posterior predictive loss approach", *Biometrika*, 85, 1–11.

Gelfand, A. E., Sahu, S. K., and Carlin, B. P. (1995), "Efficient parameterisations for normal linear mixed models", *Biometrika*, 82, 479–488.

Gelfand, A. E., Silander Jr., J. A., Wu, S., Latimer, A., Lewis, P. O., Rebelo, A. G., and Holder, M. (2006), "Explaining species distribution patterns through hierarchical modeling", *Bayesian Analysis*, 1, 41–92.

Gelfand, A. E. and Smith, A. F. M. (1990), "Sampling-based approaches to calculating marginal densities", *Journal of the American Statistical Association*, 85, 398–409.

Gelman, A. (2006), "Prior distributions for variance parameters in hierarchical models (Comment on article by Browne and Draper)", *Bayesian Analysis*, 1, 515–534.

Gelman, A., Carlin, J. B., Stern, H. S., and Rubin, D. B. (2004), *Bayesian Data Analysis*, second edition, Boca Raton: Chapman and Hall.

Gelman, A., Meng, X. L., and Stern, H. (1996), "Posterior predictive assessment of model fitness via realized discrepancies (with discussion)", *Statistica Sinica*, 6, 733–807.

Geman, S. and Geman, D. (1984), "Stochastic relaxation", *Gibbs Distributions, and the Bayesian Restoration of Images IEEE Trans. on PAMI*, 6, 721–741.

Ghosh, J. K., Delampady, M., and Samanta, T. (2006), *An Introduction to Bayesian Analysis: Theory and Methods*, New York: Springer.

Gilks, W. R. and Roberts, G. O. (1995), "Strategies for improving MCMC", in *Markov Chain Monte Carlo in Practice*, eds. Gilks, W. R., Richardson, S., and Spiegelhalter, D. J., London: Chapman and Hall, pp. 89–114.

Gilks, W. R., Thomas, A., and Spiegelhalter, D. J. (1994), "A language and program for complex Bayesian modelling", *The Statistician*, 43, 169–178.

Gilks, W. R. and Wild, P. (1992), "Adaptive rejection sampling for Gibbs sampling", *Applied Statistics*, 41, 337–348.

Gimenez, O., Rossi, V., Choquet, R., Dehais, C., Doris, B., Varella, H., Vila, J. P., and Pradel, R. (2007), "State-space modelling of data on marked individuals", *Ecological Modelling*, 206, 431–438.

Goodman, L. A. (1960), "On the exact variance of products", *Journal of the American Statistical Association*, 55, 708–713.

Gotelli, N. J. and Ellison, A. M. (2004), *A Primer of Ecological Statistics*, Massachusetts: USA: Sinauer Associates Publishers.

Gotelli, N. J. and Colwell, R. K. (2001), "Quantifying biodiversity: procedures and pitfalls in the measurement and comparison of species richness", *Ecology Letters*, 4, 379–391.

Gu, W. and Swihart, R. K. (2004), "Absent or undetected? Effects of non-detection of species occurrence on wildlife-habitat models", *Biological Conservation*, 116, 195–203.

Guisan, A. and Thuiller, W. (2005), "Predicting species distribution: offering more than simple habitat models", *Ecology Letters*, 8, 993–1009.

Guisan, A. and Zimmermann, N. E. (2000), "Predictive habitat distribution models in ecology", *Ecological Modelling*, 135, 147–186.

Gumpertz, M. L., Wu, C. T., and Pye, J. M. (2000), "Logistic regression for southern pine beetle outbreaks with spatial and temporal autocorrelation", *Forest Science*, 46, 95–107.

Guthery, F. S., Brennan, L. A., Peterson, M. J., and Lusk, J. J. (2005), "Information theory in wildlife science: critique and viewpoint", *Journal of Wildlife Management*, 69, 457–465.

Hahn, B. A. (2007), Socially-facilitated habitat selection by two migratory forest songbirds, Ph.D. thesis, University of Michigan.

Hahn, B. A. and Silverman, E. D. (2006), "Social cues facilitate habitat selection: American redstarts establish breeding territories in response to song", *Biology Letters*, 2, 337–340.

Hahn, B. A. and Silverman, E. D. (2007), "Managing breeding forest songbirds with conspecific song playbacks", *Animal Conservation*, 10, 436–441.

Halley, J. and Inchausti, P. (2002), "Lognormality in ecological time series", *Oikos*, 99, 518–530.

Hames, R. S., Rosenberg, K. V., Lowe, J. D., Barker, S. E., and Dhondt, A. A. (2002), "Adverse effects of acid rain on the distribution of the Wood Thrush *Hylocichla mustelina* in North America", *Proceedings of the National Academy of Sciences*, 99, 11235–11240.

Hanski, I. (1994), "A practical model of metapopulation dynamics", *Journal of Animal Ecology*, 63, 151–162.

Hanski, I. (1998), "Metapopulation dynamics", *Nature*, 396, 41–49.

Hanski, I. (1999), *Metapopulation Ecology*, Oxford, UK: Oxford Univiversity Press.

Hanski, I. and Simberloff, D. (1997), "The metapopulation approach, its history, conceptual domain, and application to conservation", in *Metapopulation Biology: Ecology, Genetics, and Evolution*, eds. Hanski, I. and Gilpin, M. E., New York: Academic Press, pp. 1–11.

Hansson, L. (1969), "Home range, population structure and density estimates at removal catches with edge effect", *Acta Theriologica*, 14, 153–160.

Hayne, D. W. (1949), "An examination of the strip census method for estimating animal populations", *The Journal of Wildlife Management*, 13, 145–157.

He, F. and Gaston, K. J. (2000a), "Estimating species abundance from occurrence", *American Naturalist*, 156, 553–559.

He, F. and Gaston, K. J. (2000b), "Occupancy-abundance relationships and sampling scales", *Ecography*, 23, 503–511.

He, F. and Gaston, K. J. (2003), "Occupancy, spatial variance, and the abundance of species", *American Naturalist*, 162, 366–375.

Hedley, S. L., Buckland, S. T., and Borchers, D. L. (1999), "Spatial modelling from line transect data", *Journal of Cetacean Research and Management*, 1, 255–264.

Heikkinen, J. and Hogmander, H. (1994), "Fully bayesian approach to image restoration with an application in Biogeography", *Applied Statistics*, 43, 569–582.

Hensler, G. L. and Nichols, J. D. (1981), "The Mayfield method of estimating nesting success: a model, estimators and simulation results", *Wilson Bulletin*, 93, 42–53.

Hines, J. E. (1996), SURVIV: Software to compute estimates of survival with multinomially distributed data.

Hines, J. E., Boulinier, T., Nichols, J. D., Sauer, J. R., and Pollock, K. H. (1999), "COMDYN: software to study the dynamics of animal communities using a capture-recapture approach", *Bird Study*, 46, 209–217.

Hines, J. E., Kendall, W. L., and Nichols, J. D. (2003), "On the use of the robust design with transient capture-recapture models", *The Auk*, 120, 1151–1158.

Hobert, J. P. (2000), "Hierarchical models: A current computational perspective", *Journal of the American Statistical Association*, 95, 1312–1315.

Hoeting, J. A., Leecaster, M., and Bowden, D. (2000), "An improved model for spatially correlated binary responses", *Journal of Agricultural, Biological, and Environmental Statistics*, 5, 102–114.

Holzmann, H., Munk, A., and Zucchini, W. (2006), "On identifiability in capture–recapture models", *Biometrics*, 62, 934–936.

Hooten, M. B., Larsen, D. R., and Wikle, C. K. (2003), "Predicting the spatial distribution of ground flora on large domains using a hierarchical Bayesian model", *Landscape Ecology*, 18, 487–502.

Hooten, M. B. and Wikle, C. K. (2007), "Invasions, epidemics and binary data in a cellular world", Proceedings of the American Statistical Association [CD-ROM], Alexandria, VA: American Statistical Association, pp. 3999–4010.

Hubbell, S. P. (2001), *The Unified Neutral Theory of Biodiversity and Biogeography*, Princeton, New Jersey: Princeton University Press.

Huffer, F. and Wu, H. (1998), "Markov chain Monte Carlo for autologistic regression models with application to the distribution of plant species", *Biometrics*, 54, 509–524.

Huggins, R. M. (1989), "On the statistical analysis of capture experiments", *Biometrika*, 76, 133–140.

Ibrahim, J. G., Chen, M. H., Lipsitz, S. R., and Herring, A. H. (2005), "Missing-data methods for generalized linear models: A comparative review", *Journal of the American Statistical Association*, 100, 332–347.

Jennelle, C. S., Runge, M. C., and MacKenzie, D. I. (2002), "The use of photographic rates to estimate densities of tigers and other cryptic mammals: A comment of misleading conclusions", *Animal conservation*, 5, 119–120.

Jetz, W. and Rahbek, C. (2002), "Geographic range size and determinants of avian species richness", *Science*, 297, 1548–1551.

Johnson, D. H. (1979), "Estimating nest success: The mayfield method and an alternative", *The Auk*, 96, 651–661.

Johnson, N. L., Kemp, A. W., and Kotz, S. (2005), *Univariate Discrete Distributions*, Hoboken, NJ: Wiley.

Jolly, G. M. (1965), "Explicit estimates from capture-recapture data with both death and dilution-stochastic model", *Biometrika*, 52, 225–247.

Jonsen, I. D., Flemming, J. M., and Myers, R. A. (2005), "Robust state-space modeling of animal movement data", *Ecology*, 86, 2874–2880.

Jung, R. E., Royle, J. A., Sauer, J. R., Addison, C., Rau, R. D., Shirk, J. L., and Whissel, J. C. (2005), "Estimation of stream salamander (Plethodontidae, Desmognathinae and Plethodontinae) populations in Shenandoah National Park, Virginia, USA", *Alytes*, 22, 72–84.

Kadane, J. B. and Lazar, N. A. (2004), "Methods and criteria for model selection", *Journal of the American Statistical Association*, 99, 279–290.

Karanth, K. U. (1995), "Estimating tiger *Panthera tigris* populations from camera-trap data using capture–recapture models", *Biological Conservation*, 71, 333–338.

Karanth, K. U., Chundawat, R. S., Nichols, J. D., and Kumar, N. S. (2004), "Estimation of tiger densities in the tropical dry forests of Panna, Central India, using photographic capture–recapture sampling", *Animal Conservation*, 7, 285–290.

Karanth, K. U. and Nichols, J. D. (1998), "Estimation of tiger densities in India using photographic captures and recaptures", *Ecology*, 79, 2852–2862.

Karanth, K. and Nichols, J. (2000), Ecological status and conservation of tigers in India, Final Report to Division of International Conservation, Washington, DC: US Fish and Wildlife Service, *Centre for Wildlife Studies. Bangalore, India*, 123.

Karanth, K. U., Nichols, J. D., Kumar, N. S., and Hines, J. E. (2006), "Assessing tiger population dynamics using photographic capture-recapture sampling", *Ecology*, 87, 2925–37.

Kass, R. E. and Raftery, A. E. (1995), "Bayes factors", *Journal of the American Statistical Association*, 90, 773–795.

Kawanishi, K. and Sunquist, M. (2004), "Conservation status of tigers in a primary rainforest of Peninsular Malaysia", *Biological Conservation*, 120, 329–344.

Kendall, W. L. (1999), "Robustness of closed capture-recapture methods to violations of the closure assumption", *Ecology*, 80, 2517–2525.

Kendall, W. L. and Nichols, J. D. (2002), "Estimating state-transition probabilities for unobservable states using capture-recapture/resighting data", *Ecology*, 83, 3276–3284.

Kendall, W. L., Nichols, J. D., and Hines, J. E. (1997), "Estimating temporary emigration using capture-recapture data with Pollock's Robust design", *Ecology*, 78, 563–578.

Kéry, M. (2002), "Inferring the absence of a species - A case study of snakes", *Journal of Wildlife Management*, 66, 330–338.

Kéry, M. (2004), "Extinction rate estimates for plant populations in revisitation studies: The importance of detectability", *Conservation Biology*, 18, 570–574.

Kéry, M. (2008), "Estimating abundance from bird counts: binomial mixture models uncover complex covariate relationships", *The Auk* (in press).

Kéry, M. and Gregg, K. B. (2003), "Effects of life-state on detectability in a demographic study of the terrestrial orchid Cleistes bifaria", *Journal of Ecology*, 91, 265–273.

Kéry, M. and Plattner, M. (2007), "Species richness estimation and determinants of species detectability in butterfly monitoring programmes", *Ecological Entomology*, 32, 53–61.

Kéry, M., Royle, J. A., and Schmid, H. (2005), "Modeling avian abundance from replicated counts using binomial mixture models", *Ecological Applications*, 15, 1450–1461.

Kéry, M. and Royle, J. A. (2008a), Hierarchical Bayes estimation of species richness and occupancy in spatially replicated surveys, *Journal of Applied Ecology*, 45, 589–598.

Kéry, M. and Royle, J. A. (2008b), Inference about species richness and community structure using species-specific models in the national Swiss breeding bird survey MHB, *Environmental and Ecological Statistics*, 15, in press.

Kéry, M., Royle, J. A., Plattner, M. and Dorazio, R. M. (2008), Species richness and occupancy estimation in communities subject to temporary emigration, *Ecology* (in press).

Kéry, M. and Schmid, H. (2004), "Monitoring programs need to take into account imperfect species detectability", *Basic and Applied Ecology*, 5, 65–73.

Kéry, M. and Schmid, H. (2006), "Estimating species richness: Calibrating a large avian monitoring program", *Journal of Applied Ecology*, 43, 101–110.

Kéry, M., Spillmann, J. H., Truong, C., and Holderegger, R. (2006), "How biased are estimates of extinction probability in revisitation studies?", *Journal of Ecology*, 94, 980–986.

Key, J. T., Pericchi, L. R., and Smith, A. F. M. (1999), "Bayesian model choice: what and why? (with discussion)", in *Bayesian Statistics* Volume 6, eds. Bernardo, J. M., Berger, J. O., Dawid, A. P., and Smith, A. F. M., New York: Oxford University Press, pp. 343–370.

King, R., Brooks, S. P., and Coulson, T. (2008), "Analyzing complex capture-recapture data in the presence of individual and temporal covariates and model undertainty", *Biometrics* (in press).

Koneff, M. D., Royle, J. A., Otto, M. C., Wortham, J. S., and Bidwell, J. K. (2008), "A double-observer approach to estimating detection rates in aerial waterfowl surveys", *Journal of Wildlife Management*, (in press).

Krebs, C. J. (2001), *Ecology: The Experimental Analysis of Distribution and Abundance*, San Francisco, CA: Addison-Wesley Educational Publishers, Inc.

Kuo, L. and Mallick, B. (1998), "Variable selection for regression models", *Sankhya*, 60B, 65–81.

Laird, N. M. and Louis, T. A. (1987), "Empirical Bayes confidence intervals based on bootstrap samples (with discussion)", *Journal of the American Statistical Association*, 82, 739–757.

Laird, N. M. and Ware, J. H. (1982), "Random-effects models for longitudinal data", *Biometrics*, 38, 963–974.

Laplace, P. S. (1986), "Memoir on the probability of the causes of events (English translation of the 1774 French original by S. M. Stigler)", *Statistical Science*, 1, 364–378.

Laud, P. W. and Ibrahim, J. G. (1995), "Predictive model selection", *Journal of the Royal Statistical Society, B*, 57, 247–262.

Lawton, J. H. (1993), "Range, population abundance and conservation", *Trends in Ecology and Evolution*, 8, 409–413.

Le Cren, E. D. (1965), "A note on the history of Mark-Recapture population estimates", *Journal of Animal Ecology*, 34, 453–454.

Lebreton, J. D., Burnham, K. P., Clobert, J., and Anderson, D. R. (1992), "Modeling survival and testing biological hypotheses using marked animals: A unified approach with case studies", *Ecological Monographs*, 62, 67–118.

Leibold, M. A., Holyoak, M., Mouquet, N., Amarasekare, P., Chase, J. M., Hoopes, M. F., Holt, R. D., Shurin, J. B., Law, R., and Tilman, D., et al. (2004), "The metacommunity concept: a framework for multi-scale community ecology", *Ecology Letters*, 7, 601–613.

Leopold, A. (1933), *Game Management*, Madison, WI: University of Wisconsin Press.

Levins, R. (1969), "Some demographic and genetic consequences of environmental heterogeneity for biological control", *Bulletin of the Entomological Society of America*, 15, 237–240.

Lincoln, F. C. (1930), *Calculating Waterfowl Abundance on the Basis of Banding Returns, Vol. 118*, U.S. Department of Agriculture Circular, pp. 1–4.

Lindley, D. V. (1986), "Comment on why isn't everyone a Bayesian?", *The American Statistician*, 40, 6–7.

Lindley, D. V. (2006), *Understanding uncertainty*, Hoboken, New Jersey: John Wiley & Sons.

Link, W. A. (2003), "Nonidentifiability of population size from capture-recapture data with heterogeneous detection probabilities", *Biometrics*, 59, 1123–1130.

Link, W. A. (2006), "Reply to Holzmann et al", *Biometrics*, 62, 936–9.

Link, W. A. and Barker, R. J. (1994), "Density estimation using the trapping web design: A geometric analysis", *Biometrics*, 50, 733–745.

Link, W. A. and Barker, R. J. (2005), "Modeling association among demographic parameters in analysis of open population capture-recapture data", *Biometrics*, 61, 46–54.

Link, W. A. and Barker, R. J. (2006), "Model weights and the foundations of multimodel inference", *Ecology*, 87, 2626–2635.

Link, W. A., Cam, E., Nichols, J. D., and Cooch, E. G. (2002), "Of bugs and birds: Markov Chain Monte Carlo for hierarchical modeling in wildlife research", *The Journal of Wildlife Management*, 66, 277–291.

Link, W. A. and Nichols, J. D. (1994), "On the importance of sampling variance to investigations of temporal variation in animal population size", *Oikos*, 69, 539–544.

Link, W. A. and Sauer, J. R. (1999), "Controlling for varying effort in count surveys: An analysis of christmas bird count data", *Journal of Agricultural, Biological, and Environmental Statistics*, 4, 116–125.

Link, W. A. and Sauer, J. R. (2002), "A hierarchical model of population change with application to cerulean warblers", *Ecology*, 83, 2832–2840.

Little, R. J. (2004), "To model or not to model? Competing modes of inference for finite population sampling", *Journal of the American Statistical Association*, 99, 546–557.

Little, R. J. (2006), "Calibrated Bayes: a Bayes/frequentist roadmap", *American Statistician*, 60, 213–223.

Long, R. A., MacKay, P., Ray, J. C., and Zielinski, W. J. (2008), *Noninvasive Survey Methods for North American Carnivores*, Washington DC: Island Press.

Lukacs, P. M. and Burnham, K. P. (2005a), "Estimating population size from DNA-based closed capture-recapture data incorporating genotyping error", *Journal of Wildlife Management*, 69, 396–403.

Lukacs, P. M. and Burnham, K. P. (2005b), "Review of capture-recapture methods applicable to noninvasive genetic sampling", *Molecular Ecology*, 14, 3909–3919.

Lukacs, P. M., Thompson, W. L., Kendall, W. L., Gould, W. R., Doherty, P. F., Burnham, K. P., and Anderson, D. R. (2007), "Concerns regarding a call for pluralism of information theory and hypothesis testing", *Journal of Applied Ecology*, 44, 456–460.

MacArthur, R. H. and Wilson, E. O. (1967), *The Theory of Island Biogeography*, Princeton, New Jersey: Princeton University Press.

MacKenzie, D. I. and Nichols, J. D. (2004), "Occupancy as a surrogate for abundance estimation", *Animal Biodiversity and Conservation*, 27, 461–467.

MacKenzie, D. I., Nichols, J. D., Hines, J. E., Knutson, M. G., and Franklin, A. B. (2003), "Estimating site occupancy, colonization, and local extinction when a species is detected imperfectly", *Ecology*, 84, 2200–2207.

MacKenzie, D. I., Nichols, J. D., Lachman, G. B., Droege, S., Royle, J. A., and Langtimm, C. A. (2002), "Estimating site occupancy rates when detection probabilities are less than one", *Ecology*, 83, 2248–2255.

MacKenzie, D. I., Nichols, J. D., Royle, J. A., Pollock, K. H., Bailey, L. L., and Hines, J. E. (2006), *Occupancy Estimation and Modeling*, Amsterdam: Elsevier.

Maffei, L., Cuéllar, E., and Noss, A. (2004), "One thousand jaguars (*Panthera onca*) in Bolivia's Chaco? Camera trapping in the Kaa-Iya National Park", *Journal of Zoology*, 262, 295–304.

Magnusson, W. E., Caughley, G. J., and Grigg, G. C. (1978), "A double-survey estimate of population size from incomplete counts", *Journal of Wildlife Management*, 42, 174–176.

Manly, B. F. J., McDonald, L. L., Thomas, D. L., McDonald, T. L., and Erickson, W. P. (2002), *Resource Selection by Animals: Statistical Design and Analysis for field studies*, 2nd edition, Dordrecht: Kluwer Academic Publishers.

Mao, C. X. (2004), "Predicting the conditional probability of discovering a new class", *Journal of the American Statistical Association*, 99, 1108–1119.

Mao, C. X., Colwell, R. K., and Chang, J. (2005), "Estimating the species accumulation curve using mixtures", *Biometrics*, 61, 433–441.

Marzolin, G. (1988), "Polygynie du Cincle plongeur (*Cinclus cinclus*) dans les côtes de Lorraine", *Oiseau et la Revue Francaise d'Ornithologie*, 58, 277–286.

Mayfield, H. F. (1975), "Suggestions for calculating nest success", *Wilson Bulletin*, 87, 456–466.

Mazerolle, M. J., Desrochers, A., and Rochefort, L. (2005), "Landscape characteristics influence pond occupancy by frogs after accounting for detectability", *Ecological Applications*, 15, 824–834.

McCarthy, M. A. (2007), *Bayesian Methods for Ecology*, Cambridge: Cambridge University Press.

McCarthy, M. A. and Masters, P. (2005), "Profiting from prior information in Bayesian analyses of ecological data", *Journal of Applied Ecology*, 42, 1012–1019.

McKenny, H. C., Keeton, W. S., and Donovan, T. M. (2006), "Effects of structural complexity enhancement on eastern red-backed salamander (*Plethodon cinereus*) populations in northern hardwood forests", *Forest Ecology and Management*, 230, 186–196.

Miller, A. J. (2002), *Subset Selection in Regression*, 2nd edition, New York: Chapman and Hall.

Moilanen, A. (1999), "Patch occupancy models of metapopulation dynamics: Efficient parameter estimation using implicit statistical inference", *Ecology*, 80, 1031–1043.

Moilanen, A. (2002), "Implications of empirical data quality to metapopulation model parameter estimation and application", *Oikos*, 96, 516–530.

Møller, J. and Waagepetersen, R. P. (2004), *Statistical inference and simulation for spatial point processes*, UK: Chapman & Hall/CRC.

Morgan, B. J. T. (1992), *Analysis of Quantal Response Data*, UK: Chapman & Hall/CRC.

Morgan, B. J. T. and Ridout, M. (2008), Estimating N: a robust approach to recapture heterogeneity, *Environmental and Ecological Statistics*, to appear.

Morris, C. N. (1983), "Parametric empirical Bayes inference: theory and applications (with discussion)", *Journal of the American Statistical Association*, 78, 47–65.

Newman, K. B., Buckland, S. T., Lindley, S. T., Thomas, L., and Fernandez, C. (2006), "Hidden process models for animal population dynamics", *Ecological Applications*, 16, 74–86.

Nichols, J. D. and Karanth, K. U. (2002), "Statistical concepts: assessing spatial distributions", in *Monitoring Tigers and Their Prey: A Manual for Researchers, Managers and Conservationists in Tropical Asia*, eds. Karanth, K. U. and Nichols, J. D., Bangalore, India: Centre for Wildlife Studies, pp. 29–38.

Nichols, J. D., Boulinier, T., Hines, J. E., Pollock, K. H., and Sauer, J. R. (1998a), "Estimating rates of local species extinction, colonization, and turnover in animal communities", *Ecological Applications*, 8, 1213–1225.

Nichols, J. D., Boulinier, T., Hines, J. E., Pollock, K. H., and Sauer, J. R. (1998b), "Inference methods for spatial variation in species richness and community composition when not all species are detected", *Conservation Biology*, 12, 1390–1398.

Nichols, J. D., Hines, J. E., Sauer, J. R., Fallon, F. W., Fallon, J. E., and Heglund, P. J. (2000), "A double-observer approach for estimating detection probability and abundance from point counts", *The Auk*, 117, 393–408.

Nichols, J. D., Hines, J. E., MacKenzie, D. I., Seamans, M. E., and Gutierrez, R. J. (2007), "Occupancy estimation and modeling with multiple states and state uncertainty", *Ecology*, 88, 1395–1400.

Nichols, J. D., Stokes, S. L., Hines, J. E., and Conroy, M. J. (1982), "Additional comments on the assumption of homogeneous survival rates in modern bird banding estimation models", *The Journal of Wildlife Management*, 46, 953–962.

Norris III, J. L. and Pollock, K. H. (1996), "Nonparametric MLE under two closed capture-recapture models with heterogeneity", *Biometrics*, 52, 639–649.

O'Brien, T. G., Kinnaird, M. F., and Wibisono, H. T. (2003), "Crouching tigers, hidden prey: Sumatran tiger and prey populations in a tropical forest landscape", *Animal Conservation*, 6, 131–139.

O'Connell Jr., A. F., Talancy, N. W., Bailey, L. L., Sauer, J. R., Cook, R., and Gilbert, A. T. (2006), "Estimating site occupancy and detection probability parameters for mammals in a coastal ecosystem", *Journal of Wildlife Management*, 70, 1625–1633.

O'Hara, R. B., Arjas, E., Toivonen, H., and Hanski, I. (2002), "Bayesian analysis of metapopulation data", *Ecology*, 83, 2408–2415.

Olson, G. S., Anthony, R. G., Forsman, E. D., Ackers, S. H., Loschl, P. J., Reid, J. A., Dugger, K. M., Glenn, E. M., and Ripple, W. J. (2005), "Modeling of site occupancy dynamics for northern spotted owls, with emphasis on the effects of barred owls", *The Journal of Wildlife Management*, 69, 918–932.

Orme, C. D. L., Davies, R. G., Burgess, M., Eigenbrod, F., Pickup, N., Olson, V. A., Webster, A. J., Ding, T. S., Rasmussen, P. C., and Ridgely, R. S., et al. (2005), "Global hotspots of species richness are not congruent with endemism or threat", *Nature*, 436, 1016–1019.

Otis, D. L., Burnham, K. P., White, G. C., and Anderson, D. R. (1978), "Statistical inference from capture data on closed animal populations", *Wildlife Monographs*, 62, 1–135.

Parmenter, R. R., Yates, T. L., Anderson, D. R., Burnham, K. P., Dunnum, J. L., Franklin, A. B., Friggens, M. T., Lubow, B. C., Miller, M., and Olson, G. S., et al. (2003), "Small-mammal density estimation: A field comparison of grid-based vs. web-based density estimators", *Ecological Monographs*, 73, 1–26.

Pellet, J. and Schmidt, B. R. (2005), "Monitoring distributions using call surveys: estimating site occupancy, detection probabilities and inferring absence", *Biological Conservation*, 123, 27–35.

Pledger, S. (2000), "Unified maximum likelihood estimates for closed capture-recapture models using mixtures", *Biometrics*, 56, 434–442.

Pledger, S. (2005), "The performance of mixture models in heterogeneous closed population capture-recapture", *Biometrics*, 61, 868–873.

Pledger, S., Pollock, K. H., and Norris III, J. L. (2003), "Open capture-recapture models with heterogeneity: I. Cormack-Jolly-Seber model", *Biometrics*, 59, 786–794.

Pollock, K. H. (1982), "A capture-recapture design robust to unequal probability of capture", *Journal of Wildlife Management*, 46, 752–757.

Pollock, K. H. (2002), "The use of auxiliary variables in capture-recapture modelling: an overview", *Journal of Applied Statistics*, 29, 85–102.

Pollock, K. H., Hines, J. E., and Nichols, J. D. (1984), "The use of auxiliary variables in capture-recapture and removal experiments", *Biometrics*, 40, 329–340.

Pradel, R. (1996), "Utilization of capture-mark-recapture for the study of recruitment and population growth rate", *Biometrics*, 52, 703–709.

Pradel, R., Hines, J. E., Lebreton, J. D., and Nichols, J. D. (1997), "Capture-recapture survival models taking account of transients", *Biometrics*, 53, 60–72.

R Development Core Team (2004), *R: A language and environment for statistical computing*, R Foundation for Statistical Computing, Vienna, Austria, ISBN 3-900051-07-0.

Rathbun, S. L. and Cressie, N. A. C. (1994), "A space-time survival point process for a longleaf pine forest in Southern Georgia", *Journal of the American Statistical Association*, 89, 1164–1174.

Reunanen, P., Nikula, A., Monkkonen, M., Hurme, E., and Nivala, V. (2002), "Predicting occupancy for the siberian flying squirrel in old-growth forest patches", *Ecological Applications*, 12, 1188–1198.

Robbins, C. S., Bystrak, D. and Geissler, P. H. (1986), The breeding bird survey: its first fifteen years, 1965–1979, Resource Publication 157, United States Fish and Wildlife Service, Washington, DC.

Robert, C. P. and Casella, G. (2004), *Monte Carlo Statistical Methods*, second edition, New York: Springer-Verlag.

Robinson, G. K. (1991), "That BLUP is a good thing: The estimation of random effects", *Statistical Science*, 6, 15–32.

Rodenhouse, N. L., Sherry, T. W., and Holmes, R. T. (1997), "Site-dependent regulation of population size: A new synthesis", *Ecology*, 78, 2025–2042.

Rosenstock, S. S., Anderson, D. R., Giesen, K. M., Leukering, T., and Carter, M. F. (2002), "Landbird counting techniques: Current practices and an alternative", *The Auk*, 119, 46–53.

Rotella, J. J., Dinsmore, S. J., and Shaffer, T. L. (2004), "Modeling nest-survival data: a comparison of recently developed methods that can be implemented in MARK and SAS", *Animal Biodiversity and Conservation*, 27, 187–205.

Roughton, C. M. and Seddon, P. J. (2006), "Estimating site occupancy and detectability of an endangered New Zealand lizard, the Otago skink (*Oligosoma otagense*)", *Wildlife Research*, 33, 193–198.

Royle, J. A. (2004a), "Generalized estimators of avian abundance from count survey data", *Animal Biodiversity and Conservation*, 27, 375–386.

Royle, J. A. (2004b), "Modeling abundance index data from anuran calling surveys", *Conservation Biology*, 18, 1378–1385.

Royle, J. A. (2004c), "N-mixture models for estimating population size from spatially replicated counts", *Biometrics*, 60, 108–115.

Royle, J. A. (2006), "Site occupancy models with heterogeneous detection probabilities", *Biometrics*, 62, 97–102.

Royle, J. A. (2007), Inference for abundance and occupancy under functional independence between detection and local abundance, (unpublished manuscript).

Royle, J. A. (2008a), Analysis of capture-recapture models with individual covariates, *Biometrics*, in press.

Royle, J. A. (2008b), "Hierarchical modeling of cluster size in wildlife surveys", *Journal of Agricultural, Biological and Environmental Statistics*, 14, 23–36.

Royle, J. A. (2008c), Modeling individual effects in the Cormack-Jolly-Seber Model: A state-space formulation, *Biometrics*, in press.

Royle, J. A., Dawson, D. K., and Bates, S. (2004), "Modeling abundance effects in distance sampling", *Ecology*, 85, 1591–1597.

Royle, J. A. and Dorazio, R. M. (2006), "Hierarchical models of animal abundance and occurrence", *Journal of Agricultural, Biological, and Environmental Statistics*, 11, 249–263.

Royle, J. A. and Kéry, M. (2007), "A Bayesian state-space formulation of dynamic occupancy models", *Ecology*, 88, 1813–23.

Royle, J. A., Dorazio, R. M., and Link, W. A. (2007a), "Analysis of multinomial models with unknown index using data augmentation", *Journal of Computational and Graphical Statistics*, 16, 67–85.

Royle, J. A., Kéry, M., Gautier, R., and Schmid, H. (2007b), "Hierarchical spatial models of abundance and occurrence from imperfect survey data", *Ecological Monographs*, 77, 465–481.

Royle, J. A. and Link, W. A. (2005), "A general class of multinomial mixture models for anuran calling survey data", *Ecology*, 86, 2505–2512.

Royle, J. A. and Link, W. A. (2006), "Generalized site occupancy models allowing for false positive and false negative errors", *Ecology*, 87, 835–841.

Royle, J. A. and Nichols, J. D. (2003), "Estimating abundance from repeated presence-absence data or point counts", *Ecology*, 84, 777–790.

Royle, J. A., Nichols, J. D. and Karanth, K. U. (2008), Hierarchical models for density estimation from camera trap arrays, *Journal of Applied Ecology*, in review.

Royle, J. A., Nichols, J. D., and Kéry, M. (2005), "Modelling occurrence and abundance of species when detection is imperfect", *Oikos*, 110, 353–359.

Royle, J. A., Stanley, T. R., and Lukacs, P. M. (2008), "Statistical modeling and inference from carnivore survey data", in *Noninvasive Survey Methods for Carnivores*, eds. Long, R., MacKay, P., Ray, J., and Zielinski, W., California, USA: Island Press.

Royle, J. A. and Young, K. V. (2008), Hierarchical models for density estimation, *Ecology*, (in press).

Rubin, D. B. (1984), "Bayesianly justifiable and relevant frequency calculations for the applied statistician", *Annals of Statistics*, 12, 1151–1172.

Russell, K. N., Ikerd, H., and Droege, S. (2005), "The potential conservation value of unmowed powerline strips for native bees", *Biological Conservation*, 124, 133–148.

Sanathanan, L. (1972), "Estimating the size of a multinomial population", *Annals of Mathematical Statistics*, 43, 142–152.

Saracco, J. F., DeSante, D. F. and Kaschube, D. R. (2008a), Assessing landbird monitoring programs and demographic causes of population trends, *Journal of Wildlife Management*, (in press).

Saracco, J. F., Royle, J. A., DeSante, D. F. and Gardner, B. (unpublished), Modeling spatial variation in avian survival.

Sargeant, G. A., Sovada, M. A., Slivinski, C. C., and Johnson, D. H. (2005), "Markov chain Monte Carlo estimation of species distributions: a case study of the swift fox in western Kansas", *Journal of Wildlife Management*, 69, 483–497.

Sauer, J. R., Fallon, J. E., and Johnson, R. (2003), "Use of north American breeding bird survey data to estimate population change for bird conservation regions", *The Journal of Wildlife Management*, 67, 372–389.

Sauer, J. R., Peterjohn, B. G., and Link, W. A. (1994), "Observer differences in the North American Breeding Bird Survey", *The Auk*, 111, 50–62.

Schaub, M., Pradel, R., Jenni, L., and Lebreton, J. D. (2001), "Migrating birds stop over longer than usually thought: An improved capture-recapture analysis", *Ecology*, 82, 852–859.

Schmid, H., Luder, R., Naef-Daenzer, B., Graf, R. and Zbinden, N. (1998), Schweizer Brutvogelatlas. Verbreitung der Brutvögel in der Schweiz und im Fürstentum Liechtenstein 1993–1996, Schweizerische Vogelwarte, Sempach, Switzerland.

Schmid, H., Zbinden, N. and Keller, V. (2004), Überwachung der Bestandsentwicklung häufiger Brutvögel in der Schweiz, Tech. rep., Swiss Ornithological Institute, Sempach, Switzerland.

Schmidt, B. R. (2005), "Monitoring the distribution of pond-breeding amphibians when species are detected imperfectly", *Aquatic Conservation: Marine and Freshwater Ecosystems*, 15, 681–692.

Schmidt, B. R. and Pellet, J. (2005), "Relative importance of population processes and habitat characteristics in determining site occupancy of two anurans", *The Journal of Wildlife Management*, 69, 884–893.

Schwarz, C. J. (2001), "The Jolly-Seber model: More than just abundance", *Journal of Agricultural, Biological, and Environmental Statistics*, 6, 195–205.

Schwarz, C. J. and Arnason, A. N. (1996), "A general methodology for the analysis of capture-recapture experiments in open populations", *Biometrics*, 52, 860–873.

Schwarz, C. J. and Arnason, A. N. (2005), "Jolly-Seber models in MARK", in *Program MARK: A Gentle Introduction*, eds. Cooch, E. and White, G., 5th ed., On-line manual available at http://www.phidot.org/softeware/mark/docs/book.

Schwarz, C. J. and Seber, G. A. F. (1999), "Estimating animal abundance: review III", *Statistical Science*, 14, 427–456.

Searle, S. R., Casella, G., and McCulloch, C. E. (1992), *Variance components*, New York: John Wiley & Sons.

Seber, G. A. F. (1965), "A note on the multiple–recapture census", *Biometrika*, 52, 249–259.

Seber, G. A. F. (1982), *The Estimation of Animal Abundance and Related Parameters*, 2nd ed., London: Charles Griffin.

Seber, G. A. F. (1992), "A review of estimating animal abundance II", *International Statistical Review*, 60, 129–166.

Serfling, R. J. (1980), *Approximation Theorems of Mathematical Statistics*, New York: Wiley-Interscience.

Sergio, F. and Newton, I. (2003), "Occupancy as a measure of territory quality", *Ecology*, 72, 857–865.

Service, P. M. (2000), "Heterogeneity in individual mortality risk and its importance for evolutionary studies of senescence", *American Naturalist*, 156, 1–13.

Sileshi, G., Hailu, G., and Mafongoya, P. (2006), "Occupancy-abundance models for predicting densities of three leaf beetles damaging the multipurpose tree *Sesbania sesban* in eastern and southern Africa", *Bulletin of Entomological Research*, 96, 61–69.

Silver, S. C., Ostro, L. E. T., Marsh, L. K., Maffei, L., Noss, A. J., Kelly, M. J., Wallace, R. B., Gómez, H., and Ayala, G. (2004), "The use of camera traps for estimating jaguar *Panthera onca* abundance and density using capture/recapture analysis", *Oryx*, 38, 148–154.

Silverman, E. D. (2004), "A group movement model for waterfowl aggregation", *Ecological Modelling*, 175, 411–424.

Silverman, E. D., Kot, M., and Thompson, E. (2001), "Testing a simple stochastic model for the dynamics of waterfowl aggregations", *Oecologia*, 128, 608–617.

Smith, G. W. (1995), A Critical Review of the Aerial and Ground Surveys of Breeding Waterfowl in North America, US Dept. of the Interior, National Biological Service; National Technical Information Service distributor.

Soisalo, M. K. and Cavalcanti, S. M. C. (2006), "Estimating the density of a jaguar population in the Brazilian Pantanal using camera-traps and capture-recapture sampling in combination with GPS radio-telemetry", *Biological Conservation*, 129, 487–496.

Spiegelhalter, D. J. (1995), "Discussion of Assessment and propagation of model uncertainty by D. Draper", *Journal of the Royal Statistical Society, Series B*, 57, 71–73.

Spiegelhalter, D. J., Best, N. G., Carlin, B. P., and van der Linde, A. (2002), "Bayesian measures of model complexity and fit (with discussion)", *Journal of the Royal Statistical Society, B*, 64, 583–639.

Stanley, T. R. (2000), "Modeling and estimation of stage-specific daily survival probabilities of nests", *Ecology*, 81, 2048–2053.

Stanley, T. R. (2004), "Estimating stage-specific daily survival probabilities of nests when nest age is unknown", *Auk*, 121, 134–147.

Stanley, T. R. and Royle, J. A. (2005), "Estimating site occupancy and abundance using indirect detection indices", *The Journal of Wildlife Management*, 69, 874–883.

Stephens, P. A., Buskirk, S. W., Hayward, G. D., and Del Rio, C. M. (2005), "Information theory and hypothesis testing: a call for pluralism", *Journal of Applied Ecology*, 42, 4–12.

Stephens, P. A., Buskirk, S. W., Hayward, G. D., and Del Rio, C. M. (2007), "A call for statistical pluralism answered", *Journal of Applied Ecology*, 44, 461–463.

Sturtz, S., Ligges, U., and Gelman, A. (2005), "R2WinBUGS: A Package for Running WinBUGS from R", *Journal of Statistical Software*, 12, 1–16.

Tanner, M. A. (1996), *Tools for Statistical Inference: Methods for the Exploration of Posterior Distributions and Likelihood Functions*, third edition, New York: Springer-Verlag.

Ter Braak, C. J. E. and Etienne, R. S. (2003), "Improved Bayesian analysis of metapopulation data with an application to a tree frog metapopulation", *Ecology*, 84, 231–241.

Thompson, S. K. (2002), *Sampling*, New York: Wiley.

Thompson, W. L. (2004), *Sampling Rare Or Elusive Species: Concepts, Designs, and Techniques for Estimating Population Parameters*, Island Press.

Trolle, M. and Kéry, M. (2003), "Estimation of ocelot density in the Pantanal using capture-recapture analysis of camera-trapping data", *Journal of Mammalogy*, 84, 607–614.

Trolle, M. and Kéry, M. (2005), "Camera-trap study of ocelot and other secretive mammals in the northern Pantanal", *Mammalia*, 69, 409–416.

Tyre, A. J., Tenhumberg, B., Field, S. A., Niejalke, D., Parris, K., and Possingham, H. P. (2003), "Improving precision and reducing bias in biological surveys: estimating false-negative error rates", *Ecological Applications*, 13, 1790–1801.

Ugland, K. I., Gray, J. S., and Ellingsen, K. E. (2003), "The species-accumulation curve and estimation of species richness", *Journal of Animal Ecology*, 72, 888–897.

van der Linde, A (2005), "DIC in variable selection", *Statistica Neerlandica*, 59, 45–56.

Ver Hoef, J. M. and Frost, K. J. (2003), "A Bayesian hierarchical model for monitoring harbor seal changes in Prince William Sound, Alaska", *Environmental and Ecological Statistics*, 10, 201–219.

Viljugrein, H., Stenseth, N. C., Smith, G. W., and Steinbakk, G. H. (2005), "Density dependence in North American Ducks", *Ecology*, 86, 245–254.

Wallace, R. B., Gomez, H., Ayala, G., and Espinoza, F. (2003), "Camera trapping for jaguar (*Panthera onca*) in the Tuichi Valley, Bolivia", *Journal of Neotropical Mammalogy*, 10, 133–139.

Wegge, P., Pokheral, C. P., and Jnawali, S. R. (2004), "Effects of trapping effort and trap shyness on estimates of tiger abundance from camera trap studies", *Animal Conservation*, 7, 251–256.

Weir, L. A. and Mossman, M. J. (2005), North American Amphibian Monitoring Program (NAAMP), *Amphibian Declines: the conservation status of United States species. University of California Press, Berkeley, California, USA*, 307–313.

Weir, L. A., Royle, J. A., Nanjappa, P., and Jung, R. E. (2005), "Modeling anuran detection and site occupancy on north american amphibian monitoring program (NAAMP) routes in Maryland", *Journal of Herpetology*, 39, 627–639.

White, G. C. (1983), "Numerical estimation of survival rates from band-recovery and biotelemetry data", *The Journal of Wildlife Management*, 47, 716–728.

White, G. C. and Burnham, K. P. (1999), "Program MARK: survival estimation from populations of marked animals", *Bird Study*, 46, 120–138.

White, G. C. and Garrott, R. A. (1990), *Analysis of wildlife radio-tracking data*, Academic Press.

White, G. C. and Shenk, T. M. (2001), "Population estimation with radio-marked animals", in *Design and Analysis of Wildlife Radiotelemetry Studies*, San Diego, CA: Academic Press, pp. 329–350.

White, G. C. (2005), "Correcting wildlife counts using detection probabilities", *Wildlife Research*, 32, 211–216.

Wikle, C. K. (2003), "Hierarchical Bayesian models for predicting the spread of ecological processes", *Ecology*, 84, 1382–1394.

Wikle, C. K. and Royle, J. A. (2005), "Dynamic design of ecological monitoring networks for non-Gaussian spatio-temporal data", *Environmetrics*, 16, 507–522.

Williams, B. K., Nichols, J. D., and Conroy, M. J. (2002), *Analysis and Management of Animal Populations*, San Diego, California: Academic Press.

Wilson, K. R. and Anderson, D. R. (1985a), "Evaluation of a density estimator based on a trapping web and distance sampling theory", *Ecology*, 66, 1185–1194.

Wilson, K. R. and Anderson, D. R. (1985b), "Evaluation of a nested grid approach for estimating density", *Journal of Wildlife Management*, 49, 675–678.

Wilson, K. R. and Anderson, D. R. (1985c), "Evaluation of two density estimators of small mammal population size", *Journal of Mammalogy*, 66, 13–21.

Winchell, C. S. and Doherty Jr., P. F. (2008), "Testing underlying models used in habitat conservation plans: Coastal California gnatcatcher as an example", *Journal of Wildlife Management*, (in press).

Wintle, B. A. and Bardos, D. C. (2006), "Modeling species-habitat relationships with spatially autocorrelated observation data", *Ecological Applications*, 16, 1945–1958.

Wintle, B. A., McCarthy, M. A., Parris, K. M., and Burgman, M. A. (2004), "Precision and bias of methods for estimating point survey detection probabilities", *Ecological Applications*, 14, 703–712.

Woods, J. G., Paetkau, D., Lewis, D., McLellan, B. N., Proctor, M., and Strobeck, C. (1999), "Genetic tagging of free-ranging black and brown bears", *Wildlife Society Bulletin*, 27, 616–627.

Wu, H. and Huffer, F. (1997), "Modelling the distribution of plant species using the autologistic regression model", *Environmental and Ecological Statistics*, 4, 31–48.

Yang, H. and Chao, A. (2005), "Modeling animals behavioral response by Markov chain models for capture–recapture experiments", *Biometrics*, 61, 1010–1017.

Yoshizaki, J. (2007), Use of natural tags in closed population capture-recapture studies: modeling misidentification, Ph.D. thesis, North Carolina State University.

Young, K. V. and Royle, J. A. (2006), Abundance and site occupancy of flat-tailed horned lizard (*Phrynosoma mcallii*) populations in Arizona and California, Report to: U.S. Navy and U.S. Bureau of Reclamation. 21 pp.

Zellner, A., Keuzenkamp, H. A., and McAleer, M. (2001), *Simplicity, inference, and modelling*, Cambridge: Cambridge University Press.

Zhu, J., Huang, H., and Wu, J. (2005), "Modeling spatial-temporal binary data using Markov random fields", *Journal of Agricultural, Biological, and Environmental Statistics*, 10, 212–225.

Zielinski, W. J. (1995), "Baited track-plate stations", *A detection manual for wolverine, fisher, lynx and marten in the western United States (Zielinski, W.J., and Kucera,*

T.E., eds.). United States Department of Agriculture, Forest Service, Pacific Southwest Research Station, General Technical Report PSW-GTR-157, 1–163.

Zielinski, W. J., Truex, R. L., Schlexer, F. V., Campbell, L. A., and Carroll, C. (2005), "Historical and contemporary distributions of carnivores in forests of the Sierra Nevada, California, USA", *Journal of Biogeography*, 32, 1385–1407.

Zippin, C. (1956), "An evaluation of the removal method of estimating animal populations", *Biometrics*, 12, 163–189.

INDEX

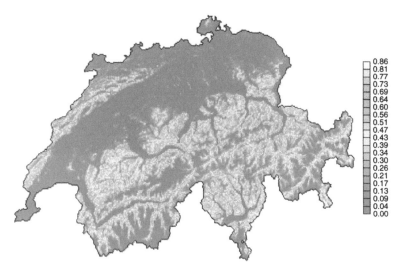

Figure 3.4. Estimated probabilities of occurrence of the willow tit for every 1 km quadrat in Switzerland.

Figure 5.1. Left: Flat-tailed horned lizard (*Phrynosoma mcallii*) in all its glory. Right: Quality habitat of the flat-tailed horned lizard. *Photos courtesy of K.V. Young.*

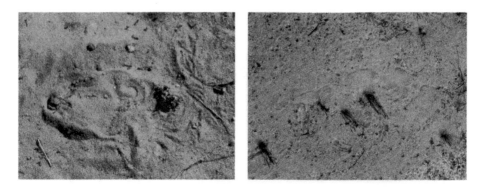

Figure 5.2. Figures illustrating FTHL elusiveness. Left: typical lizard 'dug-in'. Right: this area was thoroughly searched and lizard in the middle was missed by surveyors who nearly stepped on it (note footprints) before eventually noticing it. *Photos courtesy of K.V. Young.*

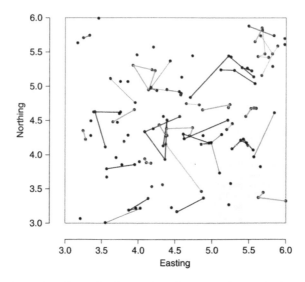

Figure 7.1. Locations of 68 flat-tailed horned lizards captured a total of 134 times on a 9 ha plot in southwestern Arizona. Captures of the same individual are connected by lines of the same color and thickness.

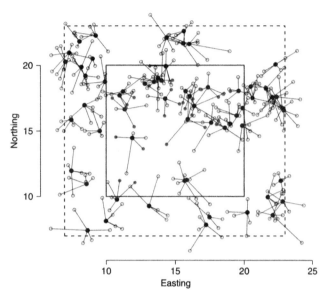

Figure 7.4. Simulated spatial capture–recapture data set for an 'area search' type of design. The surveyed area is the 10×10 sample plot nested within 16×16 quadrat, containing 60 individual activity centers (solid black circles) of which 24 are contained within the sample plot. All locations of each individual are marked with open black circles. Captures are indicated with red.

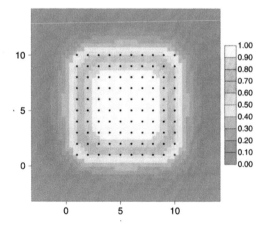

Figure 7.6. Probability of exposure of individuals to a 10 by 10 grid of traps as a function of their center of activity, s.

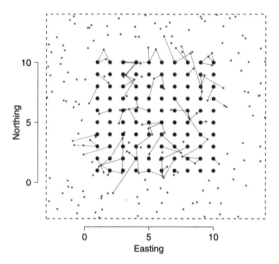

Figure 7.7. Simulated realization of trapping grid with captured individuals. Simulated captures of individuals (red dots) were made by 10×10 grid of traps (black dots). The trap(s) in which each individual was captured are indicated with blue lines.

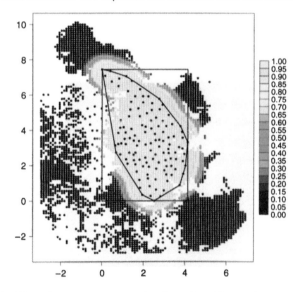

Figure 7.9. Probability of exposure to trapping in a $J = 12$ sample camera trapping study. The color of each pixel is the Pr(exposure to sampling) of a tiger with activity center at that pixel. Only pixels judged to be suitable habitat are included. The sum of all pixels is the effective sample area of the trapping array.

Figure 8.3. The West Indian manatee.

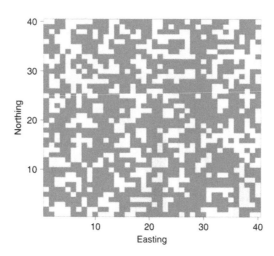

Figure 9.4. Binary image with auto-logistic parameters $\alpha = -0.5$ and $\beta = 1.5$ yielding an occupancy process with $\psi \approx 0.60$. The white cells are unoccupied and the colored cells are occupied.

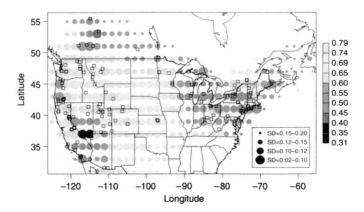

Figure 11.2. Map depicting spatial variation in yellow warbler survival. Values are point-wise posterior means. The shading reflects the magnitude of predicted survival probability whereas the size of the plot symbol is proportional to the inverse of the posterior standard deviation. Larger circles indicate smaller posterior standard deviations. The 139 Monitoring Avian Productivity and Survivorship (MAPS) stations used in the analysis are indicated with square symbols.

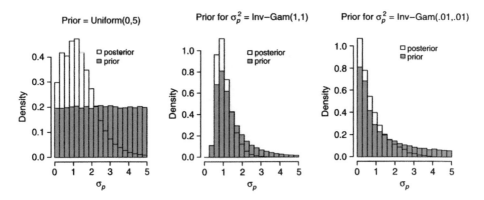

Figure 11.3. Estimated posterior distributions of σ_p and corresponding histograms obtained under several prior distributions. This figure appears in color in the electronic version of this article. Reproduced from Royle (2007b).

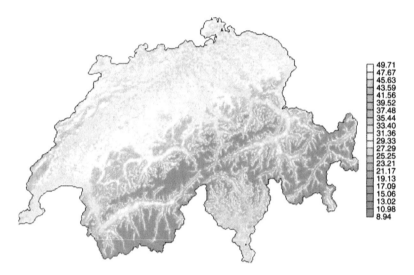

	49.71
	47.67
	45.63
	43.59
	41.56
	39.52
	37.48
	35.44
	33.40
	31.36
	29.33
	27.29
	25.25
	23.21
	21.17
	19.13
	17.09
	15.06
	13.02
	10.98
	8.94

Figure 12.7. Geographic distribution of the posterior-predictive mean number of avian species present at each of 41,365 locations in Switzerland.

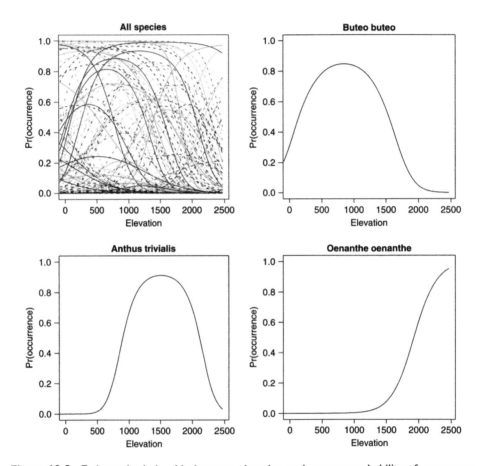

Figure 12.8. Estimated relationship between elevation and average probability of occurrence for each of the 134 species observed in the 2001 survey of Switzerland and for each of 3 individual species.

Printed and bound by CPI Group (UK) Ltd, Croydon, CR0 4YY

03/10/2024

01040316-0004